Adhesion and Friction in Biological and Bioinspired Systems

Adhesion and Friction in Biological and Bioinspired Systems

Guest Editors

Thies H. Büscher
Stanislav N. Gorb

MDPI

Basel • Beijing • Wuhan • Barcelona • Belgrade • Novi Sad • Cluj • Manchester

Guest Editors
Thies H. Büscher
Department of Functional Morphology and Biomechanics
Kiel University
Kiel
Germany

Stanislav N. Gorb
Department of Functional Morphology and Biomechanics
Kiel University
Kiel
Germany

Editorial Office
MDPI AG
Grosspeteranlage 5
4052 Basel, Switzerland

This is a reprint of the Special Issue, published open access by the journal *Biomimetics* (ISSN 2313-7673), freely accessible at: https://www.mdpi.com/journal/biomimetics/special_issues/112X4XL3J2.

For citation purposes, cite each article independently as indicated on the article page online and as indicated below:

Lastname, A.A.; Lastname, B.B. Article Title. *Journal Name* **Year**, *Volume Number*, Page Range.

ISBN 978-3-7258-7761-4 (Hbk)
ISBN 978-3-7258-7762-1 (PDF)
https://doi.org/10.3390/books978-3-7258-7762-1

Cover image courtesy of Thies H. Büscher

Contents

About the Editors

Thies H. Büscher

Thies Büscher is a researcher in the Department of Functional Morphology and Biomechanics at the Zoological Institute of Kiel University, Germany. He studied chemistry, biology, and education, followed by a doctoral degree in biology with a focus on the functional morphology and biomechanics of insects at Kiel University. Büscher's research centers on the biomechanics and evolution of biological attachment systems, particularly the tarsal structures of Phasmatodea (stick insects), next to functional systems in other arthropods. His work employs imaging techniques including confocal laser scanning microscopy, cryo-scanning electron microscopy, 3D reconstruction and methods to analyze attachment performance, such as force-measurements, to understand the functionality and performance of adaptive traits and their biomimetic potential. He serves as Editor in Chief for *Physiological Entomology* (Royal Entomological Society) and is an Associate Editor for *Zootaxa*, *Frontiers in Zoology*, and *Biodiversity Data Journal*. He has published extensively in peer-reviewed journals, contributing to phasmid taxonomy, systematics, and biomechanical studies. Büscher is actively involved in teaching zoology and mentoring students at Kiel University, as well as being a speaker for the Morphological Section of the German Zoological Society. Thies Büscher has received different awards, including the Frontiers in Zoology Award of the German Zoological Society, the Ian Abercrombie Award of the Phasmid Study Group (NHMUK, London) and the Biodiversity Award of the Academy of Sciences and Literature, Mainz.

Stanislav N. Gorb

Stanislav Gorb is a Professor and Director at the Zoological Institute of the Kiel University, Germany. He received his PhD in zoology and entomology at the Schmalhausen Institute of Zoology of the Ukrainian Academy of Sciences in Kiev (Ukraine). Gorb was a postdoctoral researcher at the University of Vienna (Austria), a research assistant at the University of Jena, and a group leader at the Max Planck Institutes for Developmental Biology in Tübingen and for Metals Research in Stuttgart (Germany). Gorb's research focuses on the morphology, structure, biomechanics, physiology, and evolution of surface-related functional systems in animals and plants, as well as the development of biologically inspired technological surfaces and systems. He received the Schlossmann Award in Biology and Materials Science in 1995, International Forum Design Gold Award in 2011 and Materialica "Best of" Award in 2011. In 1998, he was the BioFuture Competition winner for his works on biological attachment devices as possible sources for biomimetics. In 2018, he received Karl-Ritter-von-Frisch Medal of German Zoological Society and in 2026 the President´s Medal of the Royal Entomological Society, UK. In 2018, he received Friendship Award (China). Gorb is Corresponding member of Academy of the Science and Literature Mainz, Germany (since 2010) and Member of the National Academy of Sciences Leopoldina, Germany (since 2011). Gorb has authored several books and more than 800 papers in peer-reviewed journals and has five patents.

biomimetics

MDPI

Editorial

Adhesion and Friction in Biological and Bioinspired Systems

Thies H. Büscher * and Stanislav N. Gorb

Functional Morphology and Biomechanics, Zoological Institute, Kiel University, D-24118 Kiel, Germany; sgorb@zoologie.uni-kiel.de
* Correspondence: tbuescher@zoologie.uni-kiel.de

In nature, biological systems employ adhesion and friction in various contexts. Organisms have evolved a variety of specialized biological attachment systems that enhance adhesion and friction, enabling them to attach to diverse substrates firmly or to interconnect different parts of their own bodies [1]. Such systems are widespread in nature and enable organisms to attach temporarily, facilitating locomotion and mating [2], or permanently aiding in settlement [3] to various substrates. Conversely, in other scenarios, such as within joints or on animal surfaces exposed to aquatic environments, contact forces are minimized through various mechanisms to reduce adhesion and friction, for example, to reduce abrasive wear [4], drag [5], or contamination [6]. The functional principles underlying these systems are based on multiple mechanisms hierarchically integrated across different length scales, and both phenomena are tailored through an interplay of (1) surface chemistry, (2) specific mechanical material properties, (3) fluid or solid secretions, and (4) sophisticated micro- and nanostructures.

Numerous adhesion- and friction-related mechanisms that maximize or minimize contact forces have independently evolved across various taxa, since different organisms must meet unique functional requirements shaped by their specific environmental demands and selective pressures [7]. The link between specialized problem-solving strategies and the application of universal physical principles makes these systems an especially promising subject of study in the field of biomimetics [8]. Examining how these distinct biological systems deal with comparable challenges can help to identify functional constraints, while the wide range of biological strategies offers valuable inspiration for biomimetic design. To date, such phenomena have been investigated in detail only for a limited number of selected biological systems and have received comparatively little attention in biomimetic research.

The aim of this Special Issue is to connect interdisciplinary research on mechanisms of adhesion and friction across biological and bioinspired systems. It seeks to highlight how organisms achieve adhesion enhancement, reduction, or mechanical interlocking through structural, chemical, and mechanical adaptations, and to explore how these principles can inform the design of functional materials and devices. By integrating perspectives from biology, physics, and engineering, the Special Issue aims to deepen our understanding of structure–property relationships and foster innovation in biomimetic applications. The completed Special Issue covers a diverse collection of studies dealing with topics on adhesion and friction, from cell adhesion phenomena via non-specific adhesion to the mechanical interlocking of structures. Additionally, several contributions addressed friction and adhesion reduction. Some contributions provide insight into how surface structure, contact mechanics, and surface chemistry contribute to biological attachment, while others explore bioinspired engineered developments and structure–property relationships in artificial systems.

Biomimetics **2026**, *11*, 295 https://doi.org/10.3390/biomimetics11050295

Together, these contributions showcase research from natural systems and biological insight to applied biomimetic design and engineering realization. The collection underscores how studying the structural and mechanical principles of adhesion, friction, and surface interaction across living organisms can inspire innovative technological solutions. Ranging from biological phenomena, through experimental and simulation-driven understanding, to applied studies and material design concepts, the Special Issue also bridges different disciplines. This line—from biological observation to bioinspired application—reflects the overarching aim of biomimetics: transforming natural mechanisms into functional technologies for diverse industrial and biomedical contexts.

In summary, the reader can expect an overview of recent activities in the field of adhesion- and friction-related phenomena in biomimetics. On the side of property characterization of biological and bioinspired systems, Zhou et al. [9] investigated the role of substrate compliance on the jumping performance in a tree frog species with adhesive toepads. They explored the functional adaptation of tree frog adhesive devices in locomotion and surface interaction between this biological system and its substrates. One step further in the progress of generalization of functional principles is the characterization of properties of functional surfaces that are inspired by a biological template. In this setting, Saccardi et al. [10] investigated the anti-adhesive surfaces they produced based on the mandible microstructures of bees. The mandibles of honeybees are confronted with naturally strong sticking substances (propolis) and avoid persistent adhesion by microstructures and an anti-adhesive coating on the inner side of the mandibles. The authors conducted experiments on a range of artificial replicas of this system to assess the influence of microstructure and surface chemistry on propolis adhesion in this context. Lastly, Ji et al. [11] present a study on bioinspired footwear that is inspired by reptilian adhesive systems. Their development is intended to provide stabilization for foot placement under microgravity in space application contexts and, furthermore, aids in the physiological protection of muscle deterioration in weightless environments.

Two papers present studies on experimental and methodological advances in the field of adhesion- and friction-related biomimetics. Tabraue-Rubio et al. [12] developed and compared methods for binding cells to functionalized AFM probes, advancing nanoscale measurement and biointerface characterization methodologies on the cellular level. Ferreira et al. [13] studied adhesive performance at the macroscale in horseshoe bonding through mechanical and numerical analysis. Combining engineering testing with numerical modeling to understand composite joining, the authors compared different adhesives used in the equine industry to establish methods for reliable and less invasive horseshoeing procedures.

In the field of computational and simulation studies, the paper by Liu et al. [14] offers insight into surface-fluid interactions that minimize energy loss through numerical analyses of drag reduction in biomimetic microstructured surfaces.

Finally, contributions on applied biomimetics focused on advances in concepts in engineering design. Mishnaevsky et al. [15] propose a bio-inspired concept for sustainable wind turbine blades, integrating natural adhesive strategies into large-scale composite design. Among the main challenges of sustainable wind energy technology is the contrasting demand for the durability of wind turbine blades versus the recyclability of their components. By developing a bio-inspired mechanism combining chemical adhesion and mechanically interlocking fibers, the authors propose a concept to improve the operational duration of wind turbine blades while facilitating the separation of the components for recycling. For application in various contexts, Yang et al. [16] introduced phenolic intervention as a method to tune the properties of food-grade kappa-carrageenan hydrogels. They show that different plant polyphenols are capable of modifying the properties of such hydrogels

https://doi.org/10.3390/biomimetics11050295

and present a route for targeted property manipulation via tannic acid and pyrogallol for selective increase in adhesive properties, thermal behavior, and premediated disintegration. In their review article, Kumar et al. [17] provide a comprehensive review of bioinspired tribological materials for sliding, erosive, machining, and energy-absorbing applications, highlighting multifunctionality in performance as a strength for bioinspired engineering.

The contributions in this Special Issue from both biological and engineering perspectives offer valuable interdisciplinary connections, reinforcing the strong link between natural principles and their biomimetic applications. They showcase the breadth of current research in biological and bioinspired systems and highlight promising directions for future investigations. We are grateful for the wide range of contributions to this Special Issue and sincerely thank all colleagues who participated in it by writing their contributions and/or by reviewing submitted manuscripts. Their commitment made it possible to assemble this collection of articles addressing biomimetic questions on adhesion and friction from multiple scientific viewpoints.

Author Contributions: T.H.B.: conceptualization; writing—original draft; writing—review & editing. S.N.G.: conceptualization; writing—review & editing. All authors have read and agreed to the published version of the manuscript.

Funding: This research received no external funding.

Conflicts of Interest: The authors declare no conflicts of interest.

References

1. Gorb, S.N. *Attachment Devices of Insect Cuticle*; Springer Science & Business Media: Dordrecht, The Netherlands, 2001; 305p. [CrossRef]
2. Büscher, T.H.; Gorb, S.N. Physical constraints lead to parallel evolution of micro- and nanostructures of animal adhesive pads: A review. *Beilstein J. Nanotechnol.* **2021**, *12*, 725–743. [CrossRef] [PubMed]
3. Aldred, N.; Clare, A.S. Mechanisms and principles underlying temporary adhesion, surface exploration and settlement site selection by barnacle cyprids: A short review. In *Functional Surfaces in Biology*; Gorb, S.N., Ed.; Springer: Dordrecht, The Netherlands, 2009; pp. 43–65. [CrossRef]
4. Nadein, K.; Kovalev, A.; Thøgersen, J.; Weidner, T.; Gorb, S.N. Insects use lubricants to minimize friction and wear in leg joints. *Proc. Biol. Sci.* **2021**, *288*, 20211065. [CrossRef] [PubMed]
5. Bushnell, D.M.; Moore, K.J. Drag reduction in nature. *Annu. Rev. Fluid Mech.* **1991**, *23*, 65–79. [CrossRef]
6. Peisker, H.; Gorb, S.N. Always on the bright side of life: Anti-adhesive properties of insect ommatidia grating. *J. Exp. Biol.* **2010**, *213*, 3457–3462. [CrossRef] [PubMed]
7. Büscher, T.H.; Gorb, S.N. Convergent evolution of animal adhesive pads. In *Convergent Evolution. Fascinating Life Sciences*; Bels, V.L., Russell, A.P., Eds.; Springer: Cham, Switzerland, 2023; pp. 257–287. [CrossRef]
8. Büscher, T.H.; Gorb, S.N. Biological attachment systems and biomimetics—In memory of William Jon P. Barnes. *Biomimetics* **2025**, *10*, 220. [CrossRef] [PubMed]
9. Zhou, R.; Zhang, B.; Wang, Z.; Dai, Z. Effect of substrate compliance on the jumping mechanism of the tree frog (*Polypedates dennysii*). *Biomimetics* **2025**, *10*, 604. [CrossRef] [PubMed]
10. Saccardi, L.; Schiebl, J.; Balluff, F.; Christ, U.; Gorb, S.N.; Kovalev, A.; Schwarz, O. Anti-adhesive surfaces inspired by bee mandible surfaces. *Biomimetics* **2023**, *8*, 579. [CrossRef] [PubMed]
11. Ji, Y.; Li, Z.; Zou, P.; Li, C.; Wang, X.; Yang, X.; Dai, Z.; Ji, K. Anti-weightlessness physiological protection for the lower limb muscle system based on biomimetic adhesive force stimulation. *Biomimetics* **2025**, *10*, 800. [CrossRef] [PubMed]
12. Tabraue-Rubio, R.; Yuste Muñoz, L.; Vázquez, M.; Daza, R.; Colchero, L.; Fernández-Santos, M.E.; Elices, M.; Panetsos, F.; Guinea, G.V.; Pérez-Rigueiro, J. Development of three alternative strategies for the binding of cells to functionalized DeepTip^{TM} AFM probes. *Biomimetics* **2026**, *11*, 95. [CrossRef] [PubMed]
13. Ferreira, C.M.C.; Simões, B.D.; Marques, E.A.S.; Carbas, R.J.C.; da Silva, L.F.M. Exploring adhesive performance in horseshoe bonding through advanced mechanical and numerical analysis. *Biomimetics* **2025**, *10*, 2. [CrossRef] [PubMed]
14. Liu, J.; Xu, J.; Ding, C.; Shan, D.; Guo, B. Numerical investigation on drag reduction mechanisms of biomimetic microstructure surfaces. *Biomimetics* **2026**, *11*, 77. [CrossRef] [PubMed]

https://doi.org/10.3390/biomimetics11050295

15. Mishnaevsky, L., Jr.; Jafarpour, M.; Krüger, J.; Gorb, S.N. A new concept of sustainable wind turbine blades: Bio-inspired design with engineered adhesives. *Biomimetics* **2023**, *8*, 448. [CrossRef] [PubMed]
16. Yang, H.-Y.; Seo, J.; Choi, W.; Kim, E.; Yeo, S.; Park, S.; Lee, H. Adhesive κ-Carrageenan hydrogels by polyphenol intervention. *Biomimetics* **2026**, *11*, 290. [CrossRef]
17. Kumar, R.; Rezapourian, M.; Rahmani, R.; Maurya, H.S.; Kamboj, N.; Hussainova, I. Bioinspired and multifunctional tribological materials for sliding, erosive, machining, and energy-absorbing conditions: A review. *Biomimetics* **2024**, *9*, 209. [CrossRef] [PubMed]

https://doi.org/10.3390/biomimetics11050295

biomimetics

Article

Effect of Substrate Compliance on the Jumping Mechanism of the Tree Frog (*Polypedates dennys*)

Rui Zhou [1,†], Baowen Zhang [1,2,†], Zhouyi Wang [1,3,*] and Zhendong Dai [1]

1 College of Mechanical and Electrical Engineering, Nanjing University of Aeronautics and Astronautics, Nanjing 210016, China; zhourui@nuaa.edu.cn (R.Z.); baowen.zhang@wur.nl (B.Z.); zddai@nuaa.edu.cn (Z.D.)
2 Experimental Zoology Group, Wageningen University & Research, 6708 WD Wageningen, The Netherlands
3 Shenzhen Research Institute, Nanjing University of Aeronautics and Astronautics, Shenzhen 518063, China
* Correspondence: wzyxml@nuaa.edu.cn
† These authors contributed equally to this work.

Abstract

Animal locomotion in complex environments depends on the ability to adaptively regulate movement in response to substrate mechanics. Tree frogs (*Polypedates dennysi*), which combine jumping and adhesive capabilities, inhabit arboreal habitats with a wide range of compliant substrates. While previous studies have offered preliminary insights into their locomotion, the biomechanical mechanisms underlying their adaptability remain poorly characterized. In this study, we developed a stiffness-adjustable takeoff substrate supported by four springs, and combined it with a 3D motion capture system to analyze the jumping dynamics and kinematics of frogs across a broader range of compliant substrates. We found that energy recovery from the substrate was influenced by compliance. On the stiffest substrate, up to 50% of the stored energy was recovered during takeoff, whereas highly compliant substrates caused nonlinear damping, energy dissipation, and even takeoff failure. During takeoff, frogs generated peak normal forces up to 6 times their body weight and fore–aft forces up to 4.5 times their body weight. However, force generation showed limited adaptability to substrate mechanics, while takeoff velocity exhibited stronger adaptability to changes in compliance. These findings reveal a trade-off between substrate mechanics and jump performance. This work provides biomechanical insight into substrate preference and informs the design of bioinspired systems capable of efficient locomotion on compliant substrates.

Keywords: compliant substrate; jumping mechanism; tree frog; energy recovery; kinematics

1. Introduction

The behavioral and morphological adaptations of species to their physical environment play a crucial role in their survival and reproduction. In terms of locomotion, numerous species have evolved movement patterns and morphological characteristics that are adapted to their environments over long evolutionary timescales [1–4]. For instance, sand lizards (*Uma scoparia* and *Callisaurus draconoides*) have evolved elongated feet and specialized toe fringes to enhance mobility on loose, granular substrates [5]. Water striders exploit superhydrophobic structures formed by nanoscale setae on their legs to harness surface tension, enabling rapid gliding and jumping on the water surface [6]. Basilisk lizards utilize a combination of unique foot morphology and rapid slapping motions to generate stable air pockets beneath their feet, allowing them to run across water surfaces, a phenomenon often referred to as "water walking" [7,8]. Geckos, on the other hand, achieve reliable adhesion even on violently oscillating leaves by means of van der Waals forces generated through dense arrays of nanoscale setae on their toe pads [9]. These examples illustrate how organisms develop morphological and behavioral adaptations to meet locomotor challenges posed by substrate mechanical properties. Such adaptive strategies are especially critical when animals interact with dynamic substrates, including those that are elastic or whose effective stiffness changes during locomotion (e.g., through variations in substrate diameter or unpredictable energy return [10–12]). In these contexts, effective locomotion requires animals to synchronize their mechanical output with the phase-specific feedback of the substrate in order to achieve efficient propulsion.

Among various locomotor modes, jumping is particularly sensitive to changes in substrate mechanical properties due to its requirement for high instantaneous power output [13]. This sensitivity is especially evident when the takeoff substrate compliance increases. For example, in click beetles (Elateridae), jump height decreases significantly on more compliant substrates, and a 50% reduction in substrate compliance can significantly limit their ability to jump, leaving them capable only of a weak flipping motion [14]. Similarly, in the diamond dove (*Geopelia cuneata*), takeoff velocity is substantially reduced on compliant perches, which explains their natural preference for stiffer substrates in the wild [15]. In lizards (*Anolis carolinensis*), jumping performance is correlated with substrate compliance, with both takeoff velocity and jump distance declining significantly as perch compliance increases [16]. Mammals also exhibit such sensitivity: tree squirrels display divergent mechanical strategies according to the type of substrate, prioritizing force production on flat ground versus center of mass (COM) displacement on narrower poles. This adjustment is likely related to the coupling of substrate diameter and compliance in natural supports, where compliant substrates may compromise jumping performance by absorbing mechanical work [17]. Adult migratory locusts (*Locusta migratoria*) exhibit greater adaptability, they optimize their takeoff strategy by adjusting the timing of leg muscle activation and modifying the angle of leg extension to match substrate resonance frequencies, thereby not only adjusting their jumping kinematics in response to substrate mechanics but also partially recovering energy from substrate deformation [18]. These interspecific differences highlight the diverse adaptive strategies that have evolved in response to varying substrate compliance conditions.

Tree frogs, as highly specialized three-dimensional jumpers and exceptional adhesive climbers [19–24], inhabit a wide range of plant based substrates that vary from soft leaves (Figure 1) to stiff branches. These substrates differ by orders of magnitude in terms of compliance, surface curvature [20], and dynamic stability. Such complex environments present unique locomotor challenges for tree frogs. They must adapt to rapid changes in substrate compliance, achieve precise landings on vertical or even inverted surfaces [25], and cope with humidity-induced variations in adhesive performance [26]. Although previous studies have indicated that tree frogs can recover substrate energy during takeoff by modulating limb extension [27,28], these studies have primarily focused on behavioral observations, lacking direct mechanical measurements, and have been constrained by a narrow range of substrate compliance. These limitations have restricted a deeper understanding of the underlying mechanisms of takeoff on compliant substrates. To address this gap, we developed a takeoff platform with a broader adjustable stiffness range, integrated with synchronized motion capture, to quantify substrate deformation, elastic energy storage, and associated kinematic and kinetic variables. Specifically, we sought to determine whether tree frogs are still able to recover energy on more compliant substrates and whether peak takeoff forces decrease as substrate compliance increases. This experimental framework provides new insights into the biomechanical strategies underlying adhesive locomotion on compliant substrates and carries important implications for the design of bioinspired climbing and jumping robots.

Figure 1. Tree frog locomotion on compliant substrates, including foliage. (Image source: www.quanjing.com, accessed on 20 June 2021).

2. Materials and Methods

2.1. Performance Testing Apparatus

Six active individuals of the large tree frog (*Polypedates dennysi*) were obtained from commercial suppliers and used in this study. Their average body mass was 71.9 ± 10.6 g and a snout–vent length (SVL) of 9.2 ± 0.5 cm. Each frog was housed individually in a glass enclosure measuring 0.4 m (length) × 0.4 m (width) × 0.5 m (height), equipped with a water source. The animals were fed three times per week with *Gryllus bimaculatus*. The husbandry conditions were maintained at a temperature of 26–29 °C, with a 12-h light/dark cycle and relative humidity ranging from 70% to 90%.

All research on animals was performed under the guidelines of the Chinese Regulations for the Management of Laboratory Animals. The laboratory has obtained long-term ethical approval for animal-related experiments from the Jiangsu Provincial Society for Laboratory Animals Science (Approval No. 2010012, Date: 2010), which remained valid during the conduct of this study. Specifically, the experiments in this study were carried out from July to September 2022, and all operations strictly adhered to the latest ethical regulations and the requirements specified in the above approval.

2.2. Experiment Setup

Figure 2A illustrates the experimental setup employed in this study. The system consists of three main modules: a variable-stiffness takeoff substrate, a multi-camera motion capture system, and a synchronized data acquisition and processing unit. The variable-stiffness substrate is composed of a 12 cm × 13.6 cm carbon fiber plate constrained by four springs of different stiffness values. By adjusting the stiffness of individual springs, the substrate can simulate substrates of varying compliance. In the initial state, the substrate remains in a mechanically balanced state without external forces. When a tree frog initiates a jump from the substrate, the forces applied by its hindlimb cause displacement of the substrate. Owing to the linear mechanical properties of the springs, namely, that force is proportional to deformation, the force exerted on each spring can be calculated by tracking the spatial displacement of its two endpoints in real time. Since the substrate is only subject to the frog's forces and the reactive forces from the springs, the net reaction force from all four springs corresponds to the ground reaction force generated by the frog's feet. Based on this mechanical principle, the system enables real time quantification of the interaction forces between the frog's feet and the compliant substrate during takeoff. Let K denote the stiffness coefficient of the springs, and let A_1, A_2, A_3, and A_4 represent the spatial coordinates of the fixed ends of the four springs, which are attached to the base frame, while the other ends are connected to the corners of the compliant substrate. The corresponding coordinates of the movable marker points located at the four corners of the compliant substrate are denoted as M_1, M_2, M_3, and M_4. The force vectors for the four springs are then given by:

$$\begin{cases} \vec{F_1} = K(A_1 - M_1) \\ \vec{F_2} = K(A_2 - M_2) \\ \vec{F_3} = K(A_3 - M_3) \\ \vec{F_4} = K(A_4 - M_4) \end{cases} \quad (1)$$

Therefore, substrate reaction force can be defined as:

$$F = \vec{F_1} + \vec{F_2} + \vec{F_3} + \vec{F_4} - G \quad (2)$$

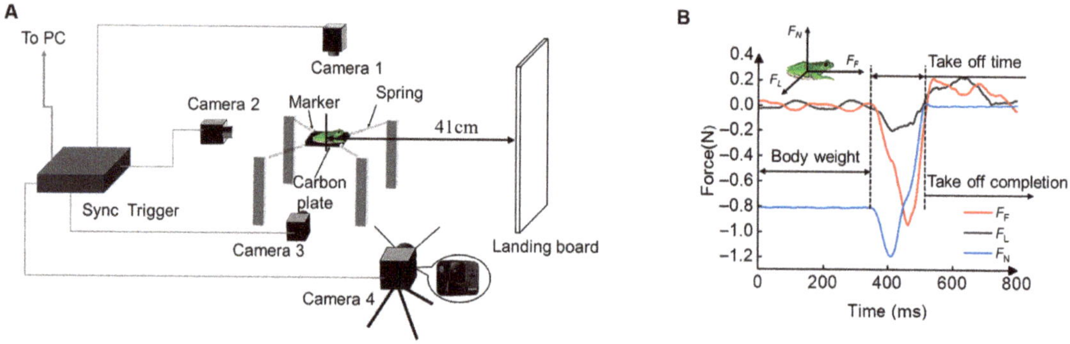

Figure 2. Measurement of jumping performance in tree frogs from substrates of various compliance. (**A**) Diagram of the experimental setup showing the location of the jump substrate and cameras used to record takeoff behavior. (**B**) A typical example of substrate reaction force, calculated using the force measurement system, during takeoff is shown, where F_F represents the fore–aft force, F_L denotes the lateral force, and F_N indicates the normal force. The initial normal force corresponds to the body weight of the tree frog, and after takeoff, the compliant substrate oscillates around zero.

Here, G denotes the gravitational force vector generated by the weight of the substrate, which acts vertically downward and can be determined in advance. This approach enables not only the measurement of the resultant substrate reaction force exerted by the frog's feet, but also the real-time determination of its direction.

In this study, to systematically modulate the stiffness of the takeoff substrate, springs with varying specifications were selected to achieve the target stiffness coefficients. We initially selected six types of commercially available springs, whose wire diameter multiplied by outer diameter yielded the following combinations: 0.7 mm × 3 mm, 0.6 mm × 3 mm, 0.5 mm × 3 mm, 0.5 mm × 4 mm, 0.3 mm × 5 mm, and 0.3 mm × 2 mm, all maintaining a consistent length of 27 mm. These springs were arranged in descending order of stiffness, with calculated stiffness coefficients of 6.34 N/cm, 2.58 N/cm, 0.87 N/cm, 0.32 N/cm, 0.11 N/cm, and 0.03 N/cm, respectively, demonstrating an approximately threefold difference between adjacent groups. In our preliminary experiments, we observed a critical threshold: when spring stiffness was ≤0.11 N/cm, tree frogs were unable to execute successful jumps. Although the frogs initiated takeoff movements, they consistently failed to detach their toe pads from the substrate. This phenomenon resulted from excessive substrate compliance, which prevented the generation of sufficient propulsive force for complete detachment. Therefore, although all six spring configurations were initially tested, only those with stiffness coefficients $\geq$0.32 N/cm were included in the final experimental protocol to ensure reliable jump performance. The overall stiffness of the takeoff substrate was then calculated accordingly, based on the combined stiffness of the selected springs.

The vertical stiffness calculation model is illustrated in Figure 3A. Suppose the substrate is subjected to a vertical load $F_{Vertical}$, resulting in a vertical displacement Δx. Let the original length of the spring be l_0, and its length after deformation be l_1. Then, the following relationship holds:

$$l_1 = \sqrt{\Delta x^2 + l_0^2} \tag{3}$$

In addition:

$$\begin{cases} F_{Spring} = K(l_1 - l_0) \\ F_{Spring}\cos\theta = F_{Vertical}/4 \\ \cos\theta = \Delta x / l_1 \end{cases} \tag{4}$$

According to Equations (3) and (4):

$$F_{Vertical} = 4K\Delta x\left(1 - \frac{l_0}{\sqrt{\Delta x^2 + l_0^2}}\right) \tag{5}$$

So,

$$K_{Vertical} = F_{Vertical}/\Delta x = 4K\left(1 - \frac{l_0}{\sqrt{\Delta x^2 + l_0^2}}\right) \tag{6}$$

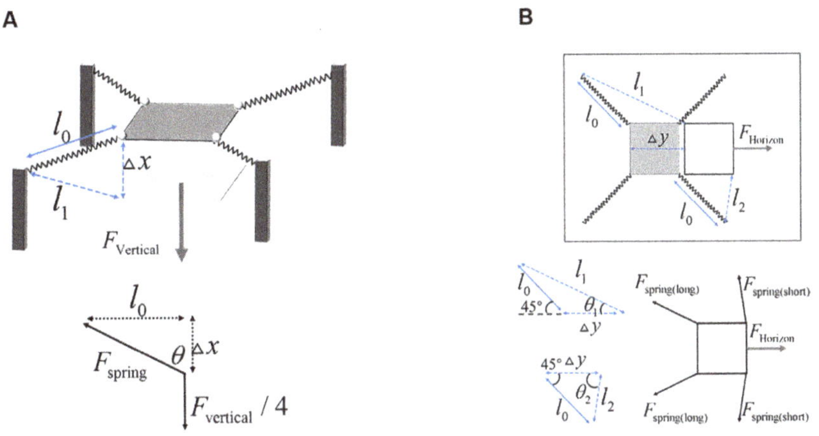

Figure 3. Stiffness calculation of the compliant takeoff substrate. (**A**) Illustrates the deformation of the substrate under vertical loading. The relationship between the deformed spring length and the vertical displacement can be derived based on geometric principles, allowing the calculation of spring force as a function of vertical displacement. Force analysis yields the relationship between the applied vertical load and the resulting vertical displacement. (**B**) Top view deformation of the substrate under horizontal loading. The gray square represents the initial (unloaded) position of the substrate, while the transparent square indicates its deformed state under load. Due to asymmetrical deformation, the springs on the side closer to the direction of applied force deform differently from those on the opposite side. The deformed lengths of these springs are computed separately based on geometric constraints. Force analysis is then used to establish the relationship between horizontal load and horizontal displacement.

The horizontal stiffness calculation model is shown in Figure 3B. When a horizontal force is applied to the substrate, it induces a horizontal displacement Δy. Under this loading condition, the springs on the side closer to the direction of the applied force undergo compression, while those on the opposite side experience tension. Therefore, the geometric relationships for spring deformation must be calculated separately for the two cases. Let the original (unloaded) length of each spring be l_0. After loading, the length of the spring on the side closer to the applied force is l_1, and the length of the spring on the far side is l_2. The following relationships then hold:

$$\begin{cases} l_1 = \sqrt{\frac{l_0^2}{2} + (\Delta y + \frac{l_0}{\sqrt{2}})^2} \\ l_2 = \sqrt{\frac{l_0^2}{2} + (\Delta y - \frac{l_0}{\sqrt{2}})^2} \end{cases} \quad (7)$$

Let the force exerted on the spring located on the side closer to the direction of loading be denoted as $F_{Spring(long)}$, and the force on the spring on the opposite side as $F_{Spring(short)}$. Then, the relationship between spring deformation and the corresponding forces can be expressed as follows:

$$\begin{cases} F_{Spring(long)} = K(l_1 - l_0) \\ F_{Spring(short)} = K(l_2 - l_0) \\ \tan\theta_1 = \frac{l_0}{\sqrt{2}\Delta y + l_0} \\ \tan\theta_2 = \frac{l_0}{\sqrt{2}\Delta y - l_0} \\ 2F_{Spring(long)}\cos\theta_1 + 2F_{Spring(short)}\cos\theta_2 = F_{Horizon} \end{cases} \quad (8)$$

The compliance characteristics of the substrate, as determined through the model calculations, are summarized in Table 1.

Table 1. Compliance calculation results of the substrate.

Spring Stiffness (N/cm)	Horizontal Stiffness (N/cm)	Vertical Stiffness (N/cm)
6.34	11.41	2.33
2.58	4.64	0.96
0.87	1.56	0.37
0.32	0.57	0.18
0.11	0.19	0.11
0.03	0.04	0.08

To ensure that the performance testing apparatus allowed us to accurately estimate the substrate reaction forces generated during tree frog jumps, its measurement accuracy was evaluated before experimentation. First, the original positions of the marker points were recorded using the multi-camera system. Then, a tree frog with a known body weight was gently placed at the center of the force measurement substrate. After the frog remained stationary and the substrate stabilized, the positions of the marker points were captured again using the camera system. Based on the previously described force calculation method, the average reaction force exerted on the substrate over a short period of time was computed. This procedure was repeated three times for each substrate configuration with different spring stiffness combinations. The vertical ground reaction forces measured in the three trials were averaged, and the standard deviation was calculated. Finally, the measured forces were normalized by the body weight of the frog to obtain unitized values. The summarized results are presented in Table 2.

Three high-speed cameras (Cameras 1–3; Blackfly, Teledyne FLIR, Wilsonville, OR, USA) with a frame rate of 425.17 FPS were used to synchronously record substrate motion, while Camera 4 (GoPro Hero 8; GoPro, Inc., San Mateo, CA, USA) recorded the jumping behavior of the tree frogs at 239.76 FPS. In the experiments, a soft sponge board (1 m × 0.6 m × 0.1 m), positioned 41 cm from the center of the takeoff substrate, was vertically fixed to the ground as the landing substrate to reduce impact forces and minimize the risk of injury during tree frog landings (Figure 2A). Preliminary tests revealed that tree frogs prefer to land on darker surfaces. Therefore, black paper of equal size was affixed to the sponge surface to guide the frogs' landing behavior.

Table 2. Measurement accuracy of substrate reaction forces.

Spring Stiffness (N/cm)	Load (BW)	Measurement Result (BW)	Relative Error
6.34	1	0.89 ± 0.02	11%
2.58	1	0.96 ± 0.04	4%
0.87	1	0.91 ± 0.01	9%
0.32	1	0.87 ± 0.01	13%

Note: BW represents the body weight of the tree frog. The measurement results are presented as the mean $\pm$ standard deviation of three repeated trials.

Each frog was subjected to a maximum of three trials every 2 days. The frogs were placed on the takeoff substrate and then encouraged to jump by slightly touching their anus with a writing brush. If a tree frog's jump trajectory deviates from the camera's tracking plane, the corresponding experimental data will be excluded from analysis. To ensure data reliability, each frog was required to complete at least seven valid jumps on each compliant substrate.

2.3. Analyses

Video Data Processing: The recorded videos were used to calculate the reaction forces of the takeoff substrate and to capture the complete jumping sequence of the tree frogs. Accurate tracking was achieved by placing markers at the four corners of the substrate, both ends of each spring, and near the frog's COM. First, the motion trajectories of the marker points were tracked using the DLTdv8 (version 8.2.0) in MATLAB R2020a (MathWorks, Natick, MA, USA) [29]. A combination of automated and manual tracking was employed: when markers were occluded or located at the edge of the video frame, manual frame correction was applied. The processed trajectory data were then imported into MATLAB (R2020a, MathWorks, Natick, MA, USA) for kinematic and dynamic analysis.

Force estimation and processing based on marker trajectories: Before the experiments, the spatial coordinates of the marker points and the fixed ends of the springs were measured in the unloaded state of the substrate. These data were used to construct a physical model of the substrate in a global coordinate system. Camera calibration was also performed to establish the relationship between pixel coordinates and real-world distances. During substrate reaction force calculation, the 3D positions of the four marker points on the carbon fiber plate were determined based on their displacements from their initial positions. Using Equations (1) and (2), the resulting force vectors acting on the substrate were decomposed into fore–aft, normal, and lateral components (Figure 2B). The time window for substrate reaction force analysis was defined from the onset of marker movement to the moment the frog's feet completely left the carbon fiber plate. The raw force data were processed using a Butterworth low-pass filter to eliminate high-frequency noise. The time-varying profile of the takeoff forces on compliant substrates is illustrated in Figure 2B. It is worth noting that although small peaks in lateral force were occasionally observed—likely due to asynchronous hindlimb extension or non-linear jumping trajectories—these were relatively minor compared to the fore–aft and normal forces. Therefore, lateral force peaks were not included in the subsequent peak force comparison analysis.

Kinematic Data Processing: In studies of tree frog jumping, the hip (or vent) is commonly used as a proxy for the COM [30,31]. Based on this approximation, we calculated takeoff velocity, takeoff duration, and takeoff angle; Takeoff duration was defined as the time interval from the initiation of movement to the complete detachment of the hindlimbs from the substrate; Takeoff velocity was calculated as the velocity of the COM at the moment when the frog's toes lost contact with the substrate, calculated as the displacement between adjacent frames divided by the time interval; Takeoff angle was calculated as the angle between the linear fit of the COM trajectory over the last 10 frames of takeoff (approximately 27 ms) and the horizontal plane. Kinematic trajectories were smoothed using polynomial spline functions. To eliminate potential confounding effects of the landing substrate on jumping performance, only trials in which the frog landed above the substrate height were included in the final analysis. The push-off phase is defined as the period from the initiation of movement by the tree frog on the substrate to the moment when the hindlimb toes completely lose contact with the substrate. This process is further divided into two distinct phases: the loading phase and the unloading phase. The loading phase refers to the interval from the onset of movement to the point at which the substrate reaches its maximum deformation. The unloading phase spans from the moment of maximum substrate deformation to the instant the frog's toes lift off the substrate.

To quantify the energy exchange between the tree frog and the substrate, we calculated the energy components of the system.

Total System Energy:

$$E_{sys} = E_{frog} + E_{spring} + E_{substrate\ motion} \tag{9}$$

Here, E_{sys} represents the total energy of the system; E_{frog} is the sum of the tree frog's kinetic energy and gravitational potential energy; E_{spring} denotes the elastic potential energy stored in the springs; and $E_{substrate\ motion}$ represents the sum of the kinetic and gravitational potential energy of the substrate. The change in the mechanical energy of the tree frog can be calculated using the following equation:

$$E_{frog} = \frac{1}{2} m_{frog} v_{frog}^2 + m_{frog} g h_{frog} \tag{10}$$

Here, m_{frog} is the mass of the tree frog, and v_{frog} is the magnitude of the velocity in the direction of the COM motion. The takeoff velocity is defined as the instantaneous velocity measured in the first frame after the frog's hindlimbs completely leave the ground. It is calculated as the displacement of the COM between the frame before and after takeoff divided by the time interval between those frames. g is the acceleration due to gravity, and h is the vertical displacement of the COM from the onset of motion to the moment the toes leave the substrate. The elastic potential energy stored in the substrate is given by:

$$E_{spring} = \frac{1}{2} k(\Delta x_1)^2 + \frac{1}{2} k(\Delta x_2)^2 + \frac{1}{2} k(\Delta x_3)^2 + \frac{1}{2} k(\Delta x_4)^2 \tag{11}$$

Here, k is the spring stiffness, Δx_i represents the deformation of the i = 1, 2, 3, 4, and there are four springs in total. The kinetic and potential energy of the moving substrate is given by:

$$E_{substrate\ motion} = \frac{1}{2} m_{substrate} v_{substrate}^2 + m_{substrate} g h_{substrate} \tag{12}$$

Here, $m_{substrate}$ is the mass of the substrate (14 g in this study), $v_{substrate}$ is the velocity of the substrate, g is the acceleration due to gravity, and $h_{substrate}$ is the vertical displacement of the diagonal intersection point formed by the four substrate markers, measured from the onset of motion to the moment when the frog's toes leave the substrate.

The total energy of the substrate, $E_{substrate}$ is the sum of its elastic potential energy and motion-related mechanical energy:

$$\eta = \frac{E_{substrate,\max} - E_{substrate,toeoff}}{E_{substrate,\max}} \tag{13}$$

Here, $E_{substrate,\max}$ represents the energy of the substrate at the point of maximum elastic deformation, and $E_{substrate,toeoff}$ denotes the substrate energy at the moment the frog's toes leave the substrate.

At the onset of the jump, all energy values were initialized to zero. Power and energy were normalized by the mass of the frog and the substrate, forces were normalized by the frog's body weight, and substrate displacement was normalized by the frog's SVL. In the unloaded state of the substrate, the spatial positions of the four fluorescent markers attached to the corners of the carbon fiber plate and the fixed ends of the four springs were recorded to establish a physical model of the substrate in the global coordinate system. Before experimentation, the cameras were calibrated to determine the pixel-to-length ratio in the world coordinate system. The displacement of the substrate and the ground reaction forces during takeoff were calculated by tracking the positions of the four markers on the carbon fiber plate and computing their spatial displacements (Δx, Δy, Δz). The diagonal intersection point of the four markers was used to determine the overall movement of the substrate.

Statistical Analysis: A mixed-effects ANOVA model was used to compare the frogs' performance across substrates with different compliance levels, while accounting for individual variation and jump sequence as random effects. Fixed effects included substrate compliance and trial order, and random effects accounted for individual differences among frogs. For indices with significant main effects, pairwise comparisons between groups were conducted using Bonferroni-corrected post hoc tests, with the significance level set at $\alpha = 0.05$. All statistical analyses were performed using SPSS 26.0 software (IBM, Armonk, NY, USA).

3. Results

3.1. Substrate Parameter Analysis

During the push-off phase, the substrate, constrained by springs, exhibited both vertical and horizontal displacements. The vertical displacement component, generated by the vertical substrate reaction force from the limbs, ranged from 2% to 58% of the SVL, as shown in Figure 4A. Mixed-effects model analysis revealed a significant main effect of substrate compliance on vertical displacement during takeoff ($F_{(3,92)} = 12.895$, $p < 0.001$). The substrate with spring stiffness of 0.32 N/cm produced significantly greater vertical displacement (19.47 $\pm$ 6.13% SVL, $p = 0.002$) than that with spring stiffness of 6.34 N/cm. Notably, the substrate with spring stiffness of 0.87 N/cm yielded the greatest vertical displacement (27.74 $\pm$ 4.75% SVL), which was significantly greater than that at all other stiffness levels ($p \leq 0.007$). The horizontal displacement component, caused by the backward propulsive force, ranged from 0.43% to 40.57% SVL. Statistical analysis showed that substrate compliance had a highly significant effect on horizontal displacement during takeoff ($F_{(3,92)} = 35.019$, $p < 0.001$). Post hoc comparisons revealed significant differences among spring stiffness conditions. The substrate with spring stiffness of 0.32 N/cm exhibited the greatest horizontal displacement (23.15 $\pm$ 2.32% SVL, $p < 0.001$), followed by those with stiffness values of 0.87 N/cm (8.03 $\pm$ 1.80% SVL) and 2.58 N/cm (2.48 $\pm$ 2.30% SVL), while the substrate with 6.34 N/cm yielded the least displacement.

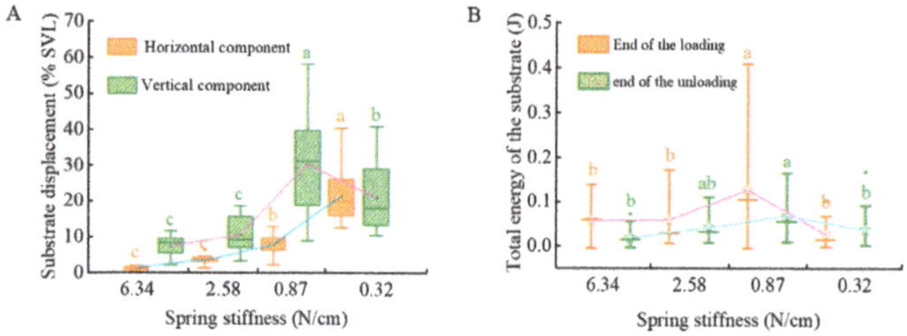

Figure 4. Effects of substrate compliance on substrate motion. (**A**) Substrate displacement; (**B**) Total substrate energy (sum of kinetic and potential energy). Spring stiffness is shown in descending order, with sample sizes of 37, 17, 28, and 18, respectively. Significant comparisons are indicated with a, b, ab and c; means not sharing the same letters are significantly different ($p < 0.05$).

The mechanical energy of the substrate is shown in Figure 4B. At the end of the loading phase, the substrate with a spring stiffness of 0.87 N/cm exhibited significantly higher mechanical energy (0.123 ± 0.013 J) compared to the substrate with spring stiffness of 0.32 N/cm (0.025 ± 0.015 J, $p < 0.001$), 2.58 N/cm (0.059 ± 0.016 J, $p = 0.015$) and 6.34 N/cm (0.058 ± 0.011 J, $p = 0.001$). No significant difference was found among the substrate with spring stiffnesses of 0.32 N/cm, 2.58 N/cm and 6.34 N/cm (all $p > 0.05$). During the unloading phase, the substrate with spring stiffness of 0.87 N/cm again exhibited the highest mechanical energy (0.067 ± 0.007 J), which was significantly greater than that of the substrates with spring stiffness of 0.32 N/cm (0.038 ± 0.008 J, $p = 0.040$) and 6.34 N/cm (0.019 ± 0.006 J, $p < 0.001$). The substrate with spring stiffness 2.58 N/cm (0.043 ± 0.008 J) did not differ significantly from the other conditions ($p > 0.05$).

3.2. Kinematic Parameter Analysis of Tree Frogs

Mixed-effects model analysis showed that substrate compliance had no significant effect on multiple kinematic parameters of tree frogs' jumping, specifically including: take-off time ($F_{(3,92)} = 0.485$, $p = 0.694$, Figure 5A), the ratio of loading time to unloading time ($F_{(3,92)} = 1.025$, $p = 0.385$, Figure 5B), and takeoff angle ($F_{(3,92)} = 1.901$, $p = 0.135$, Figure 5D).

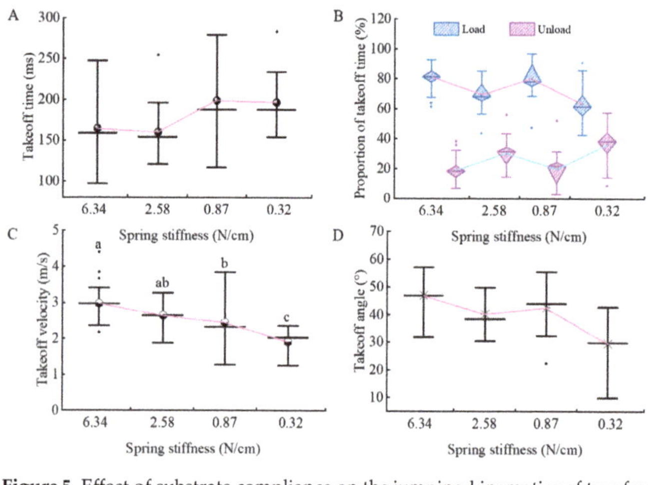

Figure 5. Effect of substrate compliance on the jumping kinematics of tree frogs. (**A**) Takeoff time; (**B**) Proportion of loading and unloading phases relative to the total takeoff time; (**C**) Takeoff velocity and (**D**) Takeoff angle. Spring stiffness is shown in descending order, with sample sizes of 37, 17, 28, and 18, respectively. Significant comparisons are indicated with a, b, ab and c; means not sharing the same letters are significantly different ($p < 0.05$).

Only takeoff velocity was significantly affected by the main effect ($F_{(3,92)} = 3.113$, $p = 0.030$, Figure 5C). Bonferroni-corrected post hoc comparisons revealed that the takeoff velocity on the 0.32 N/cm substrate (1.908 ± 0.126 m/s) was significantly lower than those on the other substrates: 0.87 N/cm (2.452 ± 0.101 m, $p = 0.007$), 2.58 N/cm (2.649 ± 0.130 m/s, $p = 0.001$), and 6.34 N/cm (2.976 ± 0.088 m/s, $p < 0.001$). Additionally, takeoff velocity on the 0.87 N/cm substrate was significantly lower than that on the 6.34 N/cm substrate ($p = 0.001$). These results indicate that as the substrate compliance increases, the takeoff speed decreases.

3.3. Dynamic Parameter Analysis of Tree Frogs

Mixed-effects model analysis revealed that substrate compliance had no significant effect on the total mechanical energy of tree frogs' jumps ($F_{(3,92)} = 2.488$, $p = 0.065$, Figure 6A), but had a significant effect on their peak power during push-off ($F_{(3,92)} = 3.851$, $p = 0.012$ Figure 6B). Compared to the 6.34 N/cm substrate, the 0.32 N/cm substrate led to a significant reduction in jump power ($p = 0.002$), and a similar reduction was observed with the 0.87 N/cm substrate ($p = 0.020$).

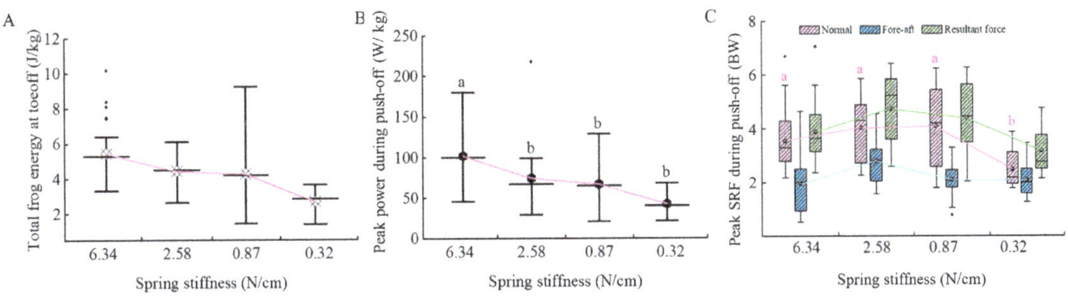

Figure 6. Dynamic parameters during takeoff in tree frogs. (**A**) Takeoff mechanical energy; (**B**) Peak power during push-off and (**C**) Peak SRF during push-off, SRF stands for Substrate Reaction Force. Spring stiffness is shown in descending order, with sample sizes of 37, 17, 28, and 18, respectively. Significant comparisons are indicated with a and b; means not sharing the same letters are significantly different ($p < 0.05$).

During push-off, tree frogs generate substrate reaction forces up to 6 times their body weight in the normal direction and approximately 4.5 times in the fore-aft direction, as illustrated in Figure 6C. Statistical analysis revealed that substrate compliance had a significant main effect on peak normal force ($F_{(3,92)} = 2.923$, $p = 0.038$). Bonferroni-corrected post hoc comparisons showed that the 0.32 N/cm substrate produced a significantly lower peak normal force (2.485 ± 0.259 BW) than the other substrates, including 0.87 N/cm (4.188 ± 0.214 BW, $p < 0.001$), 2.58 N/cm (4.030 ± 0.275 BW, $p = 0.001$) and 6.34 N/cm (3.554 ± 0.186 BW, $p = 0.007$). However, substrate compliance had no significant main effect on either peak fore-aft force ($F_{(3,92)} = 2.255$, $p = 0.087$) or peak resultant force ($F_{(3,92)} = 1.625$, $p = 0.189$).

4. Discussion

We investigated how the compliance of the horizontal launching substrate affects the jumping performance of tree frogs when landing on vertical substrate. Six types of commercially available springs with stiffness values of 6.34, 2.58, 0.87, 0.32, 0.11, and 0.03 N/cm were selected, with each adjacent pair demonstrating an approximately threefold difference in stiffness. Our preliminary experiments revealed a critical threshold: when spring stiffness was ≤0.11 N/cm, tree frogs were unable to execute successful jumps. Although the frogs initiated takeoff movements, they consistently failed to detach their toe pads from the substrate. This phenomenon resulted from excessive substrate compliance, which prevented the generation of sufficient propulsive force for complete detachment. A similar failure pattern has also been reported in click beetles (Elateridae) when attempting to jump on overly compliant substrates [14], suggesting a common biomechanical constraint among jumping animals.

Across successful takeoffs (stiffness $\geq$ 0.32 N/cm), substrate compliance clearly influenced both frog movements and substrate dynamics, yet only horizontal substrate displacement and frog takeoff velocity exhibited statistically significant changes across the full range of compliant substrates. Notably, the trend in takeoff velocity aligns with previous findings [28], highlighting the general effect of substrate compliance on amphibian jumping performance. While total mechanical energy and peak power at toe detachment showed a downward trend with increasing compliance, these changes were not statistically significant.

Although takeoff angle was not statistically significant, the takeoff angle on the 0.32 N/cm substrate was smaller than that observed on the 6.34 N/cm substrate. Mechanical modeling suggests that on rigid substrates, the direction of the substrate reaction force is consistently aligned with the COM, minimizing torque around the COM. However, on compliant substrates, deformation causes the substrate contact point to shift posteriorly, displacing the substrate reaction force vector behind the COM. This displacement generates a moment arm and induces a net clockwise torque, thereby reducing the inclination of the body's longitudinal axis (Figure 7A). The greater the compliance, the more pronounced this mechanical effect becomes. Also, takeoff time was not statistically significantly affected by substrate compliance, the takeoff time on the 0.32 N/cm substrate was longer than that on the 0.87 N/cm substrate. A plausible explanation for this is that the flexible substrate disrupts the inertial capture mechanism through which tree frogs store elastic energy in their tendons, thereby increasing the duration of takeoff [27].

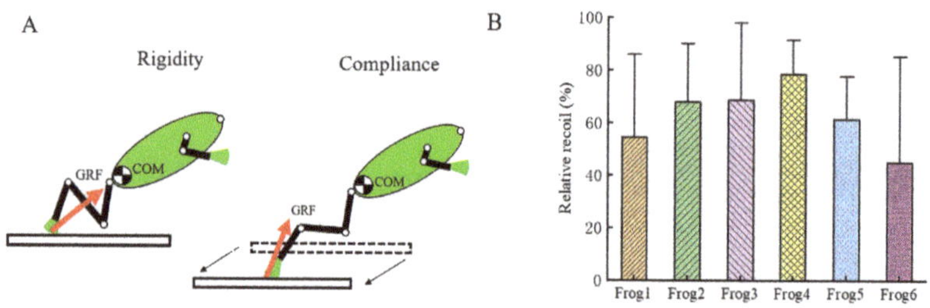

Figure 7. (**A**) Mechanical model of the effect of substrate compliance on takeoff angle and (**B**) Energy efficiency. The number of jumps for each tree frog was 6, 5, 6, 6, 6, and 7 respectively.

A distinct turning point was observed at the substrate stiffness of 0.87 N/cm. Parameters including vertical substrate displacement, total energy, as well as the frog's takeoff time, takeoff angle, and takeoff force all exhibited this shift. A plausible explanation is that, in this experiment, 0.87 N/cm may represent a critical biomechanical boundary in the interaction between the tree frog's locomotor system and the mechanical properties of the substrate. From an energy transfer perspective, substrates stiffer than 0.87 N/cm can provide sufficient rigidity to minimize energy loss due to excessive deformation, while still maintaining enough compliance to avoid harmful rebound forces.

Our study supports the findings of Astley et al. [28] that tree frogs performing static jumps on compliant substrates can recover a small portion of the energy stored in the substrate, as illustrated in Figure 7B. On a substrate with a spring stiffness of 6.34 N/cm, tree frogs recovered up to 50% of the stored elastic energy during takeoff. However, this energy recovery efficiency exhibited a pronounced dependence on substrate compliance and was only observed at the 6.34 N/cm spring stiffness level. As substrate compliance increased, the static loading caused by the frog's body weight led to a significant initial deformation of the substrate, substantially reducing the amount of energy that could be stored during the loading phase. During the unloading phase, the substrate's rebound exceeded the equilibrium position, resulting in dynamic overshoot that disrupted the energy balance. This nonlinear dynamic response indicates that traditional efficiency calculations based on energy conservation are no longer applicable under conditions of very high substrate compliance. When spring stiffness decreased to 0.11 N/cm, although the frogs could still perform the takeoff motion, the substrate was so compliant that their adhesive pads failed to fully detach from the surface. This suggests that the reactive forces generated by the substrate were insufficient to overcome gravity and fully propel the body into the air.

Force analysis further revealed that tree frogs can generate peak normal reaction forces up to 6 times their body weight and fore-aft forces of approximately 4.5 times their body weight during takeoff. This indicates that their robust musculoskeletal system has evolved to produce exceptionally high force relative to body mass, ensuring sufficient propulsion even when taking off from complex substrates. Moreover, the coordinated functions of the normal reaction force (to overcome gravity and initiate take-off) and the fore–aft force (to provide horizontal thrust and assist acceleration) enable the effective production of powerful jumps, precisely supporting their survival demands for crossing branch gaps and evading predator. Our further analysis revealed that substrate compliance had a significant effect on peak normal force, with the value at a substrate stiffness of 0.32 N/cm being significantly lower than those at all stiffer substrates. However, this reduction did not follow a progressive monotonic decrease with increasing compliance. Instead, the peak normal force was observed at a substrate stiffness of 0.87 N/cm. A possible explanation is that this intermediate stiffness provides an optimal balance between force transmission efficiency and substrate feedback, allowing the frogs to effectively exert vertical force and generate greater substrate reaction forces.

These findings offer new biomechanical insights into how adhesive jumping animals interact with compliant environments. The frogs are able to recover part of the energy from the compliant substrate, and their superior behavioral performance on substrates with intermediate to high stiffness, as indicated by higher take-off velocities, may represent an adaptive strategy that facilitates reliable detachment and effective propulsion. This helps explain the evolutionary preference of many climbing and jumping animals for stiffer substrates during locomotion [16,32–34]. Moreover, the principles uncovered in this study could be extended to the design of bioinspired systems, particularly in the context of material origami technologies. Recent advancements in bioinspired origami structures, such as origami springs and dual-morphing stretchable materials, highlight the potential for substrates with variable compliance to enhance energy efficiency and adaptive response in engineered systems [35,36]. These materials can be designed to mimic the compliance behavior observed in tree frogs, enabling the creation of multifunctional systems that respond dynamically to external forces. Furthermore, this research could play a key role in the miniaturization of bioinspired mechanical mechanisms, particularly for microrobot applications. As demonstrated in recent work on soft robotics, the interaction between compliant surfaces and adhesive mechanisms can be critical for achieving efficient locomotion and energy recovery in small scale systems [37]. By leveraging insights from tree frog locomotion, future bio-inspired microrobots could be designed to optimize force generation, energy transfer, and adaptability in complex, variable environments. These considerations open exciting avenues for the development of advanced bio-electronic systems and soft robotics, where material compliance plays a crucial role in both performance and efficiency.

5. Conclusions

Tree frogs demonstrate the ability to make behavioral decisions based on varying environmental conditions. This study systematically investigated the jumping dynamics and kinematics of frogs (*Polypedates dennysi*) on six substrate compliances to clarify the biomechanical mechanisms underlying their adaptation to variable arboreal substrates. The findings reveal that substrate compliance influences energy recovery during takeoff; on the stiffest substrates, tree frogs recovered up to 50% of the energy stored in the substrate, reflecting efficient energy utilization, while highly compliant substrates induced nonlinear damping, substantial energy dissipation, and even takeoff failure, indicating a critical threshold beyond which effective locomotion is compromised. Distinct adaptive patterns were observed between force generation and takeoff velocity. During takeoff, frogs produced peak normal forces up to six times their body weight and fore–aft forces up to 4.5 times; however, these force outputs exhibited limited adaptability to substrate compliance. By contrast, takeoff velocity showed greater resilience to compliance variation, suggesting a strategy that prioritizes maintaining movement efficiency over force regulation under substrate variability. Collectively, these results reveal a trade-off between substrate mechanics and jump performance, underscoring the evolutionary sensitivity of jumping animals to the physical properties of their environment. Furthermore, these findings provide valuable biomechanical insights for the development of bioinspired systems capable of efficient locomotion under non-ideal conditions.

Author Contributions: Conceptualization, R.Z., B.Z., Z.W. and Z.D.; methodology, R.Z. and Z.W.; software, R.Z. and B.Z.; validation, R.Z. and B.Z.; formal analysis, R.Z. and B.Z.; investigation, R.Z. and B.Z.; resources, Z.D. and Z.W.; data curation: R.Z. and B.Z.; writing—original draft preparation, R.Z.; writing—review and editing: R.Z. and B.Z.; visualization, R.Z.; supervision, Z.W.; project administration, Z.W.; funding acquisition, Z.W. and Z.D. All authors have read and agreed to the published version of the manuscript.

Funding: The National Natural Science Foundation of China (62233008), Guangdong Basic and Applied Basic Research Foundation (2023A1515010773), the Basic Research Program of Shenzhen (JCYJ20210324122812033) and the International Science and Technology Cooperation Project of the China Electronics Technology Group Corporation.

Institutional Review Board Statement: All research on animals was performed under the guidelines of the Chinese Regulations for the Management of Laboratory Animals and was approved by the Jiangsu Provincial Society for Laboratory Animals Scientific (Approval No. 2010012, Date: 2010).

Informed Consent Statement: Not applicable.

Data Availability Statement: The data generated and/or analyzed in the current study are not publicly available for legal/ethical reasons but are available from the corresponding author upon reasonable request.

Conflicts of Interest: The author declares no conflicts of interest.

Abbreviations

The following abbreviations are used in this manuscript:

COM	Center of Mass
SVL	Snout–Vent Length
BW	Body Weight

References

1. Blob, R.W.; Diamond, K.M.; Lagarde, R.; Maie, T.; Moody, K.N.; Palecek, A.M.; Ward, J.L.; Schoenfuss, H.L. Integrating Biomechanics in Evolutionary Studies, with Examples from the Amphidromous Goby Model System. *J. Exp. Biol.* **2023**, *226*, jeb244942. [CrossRef] [PubMed]
2. Gregorič, M.; Agnarsson, I.; Blackledge, T.A.; Kuntner, M. How Did the Spider Cross the River? Behavioral Adaptations for River-Bridging Webs in *Caerostris darwini* (Araneae: Araneidae). *PLoS ONE* **2011**, *6*, e26847. [CrossRef]
3. Rothier, P.S.; Fabre, A.; Benson, R.B.J.; Martinez, Q.; Fabre, P.; Hedrick, B.P.; Herrel, A. Of Flippers and Wings: The Locomotor Environment as a Driver of the Evolution of Forelimb Morphological Diversity in Mammals. *Funct. Ecol.* **2024**, *38*, 2231–2246. [CrossRef]
4. Ostridge, H.J.; Fontsere, C.; Lizano, E.; Soto, D.C.; Saxena, V.; Alvarez-Estape, M.; Barratt, C.D.; Gratton, P.; Bocksberger, G.; Lester, J.D.; et al. Local Genetic Adaptation to Habitat in Wild Chimpanzees. *Science* **2025**, *387*, eadn7954. [CrossRef]
5. Korff, W.L.; McHenry, M.J. Environmental Differences in Substrate Mechanics Do Not Affect Sprinting Performance in Sand Lizards (*Uma scoparia* and *Callisaurus draconoides*). *J. Exp. Biol.* **2011**, *214*, 122–130. [CrossRef]
6. Liu, J.; Shen, M.; Ma, Z.; Zhou, X. Research Progress on Bionic Water Strider Robots. *J. Bionic. Eng.* **2024**, *21*, 635–652. [CrossRef]
7. Yao, C.; Zhou, Y.; Wang, J.; Jiang, L. Bioinspired Universal Approaches for Cavity Regulation during Cylinder Impact Processes for Drag Reduction in Aqueous Media: Macrogeometry Vanquishing Wettability. *ACS Appl. Mater. Interfaces* **2021**, *13*, 38808–38815. [CrossRef]
8. Bush, J.W.M.; Hu, D.L. WALKING ON WATER: Biolocomotion at the Interface. *Annu. Rev. Fluid Mech.* **2006**, *38*, 339–369. [CrossRef]
9. Voigt, D.; De Souza, E.J.; Kovalev, A.; Gorb, S. Inter- and Intraspecific Differences in Leaf Beetle Attachment on Rigid and Compliant Substrates. *J. Zool.* **2019**, *307*, 1–8. [CrossRef]
10. Thorpe, S.K.S.; Crompton, R.H.; Alexander, R.M. Orangutans Use Compliant Branches to Lower the Energetic Cost of Locomotion. *Biol. Lett.* **2007**, *3*, 253–256. [CrossRef]
11. Higham, T.E.; Russell, A.P.; Niklas, K.J. Leaping Lizards Landing on Leaves: Escape-Induced Jumps in the Rainforest Canopy Challenge the Adhesive Limits of Geckos. *J. R. Soc. Interface* **2017**, *14*, 20170156. [CrossRef]
12. Byrnes, G.; Jayne, B.C. Substrate Diameter and Compliance Affect the Gripping Strategies and Locomotor Mode of Climbing Boa Constrictors. *J. Exp. Biol.* **2010**, *213*, 4249–4256. [CrossRef] [PubMed]
13. Kuo, C.-Y.; Gillis, G.B.; Irschick, D.J. Loading Effects on Jump Performance in Green Anole Lizards, *Anolis carolinensis*. *J. Exp. Biol.* **2011**, *214*, 2073–2079. [CrossRef] [PubMed]
14. Ribak, G.; Reingold, S.; Weihs, D. The Effect of Natural Substrates on Jump Height in Click-beetles. *Funct. Ecol.* **2012**, *26*, 493–499. [CrossRef]
15. Crandell, K.E.; Smith, A.F.; Crino, O.L.; Tobalske, B.W. Coping with Compliance during Take-off and Landing in the Diamond Dove (*Geopelia cuneata*). *PLoS ONE* **2018**, *13*, e0199662. [CrossRef]
16. Gilman, C.A.; Irschick, D.J. Foils of Flexion: The Effects of Perch Compliance on Lizard Locomotion and Perch Choice in the Wild. *Funct. Ecol.* **2013**, *27*, 374–381. [CrossRef]
17. Boulinguez-Ambroise, G.; Dunham, N.; Phelps, T.; Mazonas, T.; Nguyen, P.; Bradley-Cronkwright, M.; Boyer, D.M.; Yapuncich, G.S.; Zeininger, A.; Schmitt, D.; et al. Jumping Performance in Tree Squirrels: Insights into Primate Evolution. *J. Hum. Evol.* **2023**, *180*, 103386. [CrossRef]
18. Mo, X.; Romano, D.; Miraglia, M.; Ge, W.; Stefanini, C. Effect of Substrates' Compliance on the Jumping Mechanism of *Locusta migratoria*. *Front. Bioeng. Biotechnol.* **2020**, *8*, 661. [CrossRef]
19. Bijma, N.N.; Gorb, S.N.; Kleinteich, T. Landing on Branches in the Frog *Trachycephalus resinifictrix* (Anura: Hylidae). *J. Comp. Physiol. A* **2016**, *202*, 267–276. [CrossRef]
20. Song, Y.; Zhou, R.; Qiu, Y.; Chen, J.; Dai, Z.; Wu, H.; Gorb, S.N. Tree Frogs (*Polypedates dennysi*) Landing on Horizontal Perches: The Effects of Perch Diameter. *J. Exp. Biol.* **2024**, *227*, jeb246931. [CrossRef]
21. Endlein, T.; Ji, A.; Samuel, D.; Yao, N.; Wang, Z.; Barnes, W.J.P.; Federle, W.; Kappl, M.; Dai, Z. Sticking like Sticky Tape: Tree Frogs Use Friction Forces to Enhance Attachment on Overhanging Surfaces. *J. R. Soc. Interface* **2013**, *10*, 20120838. [CrossRef]
22. Endlein, T.; Ji, A.; Yuan, S.; Hill, I.; Wang, H.; Barnes, W.J.P.; Dai, Z.; Sitti, M. The Use of Clamping Grips and Friction Pads by Tree Frogs for Climbing Curved Surfaces. *Proc. R. Soc. B* **2017**, *284*, 20162867. [CrossRef] [PubMed]
23. Herrel, A.; Perrenoud, M.; Decamps, T.; Abdala, V.; Manzano, A.; Pouydebat, E. The Effect of Substrate Diameter and Incline on Locomotion in an Arboreal Frog. *J. Exp. Biol.* **2013**, *216*, 3599–3605. [CrossRef] [PubMed]
24. Federle, W.; Barnes, W.J.P.; Baumgartner, W.; Drechsler, P.; Smith, J.M. Wet but Not Slippery: Boundary Friction in Tree Frog Adhesive Toe Pads. *J. R. Soc. Interface* **2006**, *3*, 689–697. [CrossRef] [PubMed]
25. Zhou, R.; Wang, Z.; Song, Y.; Liu, S.; Dai, Z. Tree Frogs Alter Their Behavioral Strategies While Landing On Vertical Perches. *J. Exp. Zool. A Ecol. Integr. Physiol.* **2025**, *343*, 15–24. [CrossRef]

26. Meng, F.; Liu, Q.; Wang, X.; Tan, D.; Xue, L.; Barnes, W.J.P. Tree Frog Adhesion Biomimetics: Opportunities for the Development of New, Smart Adhesives That Adhere under Wet Conditions. *Phil. Trans. R. Soc. A* **2019**, *377*, 20190131. [CrossRef]
27. Reynaga, C.M.; Eaton, C.E.; Strong, G.A.; Azizi, E. Compliant Substrates Disrupt Elastic Energy Storage in Jumping Tree Frogs. *Integr. Comp. Biol.* **2019**, *59*, 1535–1545. [CrossRef]
28. Astley, H.C.; Haruta, A.; Roberts, T.J. Robust Jumping Performance and Elastic Energy Recovery from Compliant Perches in Tree Frogs. *J. Exp. Biol.* **2015**, *218*, 3360–3363. [CrossRef]
29. Hedrick, T.L. Software Techniques for Two- and Three-Dimensional Kinematic Measurements of Biological and Biomimetic Systems. *Bioinspir. Biomim.* **2008**, *3*, 034001. [CrossRef]
30. Porro, L.B.; Collings, A.J.; Eberhard, E.A.; Chadwick, K.P.; Richards, C.T. Inverse Dynamic Modelling of Jumping in the Red-Legged Running Frog *Kassina maculata*. *J. Exp. Biol.* **2017**, *220*, 1882–1893. [CrossRef]
31. Richards, C.T.; Porro, L.B.; Collings, A.J. Kinematic Control of Extreme Jump Angles in the Red Leg Running Frog (*Kassina maculata*). *J. Exp. Biol.* **2017**, *220*, 1894–1904. [CrossRef]
32. Bonser, R.H.C.; Norman, A.P.; Rayner, J.M.V. Does Substrate Quality Influence Take-off Decisions in Common Starlings? *Funct. Ecol.* **1999**, *13*, 102–105. [CrossRef]
33. Channon, A.J.; Günther, M.M.; Crompton, R.H.; D'Août, K.; Preuschoft, H.; Vereecke, E.E. The Effect of Substrate Compliance on the Biomechanics of Gibbon Leaps. *J. Exp. Biol.* **2011**, *214*, 687–696. [CrossRef] [PubMed]
34. Hunt, N.H.; Jinn, J.; Jacobs, L.F.; Full, R.J. Acrobatic Squirrels Learn to Leap and Land on Tree Branches without Falling. *Science* **2021**, *373*, 697–700. [CrossRef] [PubMed]
35. Kim, W.; Byun, J.; Kim, J.-K.; Choi, W.-Y.; Jakobsen, K.; Jakobsen, J.; Lee, D.-Y.; Cho, K.-J. Bioinspired Dual-Morphing Stretchable Origami. *Sci. Robot.* **2019**, *4*, eaay3493. [CrossRef]
36. Faber, J.A.; Arrieta, A.F.; Studart, A.R. Bioinspired Spring Origami. *Science* **2018**, *359*, 1386–1391. [CrossRef]
37. Merces, L.; Ferro, L.M.M.; Thomas, A.; Karnaushenko, D.D.; Luo, Y.; Egunov, A.I.; Zhang, W.; Bandari, V.K.; Lee, Y.; McCaskill, J.S.; et al. Bio-Inspired Dynamically Morphing Microelectronics toward High-Density Energy Applications and Intelligent Biomedical Implants. *Adv. Mater.* **2024**, *36*, 2313327. [CrossRef]

biomimetics

MDPI

Article

Anti-Adhesive Surfaces Inspired by Bee Mandible Surfaces

Leonie Saccardi [1,2,*], Jonas Schiebl [2], Franz Balluff [3], Ulrich Christ [3], Stanislav N. Gorb [4], Alexander Kovalev [4] and Oliver Schwarz [2]

1 Institute of Biomaterials and Biomolecular Systems, University of Stuttgart, 70569 Stuttgart, Germany
2 Department of Biomechatronic Systems, FraunhoferInstitute for Manufacturing Engineering and Automation IPA, 70569 Stuttgart, Germany
3 Department of Applied Coating Technology, Fraunhofer-Institute for Manufacturing Engineering and Automation (IPA), 70569 Stuttgart, Germany
4 Department Functional Morphology and Biomechanics, Zoological Institute, Kiel University, 24118 Kiel, Germany
* Correspondence: leonie.saccardi@gmx.de

Abstract: Propolis, a naturally sticky substance used by bees to secure their hives and protect the colony from pathogens, presents a fascinating challenge. Despite its adhesive nature, honeybees adeptly handle propolis with their mandibles. Previous research has shown a combination of an anti-adhesive fluid layer and scale-like microstructures on the inner surface of bee mandibles. Our aim was to deepen our understanding of how surface energy and microstructure influence the reduction in adhesion for challenging substances like propolis. To achieve this, we devised surfaces inspired by the intricate microstructure of bee mandibles, employing diverse techniques including roughening steel surfaces, creating lacquer structures using Bénard cells, and moulding resin surfaces with hexagonal patterns. These approaches generated patterns that mimicked the bee mandible structure to varying degrees. Subsequently, we assessed the adhesion of propolis on these bioinspired structured substrates. Our findings revealed that on rough steel and resin surfaces structured with hexagonal dimples, propolis adhesion was significantly reduced by over 40% compared to unstructured control surfaces. However, in the case of the lacquer surface patterned with Bénard cells, we did not observe a significant reduction in adhesion.

Keywords: adhesion; propolis; biomimetics; mandibles; honeybee

1. Introduction

In many industries, unwanted adhesion presents a significant challenge, often leading to costly and environmentally harmful cleaning procedures. One notable example of unwanted adhesion in technology pertains to the accumulation of resins on the cutting edges of woodworking tools. This issue served as the catalyst for a top-down bionic development process [1,2]. Leveraging inspiration from nature's anti-adhesive strategies can yield innovative solutions to technical challenges [3–7].

In the biomimetic development process, the initial step involves abstracting a specific problem to identify analogous challenges in the natural world [8]. The ability to prevent adhesion to surfaces or safeguard the functionality of organs from adhesion has been a selective factor in biological evolution, significantly influencing survival probabilities.

Therefore, one search strategy entails identifying adhesive substances—whether self-produced or acquired from the environment—with which certain animal surfaces regularly come into contact.

Propolis, a resin-based substance with adhesive properties, is used by honeybees to seal cracks and smooth internal hive walls [9,10]. Remarkably, honeybees appear to have no adverse reactions when in contact with propolis. They adeptly manipulate finished propolis and gather key resin components from plant buds without any issues. Propolis exhibits strong adhesion to various substrates, including glass, steel, and even PTFE, as demonstrated by our previous study [11]. Notably, in wet conditions, propolis adhesion is significantly reduced. Furthermore, our earlier findings indicate a positive correlation between temperature and propolis adhesion.

There is vast literature on Hymenoptera-mandibles [12–14]. However, a detailed anatomical investigation of honeybee mandibles, which frequently come into contact with sticky propolis and plant resins, was only recently conducted [15].

Of particular interest is the medial surface of the mandible, which plays a crucial role in propolis processing, given its frequent contact with the sticky material during biting and shaping [16,17].

Bee mandibles have a sturdy stem at the base and a concave, spoon-shaped tip at the other end, which is crucial for processing propolis (Figure 1A) [18,19]. The spoon-shaped tip tapers towards the apex and has a hairy edge on one side, with long, flexible hairs covering over half of the medial surface, while the opposite side has a sharp, hairless edge. A central ridge runs along the medial surface, fading towards the apex, adorned with stiff, grooved bristles curving toward the sharp edge. There is an elevated ledge adjacent to the hairy edge, forming a channel between the central ridge and the ledge, extending from the stem to the apex. On the other side of the central ridge, a flat area extends to the sharp edge [15].

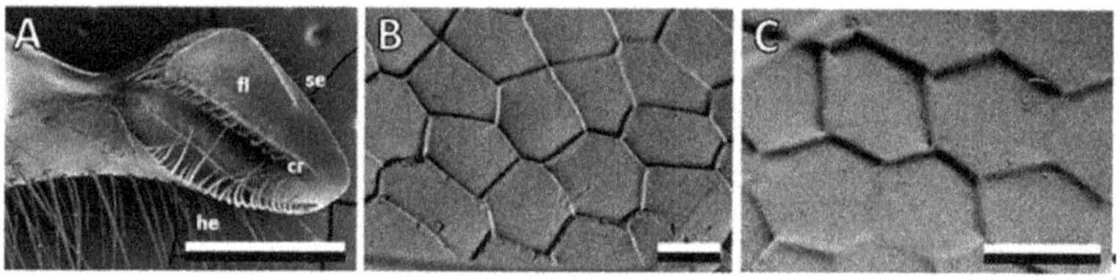

Figure 1. Bee mandibles washed with chloroform and water. (**A–C**): SEM micrographs of bee mandibles (mandible overview and close-up of flat area). cr, central ridge; fl, flat area; he, hairy edge; se, sharp edge. Scale bars: 500 µm (**A**), 10 µm (**B**,**C**).

It was observed that propolis adhesion is significantly lower on bee mandibles in comparison to other surfaces, such as glass [15]. This observation suggests that bees may have evolved an anti-adhesive strategy to facilitate propolis handling, preventing contamination with resin.

There are several hypotheses explaining the mechanisms underlying the reduction in adhesion based on principles observed in other insects. Voigt et al. [20] suggest the following potential strategies to reduce adhesion on insect cuticles: (1) specific surface chemistry; (2) surface microstructures: for example, water striders utilise complex two-level microstructures, including microtrichia, to create a super-hydrophobic surface that reduces adhesion [21] and (3) the presence of an easy-to-break solid or fluid layer preventing strong bonding through cohesion failure, like the wax layers on *Nepenthes alata* pitchers that prevent insect adhesion by breaking off [22].

Bornean stingless bees (Hymenoptera, Meliponini), like honeybees, collect resin for nest construction and protection against intruders [23]. They can remove resin from their mandibles without any residue, leading to the suggestion that these bees may use temporary lubrication to reduce resin adhesion on their mandibles [24,25].

Our recent findings indicate that a combination of surface structure and a fluid layer plays a significant role in reducing propolis adhesion on bee mandibles [15]. The removal of the liquid layer from the cuticle surface using substances like chloroform or acetone resulted in a substantial increase in propolis adhesion [15]. Consequently, we have hypothesised that this fluid serves to lubricate the mandible and minimise propolis adhesion.

The medial surface of bee mandibles also features hexagonal or pentagonal microstructures with dimensions ranging from 10–20 µm [15]. Our findings suggest that these structures alone do not independently reduce propolis adhesion; their anti-adhesive effect is achieved only in combination with fluid [15]. The functional significance of these microstructures likely lies in maintaining a specific thickness of the fluid layer and facilitating the spreading of the fluid across the surface. It is worth noting that the microstructures exhibit a slight asymmetry in shape depending on their location on the mandible, and the step height between neighbouring microscales measures 0.7 μm.

To better understand and replicate the anti-adhesive strategy of bee mandibles, we conducted a series of experiments on various substrates. These substrates were engineered to mimic the patterns found on bee mandibles, and their anti-adhesive properties were assessed through adhesion experiments. This approach aimed to gain deeper insights into the influence of material surface chemistry and structure on propolis adhesion.

2. Materials and Methods

2.1. Propolis

Raw propolis was provided by one of the co-authors (O.S., Stuttgart, Germany). The homogenisation process followed the procedure outlined in [3]. Propolis chunks were homogenised by mixing, finely ground, and subsequently stored at −20 °C. The pulverisation method employed here is akin to that commonly used for producing propolis extract [26]. To prevent contamination, all handling of the propolis was carried out using gloves that had been thoroughly cleaned with ethanol (Rotipuran®, ≥99.8%, p.a., Carl Roth GmbH & Co. KG, Karlsruhe, Germany).

2.2. Imaging of Bee Mandibles

The mandibles of honeybees (*Apis mellifera*) were prepared and examined following the procedure outlined in Saccardi et al. [15]. They were initially cleaned with a solution of chloroform and water and then examined using a scanning electron microscope (SEM) (Hitachi S-4800, Hitachi High-Technologies Corp., Tokyo, Japan) with a 3 kV accelerating voltage. Fractures of frozen bee mandibles were further analysed using cryo SEM. Additionally, a confocal 3D laser scanning microscope was used to measure the profiles of the surface structures on the mandibles.

2.3. Development and Characterisation of Bioinspired Substrates

To examine the influence of surface structures on propolis adhesion across various materials, structured substrates from steel, Spurr's resin, and UV lacquer were prepared using diverse methods for subsequent adhesion experiments. Unstructured substrates composed of the same materials served as our control group.

Rough steel: A section of an unpolished steel circular saw blade (1.2003, Fraunhofer IPA, Stuttgart, Germany) was used as a rough substrate. As a contrasting smooth control, a steel plate (1.4016, Abrams Premium Stahl®, Osnabrück, Germany) was polished using a polishing machine (Minitech 233, Presi, Eybens, France). The polishing involved the successive use of aluminium oxide suspensions with descending particle sizes (12, 3, 1, and 0.3 μm).

Bénard structures: Rayleigh–Bénard convection is a phenomenon in fluid dynamics that occurs when a layer of fluid is heated from below and cooled from above, resulting in the formation of characteristic convection cells [27,28]. These convection flows encompass Bénard cells, driven by density differences, and Marangoni cells, driven by surface tension disparities. The two effects occur in combination, with one of the two effects usually predominating. The resulting cells can form mountains and valleys, which are "frozen" by radiation curing [29]. We used this phenomenon to create structured surfaces by applying an UV lacquer (Cymel 328 with the addition of 23% isopropanol) to a metal plate. This lacquer-coated plate was structured using Rayleigh–Bénard convection, generated by heating it from the uncoated side. Once the cells reached the desired size, the lacquer was solidified using UV light (M-25-2*1-TR-SS, IST Metz, Nürtingen, Germany). For the smooth control substrate, the lacquer was applied without structuring and polymerised following application.

Hexagons: Hexagonal microstructures resembling those found on tarsal attachment pads of the bush cricket *Tettigonia viridissima* were initially described and replicated by Varenberg and Gorb [30,31]. These structures, 10 μm in diameter, closely resemble those present on bee mandibles. Negative and positive steel templates (OVD Kinegram, Zug, Switzerland) of these hexagonal structures with a 10 μm diameter were used to produce replicas of the pattern employing the two-step moulding method described by Gorb [32] and Koch et al. [33].

To ensure the success of the moulding process, some of the resulting substrates made from Spurr's epoxy resin (Spurr's low viscosity kit, Plano, Wetzlar, Germany) were sputter-coated with a 10 nm thick layer of gold–palladium and subsequently examined using a scanning electron microscope (SEM) (Hitachi S-4800, Hitachi High-Technologies Corp., Tokyo, Japan) at 3 kV accelerating voltage. The surfaces of these Spurr substrates exhibited either hexagon-shaped dimples or hexagon-shaped pins. The dimples had a contact area fraction of 25%, while the hexagonal pins had a contact area fraction of 75% [31]. As a smooth resin control surface, a resin replica of a clean glass surface was prepared following the two-step moulding method described by Gorb [32] and Koch et al. [33].

The surface free energy of all samples was characterised by measuring the contact angles of water on the substrates using the sessile drop method with a 2 μL drop volume. This was done with an optical contact angle measuring system (OCAH200, DataPhysics Instruments GmbH, Filderstadt, Germany). Between 5 and 10 contact angle measurements were conducted for each substrate.

The substrates produced through the three different methods were examined using a confocal 3D laser scanning microscope (Keyence VK-X250; Keyence Corporation, Osaka, Japan). MultiFileAnalyzer software (Version 1.2.6.106, Keyence Corporation, Osaka, Japan) was employed to measure the profiles and dimensions of surface features.

2.4. Adhesion Measurements on Structured/Bioinspired Substrates

Adhesion experiments were performed with propolis samples on structured and unstructured substrates. Prior to each adhesion experiment, a small quantity of propolis powder was defrosted and thoroughly blended by kneading to achieve a uniform consistency. Cone-shaped propolis specimens with a spherical tip were then manually crafted. The samples' surface characteristics were examined with a fast-scanning 3D measurement microscope (Keyence VR 3100; Keyence Corporation, Osaka, Japan). The samples' profiles were measured at five locations positioned in a star-like pattern extending through the apex. To approximate the radiuses at the tips of the samples, circles were matched to the profiles of the samples in five distinct orientations. The radii of these circles were measured and subsequently averaged. The effective elastic modulus and the pull-off force of propolis were measured with a microforce measurement device (Basalt-BT01; Tetra GmbH, Ilmenau, Germany [34–36]).

The setup comprised a fibre-optical sensor and micromanipulators serving as a base for securing the substrate material, along with a metal spring (utilising springs with spring constants of 618 and 539 N/m). The piezo drive was responsible for vertically displacing the sensor along with the spring, applying force during the loading phase and releasing it during the unloading of the sample.

A shortened glass capillary (5 μL micropipette Blaubrand R© Intra END, BRAND GMBH + CO KG, Wertheim, Germany) was attached to the metal spring with cyanoacrylate glue. The cone-shaped propolis sample was affixed to the end of the capillary without any additional adhesive. The test substrates were securely attached to the micromanipulator platform using double-sided adhesive tape. The propolis sample was brought into contact with the substrate, and it was withdrawn from the surface as soon as the applied force reached 5 mN. The load was chosen to approximate the load exerted by bees when handling propolis. Given the absence of specific studies on mandibular forces and pressures of honeybees, we used pressure values previously measured at the tip of the mandibles in predacious Coleoptera [37] as a reference for the load applied to the propolis sample. Our experiments were performed in a quasi-static regime (at a very slow velocity) and that is why the velocity did not play an important role in our results.

Propolis adhesion was tested on rough steel, Spurr with hexagonal pins and hexagonal dimples, UV lacquer structured with Bénard cells, and their respective unstructured control surfaces. For each propolis sample, a set of 11 single measurements was performed, each on a different spot of the substrate (N = 5 propolis samples, n = 11 measurements per sample). The maximum loading force was 5 mN. All experiments were carried out at room temperature (24 ± 0.5 °C) with a relative humidity of 37 ± 9%.

Following the 11 measurements on the structured substrate materials, we performed an additional set of five reference measurements on the unstructured control surface made from the same material.

After the adhesion experiments, the substrate material was examined under a binocular microscope (Leica M205 A, Leica Microsystems GmbH, Wetzlar, Germany) to identify any potential propolis residues or prints in the contact area.

2.5. Data Analysis and Statistics

Adhesion experiments were evaluated as described bySaccardi et al. [3] using Matlab (version R2015b, The MathWorks, Inc., Natick, MA, USA).

Statistical analysis was carried out using R software, version 3.6.1 (The R Foundation for Statistical Computing). The data were initially tested for normal distribution and variance homogeneity using the Kolmogorov–Smirnov and Levene's tests, respectively. Subsequently, the comparison of propolis adhesion on different substrates was executed with a one-way ANOVA, complemented by a pairwise multiple comparison procedure (Tukey test).

To explore potential correlations between the Young's modulus and work of adhesion obtained from the adhesion experiments, the Pearson's correlation coefficient was calculated.

2.6. Image Processing

SEM images were processed using Gimp (version 2.10.14, Spencer Kimball, Peter Mattis and the GIMP Development Team). The adjustments, like fine-tuning colour levels, contrast, and brightness, were uniformly applied to the entire image. Furthermore, scale bars and labels were added.

Profiles acquired through the confocal 3D laser scanning microscope were digitised using WebPlotDigitizer (version 4.2, https://automeris.io/WebPlotDigitizer, accessed on 25 January 2020, Ankit Rohatgi, San Francisco, CA, USA).

3. Results

3.1. Bee Mandible Surfaces

Honeybee mandibles, specifically the structures on the medial surface, served as a source of inspiration for the creation of bioinspired surfaces. These mandibles exhibit a distinct morphology, featuring a robust stem at the proximal end and a concave, spoon-shaped tip at the distal end (Figure 1A). The focal point of interest lies in the medial surface of this tip, which is covered with anisotropic scale-like micropatterns (Figure 1B,C). Most of these scales on the medial surface of the mandible are either pentagonal or hexagonal, although the specific shape and proportions vary depending on their location. In the flat area, extending from the central ridge toward the sharp edge, the scales measure 18.1 ± 2.4 μm in length and 9.55 ± 1.3 μm in width at their minimum and maximum points, respectively. In the channel area, extending from the central ridge toward the hairy edge, the structures exhibit a rounder shape, measuring 9.34 ± 0.77 μm in length and 8.21 ± 0.82 μm in width, with blunter edges. To comprehensively study the topography of these structures, we used cryo-SEM micrographs of fractured mandibles (Figure 2A) and a 3D laser scanning microscope (Figure 2C,D). The scales themselves are either flat or slightly convex, and they either overlap or form steps between them. In most areas of the mandible, the step height between the scales was consistently measured to be 0.71 ± 0.26 μm.

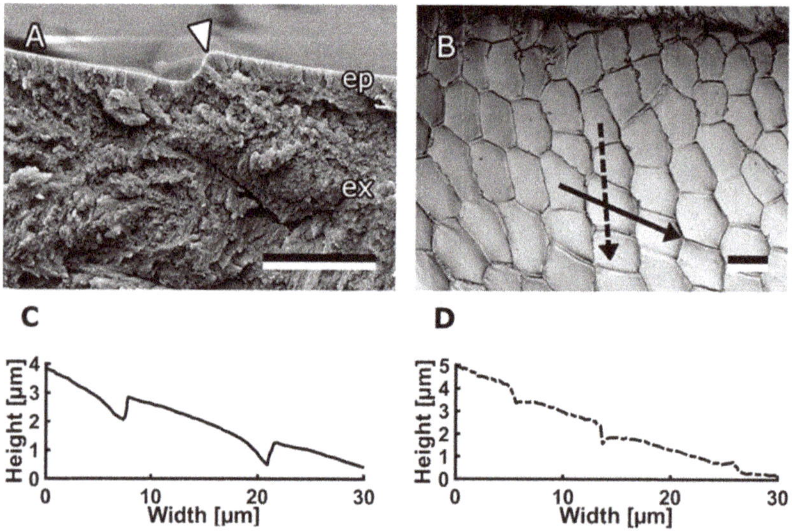

Figure 2. Profile of scales on bee mandibles. (**A**) Cryo-SEM micrograph of mandible cuticle. The arrowhead indicates the step between two scales. (**B**) 3D laser scanning microscope image of medial surface of the mandible within the channel area. (**C**) Profile of the scales along the solid arrow in (**B**). (**D**) Profile of the scales along the dashed arrow in (**B**). Scale bars: 2 μm (**A**), 10 μm (**B**). Published by Beilstein J Nanotechnol., distributed under the terms of the Creative Commons Attribution 4.0 International License, https://creativecommons.org/licenses/by/4.0 (accessed on 17 November 2023).

3.2. *Characterisation of Bioinspired Surfaces*

3.2.1. Rough Steel

As the initial candidate for a biomimetic surface with anti-adhesive characteristics, a substrate with relatively random surface structures—a sand-blasted steel plate—was chosen. This steel surface displayed an irregular pattern characterised by a series of hills and valleys (Figure 3). The surface roughness of the sand-blasted steel plate was measured to be Sa = 0.708 ± 0.155 μm. Water droplets had a contact angle of 89.7° ± 3.2° on this rough steel surface. In contrast, a polished control surface, also composed of steel, showcased significantly lower roughness, with Sa = 0.038 ± 0.013 μm and a contact angle of 82.6° ± 1.4° for water.

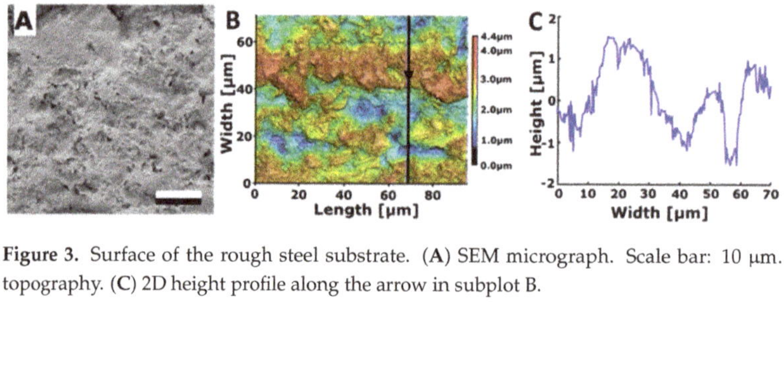

Figure 3. Surface of the rough steel substrate. (**A**) SEM micrograph. Scale bar: 10 μm. (**B**) 3D topography. (**C**) 2D height profile along the arrow in subplot B.

3.2.2. Bénard Structures

Another approach for creating a surface pattern inspired by the structures found on bee mandibles involved structuring UV lacquer using Rayleigh–Bénard convection with appropriately sized Bénard cells. To achieve this, a metal plate coated with UV lacquer was heated from below, and polymerisation was initiated once the Bénard cells attained a size of approximately 150–200 μm in width and 0.5–1.0 μm in height (Figure 4). This process of Rayleigh–Bénard convection yielded a regular pattern of hills and valleys. The surface roughness of the surface structured with Bénard cells was measured to be Sa = 0.532 ± 0.211 μm. The water contact angle on this structured surface measured 80.6° ± 3.0°.

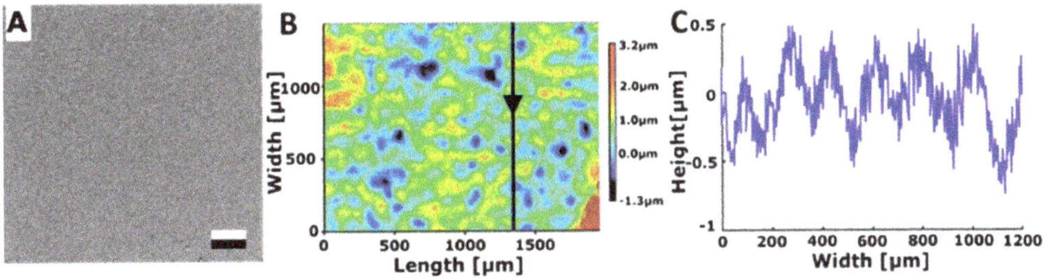

Figure 4. Surface structured with Bénard cells. (**A**) SEM micrograph. Scale bar: 10 μm. (**B**) 3D topography. (**C**) 2D height profile along the arrow in subplot B.

In contrast, for the control substrate, the UV lacquer was not structured before undergoing polymerisation and the water contact angle on this smooth surface was 77.1° ± 4.5°.

3.2.3. Hexagons

The microstructures observed on the tarsal attachment pads of the bush cricket *Tettigonia viridissima* resemble those found on bee mandibles in terms of their hexagonal shape and dimensions. These structures were previously described and replicated by Varenberg and Gorb [18,19].

Negative and positive steel templates (OVD Kinegram, Zug, Switzerland) of these hexagonal structures with a 10 μm diameter were used to produce moulds. Subsequently, the moulds were used to create structured substrates with Spurr's resin. This process resulted in the generation of two variations of hexagonal structures: hexagon-shaped dimples (Figure 5A–C) and hexagon-shaped pins (Figure 5D–F). The hexagons had an approximate diameter of 10 μm. Dimples were set apart by a border ranging from 0.5 to 1.0 μm in height, while pins were surrounded by a trench that reached a depth of 1.0 to 1.5 μm.

The surface roughness of the Spurr surface structured with hexagonal dimples was measured to be Sa = 0.771 ± 0.159 μm and for the surface with hexagonal pins it was Sa = 0.449 ± 0.137 μm. Subsequent measurements revealed that the water contact angle on the Spurr surface with hexagonal dimples was 75.9° ± 5.1°, whereas the contact angle on Spurr with hexagonal pins was 124.1° ± 4.9°. In comparison, the contact angle of water on the smooth control surface composed of Spurr resin was 88.6° ± 3.9°.

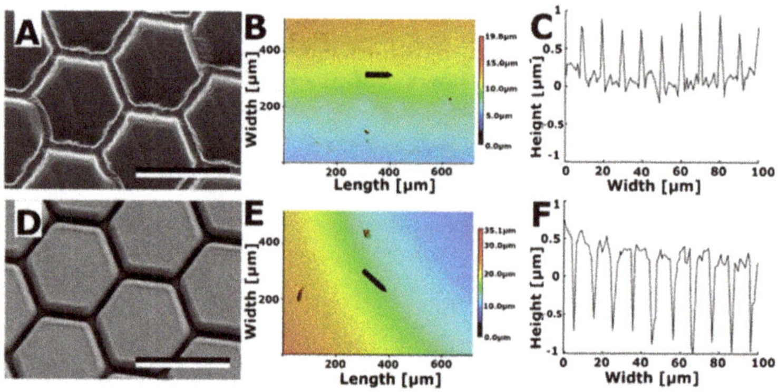

Figure 5. Hexagonal surface structures on a surface made from the Spurr's resin. (**A**) SEM image of hexagonal dimples. (**B**) 3D topography. (**C**) 2D profile of hexagonal dimples along the arrow in B. (**D**) SEM image of hexagonal pins. (**E**) 3D topography. (**F**) 2D profile of hexagonal pins along the arrow in (**E**). Scale bars: 10 µm (**A**,**D**).

3.3. Adhesion on Bioinspired Substrates

Propolis adhesion was measured on the bioinspired surfaces introduced in this study and their corresponding smooth control surfaces (Table 1). On the rough steel substrate, the measured work of adhesion was 1.26 ± 0.61 J/m^2 and there was a significant difference between the work of adhesion on the rough and polished steel (Figure 6, left section; $p < 0.0001$).

Table 1. Propolis adhesion on bioinspired surfaces. Work of adhesion and pull-off forces obtained from adhesion experiments (N = 5 propolis samples, n = 10 individual measurements per substrate). Mean values and standard deviations (s.d.) are given. Reference measurements on honeybee mandibles were reported by [15].

Substrate	N×n	Work of Adhesion [J/m^2]		Pull-off Force [mN]	
		Mean	**s.d.**	**Mean**	**s.d.**
Honeybee mandible, unwashed	150	1.01	0.21	0.71	0.33
Steel, polished	50	2.29	0.82	1.98	0.51
Steel, rough	50	1.26	0.61	1.09	0.43
Lacquer, smooth (UV lacquer)	50	1.74	0.98	4.20	0.70
Lacquer, Bénard cells (UV lacquer)	50	1.99	1.07	3.76	0.60
Spurr, smooth	50	3.61	0.95	3.35	0.87
Hexagon dimples (Spurr)	50	1.99	0.64	1.99	0.55
Hexagon pins (Spurr)	50	3.28	1.38	2.31	1.08

In the case of adhesion tests conducted with propolis on UV lacquer-coated substrates, there was no significant difference in the work of adhesion between the smooth control and the substrate structured with Bénard cells (Figure 6, middle section; $p = 0.9997$).

Additionally, we assessed propolis adhesion on structured Spurr's resin substrates and compared it to adhesion on the smooth resin surface (Figure 6, right section). Notably, there was no significant difference in the work of adhesion between the smooth Spurr substrate and the substrate with hexagonal pins ($p = 0.9920$). However, the work of adhesion was significantly lower on substrates structured with hexagonal dimples ($p < 0.0001$).

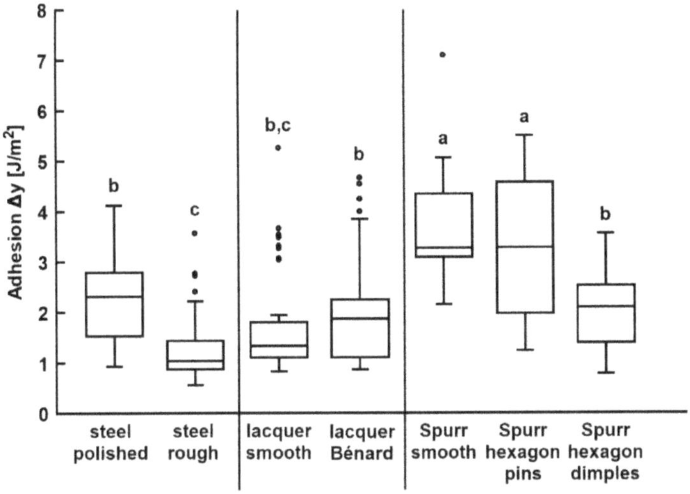

Figure 6. Propolis adhesion on structured substrates. Work of adhesion of propolis samples was tested on different substrate materials structured with bioinspired patterns and on corresponding unstructured control substrates. The ends of the boxes define the 25th and 75th percentiles, with a line at the median and error bars defining the 10th and 90th percentiles. The outlier markings represent individual measurements. Conditions and substrates marked with different letters differ significantly from each other (one-way ANOVA, $p < 0.001$ and Tukey test, $p < 0.05$).

4. Discussion

4.1. Development and Characterisation of Bioinspired Surfaces

Various approaches were employed to create structured substrates inspired by the micropatterns observed on bee mandibles. In the initial approach, the bee mandible structures were interpreted as a form of surface roughness, leading to the utilisation of a rough steel surface as a technical approximation. However, the structures on the rough steel surfaces were random in size and distribution, lacking a distinct shape akin to the scales present on the mandible. Consequently, the rough steel surface served as a rather crude approximation of the bee model.

A more regular patterned surface was attained by structuring UV lacquer with Bénard cells. These Bénard cells generated a smooth and regular pattern of valleys and hills. These structures were only visible using a confocal 3D laser scanning microscope rather than a scanning electron microscope (SEM), likely because they were relatively flat and lacked sharp edges. Notably, the lateral dimensions of these structures were approximately ten times larger than those observed on bee mandibles, and they did not possess a distinct geometric shape.

Hexagonal shapes [30,31] proved to be a better match in terms of size and shape when compared to bee patterns. Nevertheless, the hexagonal shapes still differed from the bee patterns as they were isotropic and did not overlap like the scale-like microstructures on bee mandibles. Neither hexagonal pins nor dimples provided a perfect representation of the structures on bee mandibles. However, hexagonal pins appeared to closely resemble the scales on bee mandibles, as these scales often had a relatively flat surface area. Even in instances where some scales in the mandible's channel area exhibited blunter edges, the overall profile indicated that the bulges did not significantly rise when compared to the profiles of hexagonal dimples, where clear spikes were visible between each valley.

A more promising approach for future replication of the bee mandible pattern might involve the production of a laser-engraved metal template [38–45] based on the 3D data obtained from the mandible surface. This template could then be employed to replicate the structure using a two-step moulding technique.

Additionally, it is worth noting that the chemical composition of materials differed: steel (mostly iron alloy) and epoxy resin (organic compounds) were used. Despite this distinction, both materials exhibited similar surface energies, with dispersive interactions predominating over polar interactions. A previous study conducted by Saccardi et al. [15] indicated that the substrate material did not significantly influence propolis' work of adhesion. Biological adhesives often exhibit a minimal reliance on the chemical properties of the substrate [46]. It is our hypothesis that specific fluid components within propolis, such as mono- and sesquiterpenoids, may modify the interaction between the propolis sample and the substrate, resulting in adhesion that is nearly unaffected by the substrate's characteristics.

4.2. Adhesion of Propolis on Bioinspired Surfaces

There are many recent papers on the anti-adhesive properties of various microstructures [47–56]. We wanted to investigate the influence of surface microstructures on propolis adhesion through adhesion experiments on various bioinspired substrates. Tests involving propolis on replicated mandibles and a smooth control substrate revealed that these structures did not reduce adhesion [15]. This possibly happened because, in the real mandible, the mandible surface is covered with a fluid, which was not replicated here. The role of the structures could be to evenly distribute this fluid.

Adhesion on the rough steel substrate was significantly lower compared to the smooth steel control. Irregular surface roughness likely reduced the actual contact area between the propolis sample and the substrate, as only the highest points made adhesive contact. In contrast, the Bénard cell structures on the lacquer surface did not significantly affect propolis adhesion when compared to the smooth control. These structures were larger and had a gentler slope than the previously mentioned structures, possibly allowing the propolis sample to adapt to the surface pattern, resulting in no reduction in contact area.

Hexagonal pins also did not reduce adhesion compared to the smooth Spurr control. Since these pins closely resembled the bee mandible model, it further supports the previous assumption that the structures on bee mandibles are not responsible for reduced propolis adhesion. In contrast, hexagonal dimples significantly reduced propolis adhesion. This could be due to the thin walls between the dimples hindering the formation of a tight contact between the propolis and the flat surface in sample valleys, resulting in a smaller contact area. The pins area fraction is 75% compared to only 25% of dimples [31]. Varenberg et al. [31] also found that adhesion and friction on dimples are significantly lower than on pins due to the reduced contact area. In comparison to the hexagonal dimples, the slight bulges found on the edges of some scales on the bee mandible were probably not high or sharp enough to prevent effective propolis contact with the scale surface.

It would be of interest to measure propolis adhesion on structured surfaces coated with a liquid layer that simulates the surface coating found on mandibular structures. Since propolis adhesion is significantly reduced even on smooth surfaces in wet conditions [11], we can anticipate an even more pronounced anti-adhesive effect on the microstructured surfaces covered with fluid. This expectation is based on the combination of surface structure and fluid, as previously demonstrated in water-filled silicone microstructures that prevent wetting by barnacle glue [57–60]. Furthermore, the spread of liquids on these structured substrates could be observed to confirm the assumption that the scales on bee mandibles aid in even distribution of the fluid coating the surface.

An effective anti-adhesive system based on a fluid layer evenly distributed by surface structures like those found on bee mandibles could have applications in various industries, for example in the woodworking industry, where resin-contaminated tools pose a significant problem, as cleaning is challenging and time-consuming. Similar systems involving slippery liquid-infused porous surfaces have been shown to be highly effective in reducing the adhesion of various substances such as ice [61] or biofilms [62].

Author Contributions: L.S., J.S., O.S., S.N.G. and A.K. contributed to the conception and design of the study. L.S., A.K. and S.N.G. conducted and evaluated the experiments. L.S. wrote the first draft of the manuscript. F.B. and U.C. developed and produced surfaces structured with Bénard cells. S.N.G. and A.K. developed and produced surfaces structured with hexagons. All authors have read and agreed to the published version of the manuscript.

Funding: This research received no external funding.

Institutional Review Board Statement: Not applicable.

Data Availability Statement: Data are contained within the article.

Acknowledgments: We thank Stefanie Wunder for her help in developing surfaces structured with Bénard cells.

Conflicts of Interest: The authors declare no conflict of interest.

References

1. Speck, T. *Bionik: Faszinierende Lösungen der Natur für die Technik der Zukunft*; Lavori Verlag: Freiburg, Germany, 2012.
2. Nachtigall, W. *Bionik: Grundlagen und Beispiele für Ingenieure und Naturwissenschaftler*; Springer: Berlin/Heidelberg, Germany, 2002.
3. Paz-Gómez, G.; del Caño-Ochoa, J.C.; Rodríguez-Alabanda, O.; Romero, P.E.; Cabrerizo-Vílchez, M.; Guerrero-Vaca, G.; Rodríguez-Valverde, M.A. Water-Repellent Fluoropolymer-Based Coatings. *Coatings* **2019**, *9*, 293. [CrossRef]
4. Palumbo, F.; Lo Porto, C.; Favia, P. Plasma Nano-Texturing of Polymers for Wettability Control: Why, What and How. *Coatings* **2019**, *9*, 640. [CrossRef]
5. He, Z.; Wu, C.; Hua, M.; Wu, S.; Wu, D.; Zhu, X.; Wang, J.; He, X. Bioinspired multifunctional anti-icing hydrogel. *Matter* **2020**, *2*, 723–734. [CrossRef]
6. Liu, Y.; Wang, B.; Wang, Y.; Chen, J.; Cui, B.; Yin, P.; Chen, J.; Zhang, X.; Zhang, L.; Xin, J.H. Bioinspired Superhydrophobic Surface Constructed from Hydrophilic Building Blocks: A Case Study of Core–Shell Polypyrrole-Coated Copper Nanoneedles. *Coatings* **2020**, *10*, 347. [CrossRef]
7. Volpe, A.; Gaudiuso, C.; Di Venere, L.; Licciulli, F.; Giordano, F.; Ancona, A. Direct Femtosecond Laser Fabrication of Superhydrophobic Aluminum Alloy Surfaces with Anti-icing Properties. *Coatings* **2020**, *10*, 587. [CrossRef]
8. Nachtigall, W. *Bionik als Wissenschaft: Erkennen—Abstrahieren—Umsetzen*; Springer: Berlin/Heidelberg, Germany, 2010.
9. Burdock, G.A. Review of the biological properties and toxicity of bee propolis (propolis). *Food Chem. Toxicol.* **1998**, *36*, 347–363. [CrossRef] [PubMed]
10. Bankova, V.; de Castro, S.; Marcucci, M. Propolis: Recent advances in chemistry and plant origin. *Apidologie* **2000**, *31*, 3–15. [CrossRef]
11. Saccardi, L.; Schiebl, J.; Weber, K.; Schwarz, O.; Gorb, S.N.; Kovalev, A. Adhesive Behavior of Propolis on Different Substrates. *Front. Mech. Eng.* **2021**, *7*, 1–12. [CrossRef]
12. Richter, A.; Schoeters, E.; Billen, J. Morphology and closing mechanism of the mandibular gland orifice in ants (Hymenoptera: Formicidae). *J. Morphol.* **2021**, *282*, 1127–1140. [CrossRef]
13. Klunk, C.L.; Argenta, M.A.; Casadei-Ferreira, A.; Economo, E.P.; Pie, M.R. Mandibular morphology, task specialization and bite mechanics in Pheidole ants (Hymenoptera: Formicidae). *J. R. Soc. Interface* **2021**, *18*, 20210318. [CrossRef]
14. Silveira, O.T.; dos Santos, J.N.A., Jr. Comparative morphology of the mandibles of female polistine social wasps (Hymenoptera, Vespidae, Polistinae). *Rev. Bras. Entomol.* **2011**, *55*, 479–500. [CrossRef]
15. Saccardi, L.; Brümmer, F.; Schiebl, J.; Schwarz, O.; Kovalev, A.; Gorb, S.N. Interaction between honeybee mandibles and propolis. *Beilstein J. Nanotechnol.* **2022**, *13*, 958–974. [CrossRef]
16. Nakamura, J.; Seeley, T.D. The functional organization of resin work in honeybee colonies. *Behav. Ecol. Sociobiol.* **2006**, *60*, 339–349. [CrossRef]
17. Simone-Finstrom, M.; Spivak, M. Propolis and bee health: The natural history and significance of resin use by honey bees. *Apidologie* **2010**, *41*, 295–311. [CrossRef]
18. Snodgrass, R.E. *The Anatomy of the Honey Bee*; Washington: Government Printing Office: Washington, DC, USA, 1910.
19. Carreck, N.L.; Andree, M.; Brent, C.S.; Cox-Foster, D.; Dade, H.A.; Ellis, J.D.; Hatjina, F.; van Englesdorp, D. Standard methods for Apis mellifera anatomy and dissection Métodos estandar para la disección y anatomía de Apis mellifera. *J. Apic. Res.* **2013**, *52*, 1–40. [CrossRef]
20. Voigt, D.; Gorb, S.N. An insect trap as habitat: Cohesion-failure mechanism prevents adhesion of Pameridea roridulae bugs to the sticky surface of the plant Roridula gorgonias. *J. Exp. Biol.* **2008**, *211*, 2647–2657. [CrossRef] [PubMed]
21. Perez Goodwyn, P.; De Souza, E.; Fujisaki, K.; Gorb, S.N. Moulding technique demonstrates the contribution of surface geometry to the super-hydrophobic properties of the surface of a water strider. *Acta Biomater.* **2008**, *4*, 766–770. [CrossRef]
22. Gorb, E.; Purtov, J.; Gorb, S.N. Adhesion force measurements on the two wax layers of the waxy zone in Nepenthes alata pitchers. *Sci. Rep.* **2014**, *4*, 5154. [CrossRef]
23. Leonhardt, S.D.; Blüthgen, N. A Sticky Affair: Resin Collection by Bornean Stingless Bees. *Biotropica* **2009**, *41*, 730–736. [CrossRef]

24. Gastauer, M.; Campos, L.A.O.; Wittmann, D. Handling sticky resin by stingless bees (Hymenoptera, Apidae). *Rev. Bras. De Entomol.* **2011**, *55*, 234–240. [CrossRef]
25. Gastauer, M.; Campos, L.A.O.; Wittmann, D. Handling sticky Resin by Stingless Bees: Adhesive Properties of Surface Structures. *An. Acad. Bras. Cienc.* **2013**, *85*, 1189–1196. [CrossRef]
26. Bankova, V.; Bertelli, D.; Borba, R.; Conti, B.J.; da Silva Cunha, I.B.; Danert, C.; Eberlin, M.N.; Falcão, S.I.; Isla, M.I.; Moreno, M.I.N.; et al. Standard methods for Apis mellifera propolis research. *J. Apic. Res.* **2016**, *58*, 1–49. [CrossRef]
27. Bénard, H. Les tourbillons cellulaires dans une nappe liquide. *Rev. Générale Sci. Pures Appliquées* **1900**, *11*, 1261–1271.
28. Rayleigh, L. On convection currents in a horizontal layer of fluid, when the higher temperature is on the under side. *Philos. Mag. Ser.* **1916**, *5*, 737–761. [CrossRef]
29. Seeler, F. *Numerische und Experimentelle Untersuchung des Lackfilmverlaufs*; Paderborn University: Paderborn, Germany, 2021.
30. Varenberg, M.; Gorb, S.N. Hexagonal surface micropattern for dry and wet friction. *Adv. Mater.* **2009**, *21*, 483–486. [CrossRef]
31. Varenberg, M.; Murarash, B.; Kligerman, Y.; Gorb, S.N. Geometry-controlled adhesion: Revisiting the contact splitting hypothesis. *Appl. Phys. A* **2011**, *103*, 933–938. [CrossRef]
32. Gorb, S.N. Visualisation of native surfaces by two-step moldings. *Microsc. Today* **2007**, *15*, 44–47. [CrossRef]
33. Koch, K.; Schulte, A.J.; Fischer, A.; Gorb, S.N.; Barthlott, W. A fast, precise and low-cost replication technique for nano- and high-aspect-ratio structures of biological and artificial surfaces. *Bioinspiration Biomim.* **2008**, *3*, 46002. [CrossRef] [PubMed]
34. Gorb, S.N.; Scherge, M. Biological microtribology: Anisotropy in frictional forces of orthopteran attachment pads reflects the ultrastructure of a highly deformable material. Proceedings of the Royal Society of London. *Ser. B Biol. Sci.* **2000**, *267*, 1239–1244. [CrossRef]
35. Jiao, Y.; Gorb, S.N.; Scherge, M. Adhesion measured on the attachment pads of Tettigonia viridissima (Orthoptera, insecta). *J. Exp. Biol.* **2000**, *203*, 1887–1895. [CrossRef]
36. Gorb, E.; Kastner, V.; Peressadko, A.; Arzt, E.; Gaume, L.; Rowe, N.; Gorb, S.N. Structure and properties of the glandular surface in the digestive zone of the pitcher in the carnivorous plant Nepenthes ventrata and its role in insect trapping and retention. *J. Exp. Biol.* **2004**, *207*, 2947–2963. [CrossRef]
37. Wheater, C.P.; Evans, M.E.G. The mandibular forces and pressures of some predacious Coleoptera. *J. Insect Physiol.* **1989**, *35*, 815–820. [CrossRef]
38. Švantner, M.; Kučera, M.; Smazalová, E.; Houdková, Š.; Čerstvý, R. Thermal effects of laser marking on microstructure and corrosion properties of stainless steel. *Appl. Opt.* **2016**, *55*, D35–D45. [CrossRef] [PubMed]
39. Frönd, M.; Ventzke, V.; Riekehr, S.; Kashaev, N.; Klusemann, B.; Enz, J. Microstructure and hardness evolution of laser metal deposited AA5087 wall-structures. *Procedia CIRP* **2018**, *74*, 131–135. [CrossRef]
40. Tang, M.; Shi, Y.; Zhu, W.; Zhang, N.; Zhang, L. A Convenient and High-Efficient Laser Micro-Engraving Treatment for Controllable Preparation of Microstructure on Al Alloy. *Materials* **2018**, *11*, 2297. [CrossRef] [PubMed]
41. Machida, M.; Niidome, T.; Onoe, H.; Heisterkamp, A.; Terakawa, M. Spatially-targeted laser fabrication of multi-metal microstructures inside a hydrogel. *Opt. Express* **2019**, *27*, 14657–14666. [CrossRef] [PubMed]
42. Ackerl, N.; Fisch, G.; Auerswald, J.; Wegener, K. Evolution of microstructures on stainless steel induced by ultra-short pulsed laser ablation. *SN Appl. Sci.* **2020**, *2*, 652. [CrossRef]
43. Yang, S.-N.; Liu, X.-Q.; Zheng, J.-X.; Lu, Y.-M.; Gao, B.-R. Periodic Microstructures Fabricated by Laser Interference with Subsequent Etching. *Nanomaterials* **2020**, *10*, 1313. [CrossRef]
44. Henriksen, N.G.; Andersen, O.Z.; Jellesen, M.S.; Christiansen, T.L.; Somers, M.A.J. Influence of Laser Marking on Microstructure and Corrosion Performance of Martensitic Stainless Steel Surfaces for Biomedical Applications. *HTM J. Heat Treat. Mater.* **2022**, *77*, 177–196. [CrossRef]
45. Henriksen, N.G.; Poulios, K.; Somers, M.A.J.; Christiansen, T.L. Impact of laser marking on microstructure and fatigue life of medical grade titanium. *Mater. Sci. Eng.* **2023**, *873*, 145020. [CrossRef]
46. Richter, K.; Grunwald, I.; von Byern, J. Bioadhesives. In *Handbook of Adhesion Technology*; da Silva, L.F.M., Ochsner, A., Adam, R.D., Eds.; Springer International Publishing: Cham, Switzerland, 2018; pp. 1–45. [CrossRef]
47. Liang, L.; Lu, L.; Xing, D.; Wan, Z.; Tang, Y. Preparation of superhydrophobic and anti-resin-adhesive surfaces with micro/nanoscale structures on high-speed steel via laser processing. *Surf. Coat. Technol.* **2019**, *357*, 57–68. [CrossRef]
48. Li, X.; Yang, Y. 3D Printing Collembola Cuticle Inspired Superhydrophobic Microstructures for Potential Deicing Application. In Proceedings of the 2020 International Symposium on Flexible Automation, Chicago, IL, USA, 9–10 July 2020; p. V001T07A009. [CrossRef]
49. Bing, W.; Wang, H.; Tian, L.M.; Zhao, J.; Jin, H.C.; Du, W.B.; Ren, L.Q. Small Structure, Large Effect: Functional Surfaces Inspired by Salvinia Leaves. *Small Struct.* **2021**, *2*, 2100079. [CrossRef]
50. Chen, L.; Duan, Y.; Cui, M.; Huang, R.; Su, R.; Qi, W.; He, Z. Biomimetic surface coatings for marine antifouling: Natural antifoulants, synthetic polymers and surface microtopography. *Sci. Total Environ.* **2021**, *766*, 144469. [CrossRef]
51. Lu, L.; Yao, W.; Xie, Y.; Li, K.; Wan, Z. Study on the wettability of biomimetic stainless-steel surfaces inspired by Bauhinia Linn. *Leaf Surf. Coat. Technol.* **2021**, *405*, 126721. [CrossRef]
52. Castro-Mora, M.; Vásquez-González, M.; Cordero-Guerrero, J.; Benavides-Acevedo, M.; González, J.; López-Brenes, M.J.; Vega-Baudrit, J.; Corrales-Ureña, Y. Bacterial anti-adhesive films of PDMS coated with microstructures of biogenic silica rosettes extracted from pineapple peels residues. *Surf. Interfaces* **2022**, *30*, 101881. [CrossRef]

53. Oopath, S.V.; Baji, A.; Abtahi, M. Biomimetic Rose Petal Structures Obtained Using UV-Nanoimprint Lithography. *Polymers* **2022**, *14*, 3303. [CrossRef]
54. Zhao, J.; Zhang, L.; Cheng, X.; Wang, J.; Li, Y.; You, J. Programmable Transition between Adhesive/Anti-Adhesive Performances on Porous PVDF Spheres Supported by Shape Memory PLLA. *Polymers* **2022**, *14*, 374. [CrossRef]
55. Lifka, S.; Stecher, C.; Meyer, M.; Joel, A.-C.; Heitz, J.; Baumgartner, W. Biomimetic, antiadhesive surface structure inspired by the calamistra setae of cribellate spiders for electrospun nanofiber handling. *Front. Ecol. Evol.* **2023**, *11*, 99355. [CrossRef]
56. Liu, Z.; Wu, F.; Gu, H.; Feng, J.; Huang, D.; Zheng, L.; Yao, G.; Chen, Z.; Wang, C. Adhesion failure and anti-adhesion bionic structure optimization of surgical electrodes in soft tissue cutting. *J. Manuf. Process.* **2023**, *89*, 444–457. [CrossRef]
57. Petersen, D.S.; Gorb, S.N.; Heepe, L. The influence of material and roughness on the settlement and the adhesive strength of the barnacle Balanus improvisus in the Baltic Sea. *Front. Mar. Sci.* **2020**, *7*, 1–14. [CrossRef]
58. Petersen, D.S.; Heepe, L.; Gorb, S.N. Porous silicone substrates inhibit permanent barnacle attachment under natural conditions. *Biointerphases* **2020**, *15*, 061013. [CrossRef]
59. Petersen, D.S.; Schultz, M.; Gorb, S.N.; Heepe, L. A systematic investigation into the effect of fibrillar microstructures on the settlement and attachment strength of the bay barnacle Balanus improvisus under natural conditions. *Appl. Phys. A* **2020**, *126*, 1–11. [CrossRef]
60. Petersen, D.S.; Kleinteich, T.; Gorb, S.N.; and Heepe, L. Competing with barnacle cement: Wetting resistance of a re-entrant surface reduces underwater adhesion of barnacles. *J. R. Soc. Interface* **2018**, *15*, 20180396. [CrossRef]
61. Kim, P.; Wong, T.S.; Alvarenga, J.; Kreder, M.J.; Adorno-Martinez, W.E.; Aizenberg, J. Liquid-infused nanostructured surfaces with extreme anti-ice and anti-frost performance. *ACS Nano* **2012**, *6*, 6569–6577. [CrossRef]
62. Epstein, A.K.; Wong, T.S.; Belisle, R.A.; Aizenberg, J. Liquid-infused structured surfaces with exceptional anti-biofouling performance. *Proc. Natl. Acad. Sci. USA* **2012**, *109*, 13182–13187. [CrossRef]

biomimetics

MDPI

Article

Anti-Weightlessness Physiological Protection for the Lower Limb Muscle System Based on Biomimetic Adhesive Force Stimulation

Yuanming Ji [1], Zhili Li [2,*], Peng Zou [2], Chengyang Li [1], Xipeng Wang [1], Xiyue Yang [3], Zhendong Dai [1] and Keju Ji [1,*]

1 Jiangsu Provincial Key Laboratory of Biomimetic Materials and Devices, Nanjing University of Aeronautics and Astronautics, Nanjing 210016, China; jiyuanming@nuaa.edu.cn (Y.J.); lichengyang@nuaa.edu.cn (C.L.); wxp114@nuaa.edu.cn (X.W.); zddai@nuaa.edu.cn (Z.D.)

2 State Key Laboratory of Space Medicine, China Astronaut Research and Training Center, Beijing 100094, China

3 Electric and Information Engineering, Electric and Electronic Engineering, Imperial College London, South Kensington Campus, London SW7 2AZ, UK; aryayang1017729@163.com

* Correspondence: li_zhili777@163.com (Z.L.); jikeju@nuaa.edu.cn (K.J.)

Abstract

With the advancement of crewed spaceflight, mitigating the physiological effects of microgravity, such as bone–muscle deterioration and movement instability, has become increasingly vital. Inspired by reptilian climbing mechanisms, this study presents a novel bio-inspired adhesive footwear characterized by low pre-load, strong adhesion, and controllable attachment–detachment capability. This study analyzes the adaptability of a multi-level variable modulus design to surfaces with varying curvatures and roughness. Experimental investigations were conducted to analyze the contact mechanics and interfacial mechanisms of biomimetic adhesive materials featuring microstructure arrays. Moreover, stepping exercises were performed by volunteers wearing the proposed footwear under simulated weightlessness to assess biomechanical performance. Interface contact stresses were measured using force-sensing array plates, enabling characterization of plantar adhesion under different detachment speeds and angles. Electromyographic signals from lower limb muscle groups during stepping exercises were analyzed to elucidate the mechanical stimulation patterns and effects induced by plantar adhesion forces. Results indicate that plantar adhesion forces ranging between 50 and 105 N effectively stimulate primary flexor muscles, including the biceps femoris and gastrocnemius. This biomimetic solution offers a flexible and convenient approach for stabilizing foot positioning and promoting musculoskeletal engagement in microgravity, improving astronauts' mobility and operational performance in orbit.

Keywords: biomimetic adhesion; multi-level hierarchy; microstructure; microgravity countermeasure; van der Waals force; electromyography (EMG) signals

1. Introduction

Space exploration represents the frontier of human understanding and the transformation of nature, rendering it a pursuit of paramount importance. The development of crewed space technology directly reflects a nation's comprehensive strength and enhances its international influence. Although modern spacecraft are equipped with advanced automated and intelligent systems capable of autonomously performing various tasks, such as observation, data collection, and processing, human involvement remains indispensable.

Astronauts' on-site judgment, emergency response capabilities, and adaptability to complex situations are critical for mission success. Therefore, astronauts constitute the most valuable resource in space exploration and drive the continuous advancement of human space endeavors.

In outer space, astronauts encounter challenges in daily life and work that differ drastically from those on Earth. In microgravity, foot muscles no longer support the body's weight, requiring astronauts to adjust their posture and movement to maintain balance. This adjustment requires continuous exertion of specific muscle groups. Prolonged maintenance of these postures impacts work efficiency and damages the muscular system's structure and function. In most simulated weightless environments, such as those created by suspension systems or air-cushion beds, the maximum plantar force experienced by astronauts is reduced by approximately 70–82% compared to the gravitational force on Earth, with the vertical impulse decreasing by 35–52% [1–5]. Although gait cycle times do not significantly increase compared to those in normal gravity conditions, the stance phase, which plays a crucial role, is significantly shortened. The notable reductions in maximum plantar force, stance duration, and vertical impulse indicate diminished mechanical stimulation to the lower limb muscular system, leading to muscle atrophy. Consequently, astronauts often experience impaired musculoskeletal function upon returning to Earth after extended space missions. Additionally, the plantar surface cannot generate a significant frictional cone with contact surfaces without gravity, causing astronauts to float unrestrained in microgravity. Therefore, enabling astronauts to achieve flexible and reliable position fixation, walking, and exercise in microgravity, resembling their movement on Earth, remains a primary focus in microgravity countermeasure research.

Existing exercise methods, including treadmill training, resistance training, and cycling [6,7], support the countermeasure needs in weightless environments. However, these devices are bulky and do not provide comprehensive musculoskeletal training, which can still affect muscle function. Recent studies have shown that during running exercises in space [8–11], the maximum plantar force is only 54% of that on Earth, insufficient to adequately stimulate lower limb muscles and prevent muscle degradation [12]. The existing foot fixation method using restraint bands [13] relies on a cooperative fixation approach that lacks flexibility and can cause fatigue or injury due to prolonged dorsal foot pressure. Consequently, astronauts often experience difficulty readapting to Earth's gravity upon completion of long-term space missions.

Among the existing fixation methods, magnetic adhesive nylon hook-and-loop fasteners [14] require compatible surfaces for effective adhesion, limiting astronaut mobility and task performance. Moreover, the magnetic fields generated by magnetic adhesion often interfere with onboard instruments, risking performance degradation, malfunction, or even damage. Claw-like attachments [15], although enabling fixation, are only effective on rough surfaces, restricting their applicability. Therefore, an effective device is required that can support astronauts' positional stability, walking, and exercise.

Animals such as geckos, flies, and spiders rely on the van der Waals forces between their feet and the target surfaces to achieve omnidirectional climbing ability. The attachment ability of these animals [16–19] is due to their foot adhesion functional units, which feature arrays of setae combined with multi-level adaptive structures, enabling reliable contact with the target surface. In 2014, the Russian space agency sent a gecko into space [20–23]; the gecko successfully maintained stable interface attachment and movement within the microgravity environment of the space station, demonstrating the effectiveness and adaptability of van der Waals-based mechanisms. The ability to flexibly attach to and climb vertical and ceiling surfaces in space can address astronauts' needs for secure surface fixation and enhanced stability in microgravity environments, providing a biological blueprint for assistive devices. This could enable astronauts to achieve omnidirectional mobility similar to that of climbing organisms while operating in space.

This paper presents a biomimetic adhesive shoe that integrates an array of microstructured adhesive units in its sole, composed of a biomimetic adhesion layer, a cushioning adaptive layer, and an upper wearable portion. Experiments were conducted to investigate and analyze the adhesive mechanical properties and underlying mechanics of the biomimetic adhesive material. Interface contact mechanics were evaluated under various surface conditions and attachment–detachment parameters (detachment speed, angle, preload, and substrate roughness). The feasibility of the biomimetic adhesive shoe as an assistive tool for lower limb musculoskeletal training was validated by stepping exercises performed in simulated microgravity conditions

2. Experimental Section

2.1. Fabrication of Biomimetic Adhesive Shoe

The biomimetic adhesive shoe comprises a biomimetic adhesion layer, a cushioning adaptive layer, and an upper wearable portion. The biomimetic adhesion layer was fabricated using a biomimetic adhesive material featuring an array of microstructures. Specifically, the photoresist was exposed and developed by a photolithography machine (Suss MA-6) using the photolithography process, and a mold with a microstructure array was obtained. Then, the prepolymer silica gel (No. 4130) was prepared using glue injection equipment at a rate of 2 mL/s, 0.4 MPa pressure into the microstructure mold to pour molding, which was then heated and solidified at 70 °C for 10 h; then, the mold was released to obtain the bionic adhesive layer. A porous nylon felt material (primary cushioning adaptive layer) with a tensile strength of 3.2 MPa and a tear strength of 1.3 MPa was used as a backing substrate to ensure robust bonding with the cushioning adaptive layer. The nylon felt and silicone prepolymer were simultaneously introduced during the molding process, enabling integral fabrication of the nylon felt with the biomimetic adhesion layer.

To further improve the performance of the cushioning adaptive layer, we fabricated a secondary adaptive layer comprising a micropillar array structure using a combination of 3D printing and molding techniques. This secondary adaptive layer was made from 50 A rubber and thermally bonded to the backside of the nylon felt. The adhesive sole of the biomimetic shoe was designed as discrete, three-segment adhesive units: forefoot, midfoot, and rearfoot regions. Considering the shape characteristics of plantar contact areas, circular contact zones were designated in the forefoot and rearfoot sections, ensuring that stress-concentrated regions during walking and stepping are effectively supported, thus dispersing localized stresses and enhancing comfort.

The upper wearable portion was crafted from cowhide leather and attached to the secondary cushioning adaptive layer using a combination of nylon-thread stitching and adhesive bonding. The final biomimetic adhesive shoe structure, from bottom to top, is as follows: the biomimetic adhesion layer, primary cushioning adaptive layer, secondary cushioning adaptive layer, and upper wearable portion.

2.2. Contact Mechanics and Exercise Efficacy Testing of the Biomimetic Adhesive Shoe

The experimental evaluation comprised tests on the biomimetic adhesive units and the adhesive shoe as a whole. The biomimetic adhesive unit with buffering and adhesion functions was labeled BA (Buffer and Adhesion), and its corresponding shoe was labeled BA-S (Buffer and Adhesion Shoes). The control groups included adhesive units and shoes without buffering layers, labeled N-B (No-Buffer) and N-B-S (No-Buffer Shoes), respectively. Additionally, adhesive units and shoes with buffering but without microstructured biomimetic adhesion layers were designated N-A (No-Adhesion) and N-A-S (No-Adhesion Shoes), respectively.

2.2.1. Contact Mechanics Testing of Biomimetic Adhesive Units

Normal adhesion forces were measured using a Bruker UMT tribometer (Saarbr ü cken, Germany Bruker). Samples of dimension 1 cm $\times$ 1 cm were prepared by precision cutting with a vibration knife, mounted onto the upper fixture using double-sided tape, while substrates of glass (surface roughness Sa = 5 μm) and aluminum alloy plate (Sa = 0.2 μm) were fixed onto the lower fixture. A preload of 5 N and a detachment speed of 1 mm/s were applied, and adhesion force changes over time were recorded. Tangential adhesion forces were evaluated on an aluminum alloy substrate using a self-developed 3D force sensor setup. Surface roughness levels (Sa = 0.02–5 μm) were verified by laser confocal microscopy; representative roughness maps and parameters are shown in Figure S1. A string attached to the edge of the biomimetic adhesive unit was subjected to increasing tangential pull until detachment from the substrate occurred.

The fatigue life of the biomimetic adhesive shoes was assessed using a rubber fatigue testing machine. Contact mechanics and surface morphology changes before and after fatigue testing were compared to evaluate durability and reusability.

2.2.2. Adhesion Force Testing of Biomimetic Adhesive Shoes

Adhesion performance under simulated microgravity conditions was tested using a self-developed microgravity simulation platform integrated with a force measurement platform. A constant-tension suspension system was used to emulate reduced effective body weight; the device layout is shown in Figure S2, and the statics/geometry leading to the approximately constant tensile support are detailed in Supplementary Methods S3.1 (Equations (S1)–(S4)). Ten volunteers (age: 20–35 years, mean age: 26 years; five males, five females) participated in the study. Each participant was suspended vertically using a constant-force suspension system to counterbalance body weight, thereby simulating microgravity. An aluminum alloy plate with a matte finish (roughness 0.2 μm) was mounted on the force platform. Participants performed adhesion and detachment actions between the shoe sole and aluminum alloy plate under five different detachment modes, with real-time interface contact mechanics data recorded by the platform.

Detachment methods included: vertical detachment, detachment from forefoot to rearfoot, detachment from rearfoot to forefoot, medial-to-lateral detachment, and lateral-to-medial detachment. Detachment speeds were controlled by varying stepping frequencies regulated by a metronome, enabling assessment of interface mechanics at different detachment velocities. Adhesion performance was further evaluated on substrates of varying roughness to characterize adaptability to diverse surface conditions.

2.2.3. Electromyography (EMG) Signal Testing

Surface EMG signals were acquired using the BTS FREEEMG 300 wireless electromyography system. Surface electrodes were placed on the major muscle groups of the lower limbs. Volunteers were first suspended in a simulated microgravity environment by the constant-force suspension device, performing stepping exercises without the biomimetic adhesive shoes for 5 min as a baseline control test. Subsequently, the tests were repeated with the biomimetic adhesive shoes. EMG signals from selected muscle groups were continuously recorded throughout the tests. Raw EMG signals were processed in MATLAB R2023b (9.16.0) 64-bit (MathWorks, Natick, MA, USA) using a 4th-order Butterworth bandpass filter to remove noise outside the frequency range of 20–400 Hz, followed by a root mean squared (RMS) smoothing procedure with a window size of 50 data points to obtain RMS-processed EMG values. Surface electrode locations for BF, TA, LG, GM, and VL followed standard guidelines; a placement map is provided in Figure S3 (Supplementary Materials). All anonymized paired observations (per subject and condition; loads, EMG_{RMS}, $\Delta EMG\%$, and trial indices) are available in Dataset S1.

3. Results and Discussion

Inspired by the hierarchical setal array of gecko toes, the biomimetic adhesive shoe integrates multi-level, cross-scale cushioning adaptive layers with a microstructured adhesive layer. L_0 denotes the characteristic length/compliance-path scale of the macro-buffer layer, governing energy dissipation in the load–deformation process and the redistribution of interfacial loads; L_1 denotes the characteristic microstructural contact length, primarily controlling the evolution of real contact area and the efficiency of adhesive-energy utilization. This design aims to provide astronauts with an efficient fixation and walking method based on biomimetic adhesion technology under microgravity conditions (Figure 1a). Following a Buffer→load-shaping→Adhesion (BA) chain, L_0 facilitates pressure equalization and peak attenuation, whereas L_1 affords van der Waals–dominated, direction-selective traction with reversible release under modest preload and a small tangential bias; key geometric/operating parameters (L_0, L_1, F_N, peeling angle α) are annotated in Figure 1a to enable direct mapping from biology to engineering. Figure 1b illustrates the morphology of the biomimetic adhesive shoe sole, highlighting the multi-level cushioning adaptive layers and an outer biomimetic adhesive layer. The adhesive layer comprises a hexagonal prism array with an average circums cribed diameter of approximately 40 μm, an average height of 15 μm, and an average gap of 4 μm. The prisms are arranged in an equidistant pattern, with a center-to-center spacing of 45 μm between adjacent prisms.

The macro cushioning adaptive layers significantly enhance adaptability to adhesive surfaces with varying curvature and friction coefficients. Specifically, the macro-scale cushioning layer adapts to the target surface curvature through substantial deformation under moderate preloads, while the microstructured adhesive layer further accommodates surface roughness, increasing real interfacial contact at the microscale and effectively adapting to substrates of different roughness (Figure 1c).

3.1. Contact Mechanics Behavior of Biomimetic Adhesive Units

The contact mechanical properties of biomimetic adhesive units on various target surfaces are presented in Figure 2. Figure 2a presents a schematic diagram of the biomimetic adhesive unit (BA) and comparative structural units (N-B and N-A). The experimental apparatus and testing methods used to evaluate contact mechanical behavior are illustrated in Figure 2b.

Figure 2c displays the variation in normal adhesion forces with preload ranging from 0.1 N to 8 N, at a surface roughness of 0.2 μm and detachment speed of 3 mm/s. The BA unit exhibits a significant increase in adhesion force (over tenfold), demonstrating effective adhesion and stable attachment at relatively low preload. Conversely, the adhesion force of the N-A unit is substantially lower owing to the absence of microstructures, resulting in reduced adaptability and contact area. For the BA unit, adhesion force plateaus beyond a preload of approximately 5 N, indicating that the microstructural elements achieve near-complete molecular-level contact, and additional preload yields minimal gains. The N-B unit requires a higher preload peak compared to the BA unit, as the buffer layer in the BA structure more evenly distributes stress, enabling complete contact at lower preload values (Figure 2c inset).

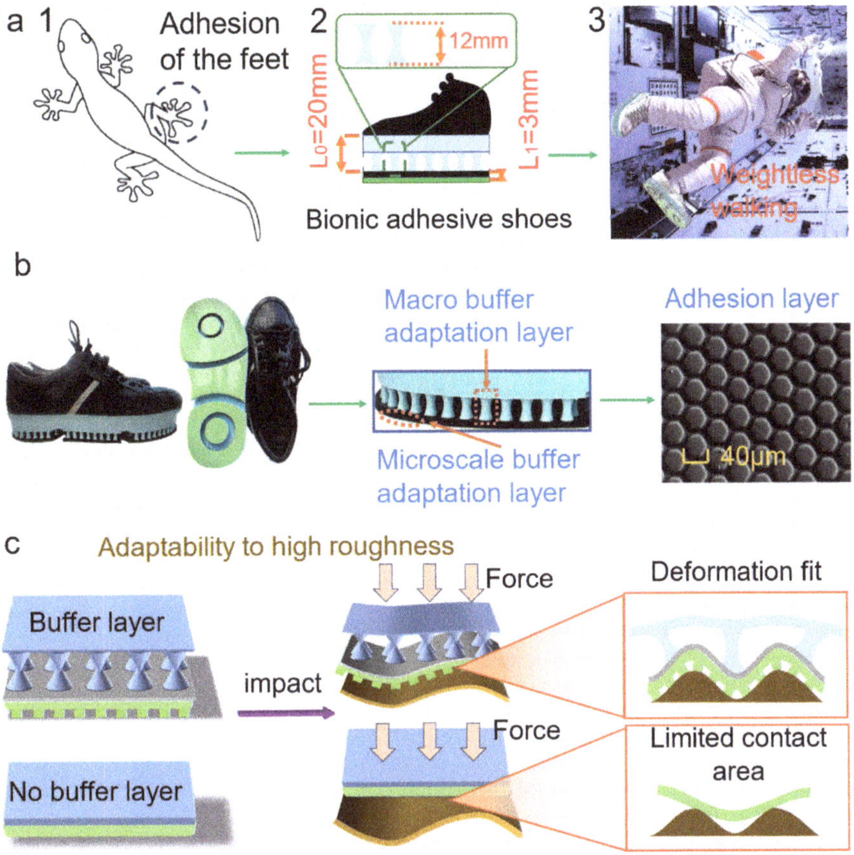

Figure 1. (**a**) (**1**) The gecko's foot morphology during climbing, (**2**) the structure of the biomimetic adhesive shoe inspired by this mechanism, and (**3**) simulated walking in a weightless space environment using the biomimetic adhesive shoe; (**b**) side-top view of biomimetic adhesive shoe, small local enlarged view, and topography of adhesive microstructure arrays; (**c**) schematic diagram of the adaptability of the multi-level buffer structure to the surface and roughness.

Figure 2d shows the adhesion force of the biomimetic adhesive units at different detachment speeds, with a fixed preload of 5 N and substrate roughness (Sa) of 0.2 μm. The adhesion force increases from approximately 7 N at 0.5 mm/s to approximately 11 N at 3 mm/s. Beyond 3 mm/s, adhesion force plateaus, indicating that the molecular adhesion formation rate has reached its limit. According to the momentum theorem (ft = mv), higher detachment speeds reduce detachment time and thus require greater force. The BA group exhibits higher adhesion than the other groups owing to the increased microscopic contact area facilitated by its microstructured adhesive layer. The lack of a buffer layer in N-B causes uneven stress distribution and a slight reduction (approximately 5%) in adhesion. The N-A group shows the lowest adhesion owing to its minimal contact area.

Figure 2e presents the adhesion performance of units tested on acrylic substrates with varying roughness levels (detachment speed of 3 mm/s, preload of 5 N). Corresponding roughness maps for the tested substrates are provided in Figure S1, which corroborate the monotonic decrease in adhesion with increasing Sa. Adhesion force decreases linearly with increasing surface roughness, as higher roughness reduces the microscopic contact area.

Notably, at very low roughness (Sa < 0.1 μm), N-A exhibits significantly higher adhesion (over 250%) due to its larger effective contact area on smoother surfaces. However, BA outperforms N-B at higher roughness values owing to its buffering layer, enabling better conformity and contact. Adhesion significantly decreases from Sa = 2.0 μm to Sa = 3.5 μm, and nearly vanishes at Sa = 5.0 μm due to reduced molecular contact probability.

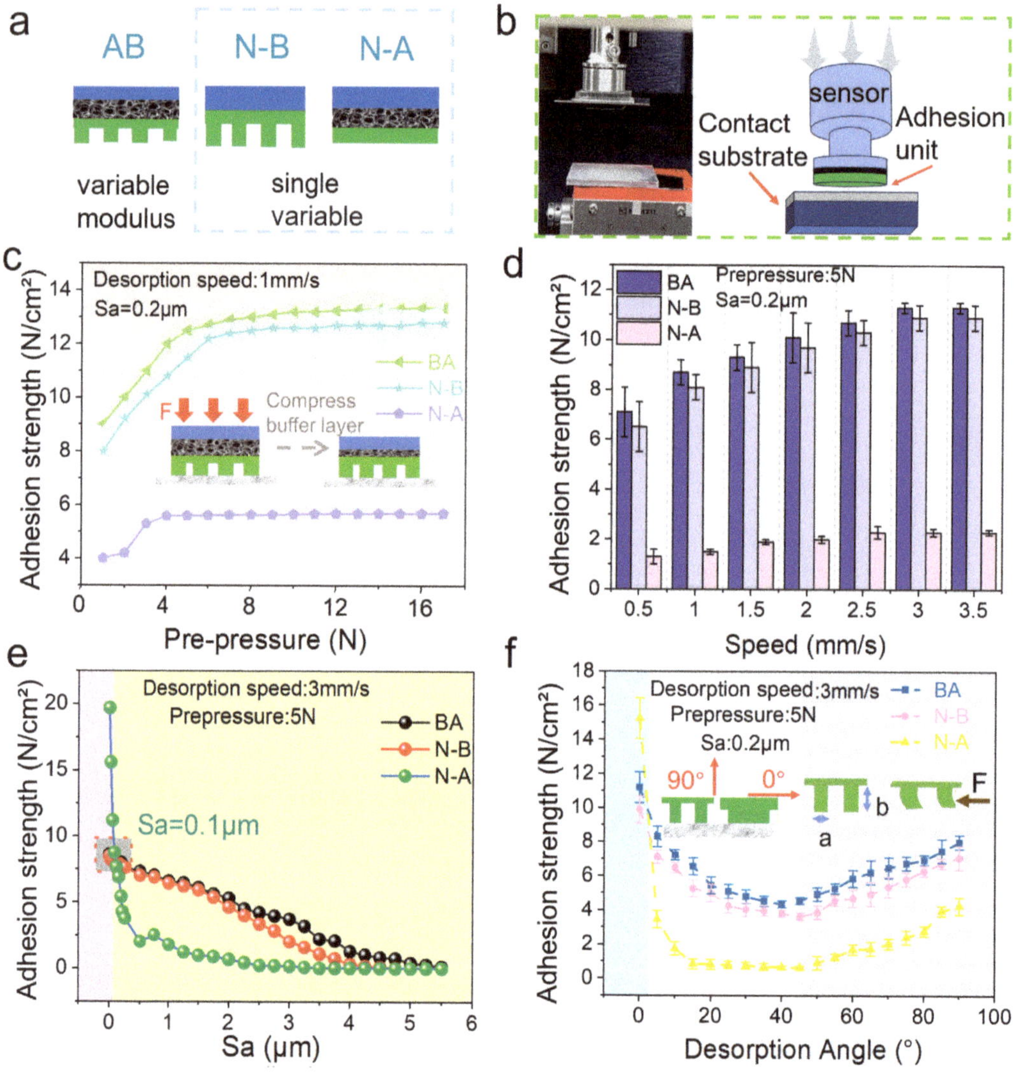

Figure 2. (**a**) Samples with buffer layer and microstructure layer (AB), samples with buffer layer and microstructure layer (N-B), and samples with buffer layer and no microstructure layer (N-A); (**b**) physical drawings and schematic diagrams of the experimental apparatus for testing biomimetic adhesive materials; (**c**) adhesion of the sample under different pre-pressures; (**d**) adhesion of the sample at different detachment speeds; (**e**) adhesion of the sample on substrates with different roughness; (**f**) adhesion of the sample at different angles, with insets showing the detachment direction and pillar height.

Figure 2f illustrates the relationship between adhesion force and detachment angle for the BA unit, measured using a custom-developed 3D force-sensing platform (detachment speed: 3 mm/s, preload: 5 N, substrate roughness: Sa = 0.2 μm). The adhesion force of the BA unit initially decreases from 8.7 N at 0° to 4.3 N at 45°, before increasing again to 7.6 N at 90°. The N-B unit experiences uneven tangential force distribution, resulting in lower detachment force and noticeable stick-slip behavior. Conversely, BA and N-A units demonstrate smoother detachment. At angles less than 3°, the adhesion of N-A is significantly higher (>30%) owing to the frictional contribution from its larger contact area. Beyond 3°, the adhesion decreases sharply. The inset on the left side of Figure 2f illustrates the direction of tensile force with or without adhesive layers. Beyond 45°, the adhesion force primarily arises from normal adhesion; hence, the BA unit performs better. The right inset of Figure 2f shows the dimensional relationship between pillar height (b) and diameter (a). With a height greater than the diameter, micropillars bend under angled tension, maintaining sufficient contact area and mitigating adhesion loss caused by angle variation.

Fatigue testing under repeated normal adhesion–detachment cycles (10,000 cycles) revealed no significant reduction in normal adhesion force, with surface morphology exhibiting only minor micro-cracks. This confirmed the long-term stability and applicability of the biomimetic adhesive units for engineering applications.

3.2. *Contact Mechanics Behavior of Biomimetic Adhesive Shoes Under Simulated Microgravity*

Participants wore biomimetic adhesive shoes and were suspended using a constant-force suspension system to simulate microgravity conditions (Figure 3a). Figure 3b illustrates the variations in plantar force recorded while participants stepped on a force-measurement platform, with stepping frequency controlled at 30 steps/min. At initial foot contact, the support force increased, peaking at approximately 100 N during the late stance phase. Subsequently, as the foot was raised, maximum plantar adhesion occurred before transitioning into the swing phase. Compared with a gravity-based load of approximately 600 N, this represents an approximately 85% reduction in force, confirming the effectiveness of microgravity simulation. The maximum normal adhesion force under simulated microgravity conditions was 75 ± 12 N, reflecting differences due to varying preload conditions during the stepping process.

Under vertical detachment conditions in a gravity environment, the adhesion performance was tested by participants lifting their feet to the same height (10 cm) at detachment speeds ranging from 30 cm/s to 150 cm/s (Figure 3c). Adhesion force increased with increasing detachment speed, with the peak force at 150 cm/s nearly double that at 30 cm/s. Additionally, the duration of force application decreased as speed increased. The detachment time at 30 cm/s was approximately five times longer than at 150 cm/s. Consequently, the adhesion force impulse decreased with increasing speed up to a certain threshold. Beyond approximately 110 cm/s, additional increases in adhesion force were minimal, indicating a limit in maximum achievable impulse given a fixed contact area.

Stepping exercises under simulated microgravity using different detachment modes (Figure 3d inset) revealed that the vertical detachment (simultaneous detachment) produced the highest adhesion force compared to the other four modes. The average adhesion force under simultaneous detachment was 67 ± 15 N (Figure 3d). This result occurs primarily because simultaneous stepping results in the heel and forefoot leaving the ground simultaneously, causing nearly all adhesive micropillars to detach simultaneously. Consequently, adhesion force predominantly acts in the normal direction, minimizing tangential friction and maximizing the observed adhesion.

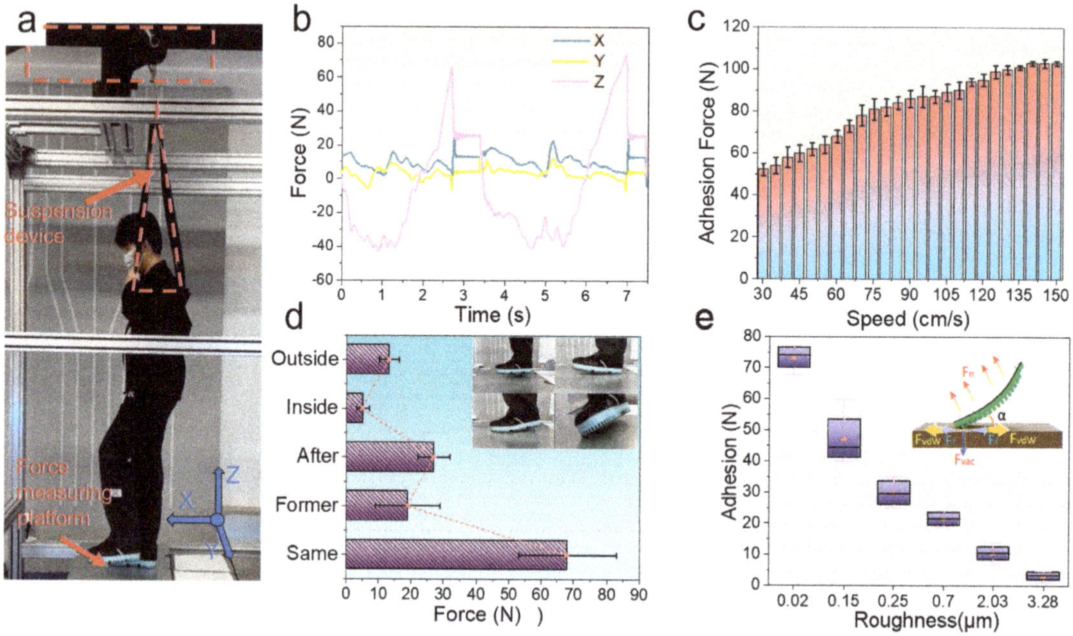

Figure 3. (**a**) Attitude diagram of the subject under the force measurement platform simulating a microgravity environment; (**b**) F_X, F_Y, and F_Z process data diagrams collected by the three-dimensional force measurement platform in the microgravity environment; (**c**) variation in adhesion of biomimetic adhesive shoes with different detachment speeds; (**d**) adhesion of the forefoot mode, hind heel mode, medial adhesion–detachment mode, and lateral detachment mode in microgravity, and the inset shows different detachment behavior patterns; (**e**) variation in the adhesion of biomimetic adhesive shoes under different roughness surfaces, and the inset shows a schematic diagram of microscopic detachment. A short demonstration of the suspension-based stepping tasks under the BA, N-A, and No-shoe conditions is provided in Video S1 (Supplementary Materials).

Compared to other detachment modes, lateral and medial detachment modes are significantly influenced by tangential friction forces. Given that the total adhesion force remains approximately constant, normal adhesion forces under these modes are substantially lower than those observed in vertical and rear-side detachment modes. Specifically, the average normal adhesion force in lateral detachment was only 18.5% and in medial detachment, 28.3% of that observed in vertical detachment. This disparity arises from the differing microstructural interactions during lateral or medial detachment. Assuming the adhesive units' micropillars contact a locally flat substrate, the adhesive interaction force (P_f) between the shoe sole and the surface [24,25] is given by

$$P_f = 6\pi r3K\omega_f \quad (1)$$

$$\omega_f = \gamma L(1 + \cos\theta) \quad (2)$$

where P_f represents the normal adhesion force (N), f denotes the effective adhesion energy (mJ/m^2), r is the radius of the micropillars (μm), K is the effective Young's modulus (MPa), and ω_f is the work of adhesion (W). γL represents liquid surface tension (N), and θ denotes the adhesion angle (°). According to Equations (1) and (2), the effective adhesion energy of the biomimetic adhesive shoe is influenced by the adhesion angle θ. During medial-side detachment, the inclination angle of the primary tensile force exerted on the lower leg relative to the z-axis ranges between 20° and 30°, with lateral forces on the coronal plane dominating the interaction. Consequently, approximately 30% of the adhesion force acts in the tangential direction and approximately 70% in the normal direction under gravitational conditions. As part of the tangential force is offset by interfacial stresses, detachment can be achieved using relatively smaller normal forces.

However, the maximum adhesion force under the forefoot detachment mode is only 8.9% of that observed during vertical detachment under identical conditions, considerably lower than other detachment modes. This significant reduction occurs because lifting the foot from the forefoot side creates a larger peeling angle approaching 90° between the forefoot area and the contact surface. A larger peeling angle facilitates crack propagation at the interface between the adhesive material and the substrate, enabling complete detachment with lower adhesion force. Conversely, the peeling angles in other detachment modes are typically smaller, rendering crack propagation more difficult and thus requiring higher forces to achieve detachment.

Surface roughness is a critical factor influencing the adhesion performance of the biomimetic adhesive shoes, particularly as contact surfaces in spacecraft exhibit varied roughness levels. Consequently, the biomimetic adhesive shoe must provide sufficient adhesion force while adapting effectively to different surface roughness conditions. The adhesion mechanism of biomimetic adhesive materials is illustrated schematically in the inset of Figure 3e. When micropillars are angled and pulled, both tangential friction forces and normal adhesion forces are generated, collectively ensuring adequate adhesion at certain inclination angles. For the biomimetic adhesive shoe with a multi-level hierarchical design under normal force, the frictional mechanism relies on dual effects: mechanical interlocking and van der Waals forces. Mechanical interlocking refers to the tangential resistance generated as the elastomer deforms and conforms to the microstructure arrays of the acrylic surface under normal force. As surface roughness increases, the effective interlocking area initially increases but decreases at higher roughness. Tangential adhesion forces resulting from van der Waals interactions can be expressed by Equation (3) [26–28]:

$$F_c \sim \sqrt{G_c}\sqrt{A/C} \quad (3)$$

where G_c represents the critical strain energy for interface separation, a constant for specific material interfaces. A denotes the actual contact area between the elastomer and the acrylic substrate, while C is the compliance of the elastomer sample in the direction of the frictional force. Increasing the interfacial contact area and decreasing the elastomer's compliance enhances the tangential adhesion force. The variable-modulus biomimetic adhesive units address the interface properties required for high adhesion forces. The high-modulus backing layer of the adhesive unit reduces the elastomer's tangential compliance, while the low-modulus contact layer increases compliance with surface roughness, thereby improving the actual contact area between the adhesive unit and the acrylic substrate. This results in a higher tangential adhesion force for the variable-modulus biomimetic adhesive unit.

As the surface roughness of the acrylic substrate increases from Sa = 0.02 μm to Sa = 3.3 μm, the adhesion force of the biomimetic adhesive shoes, tested under vertical detachment in a gravity environment, gradually decreases. A metronome was used to control the stepping frequency to 30 steps/min. Figure 3e shows that as the roughness increases, the adhesion force of the shoe decreases. At the microscale, higher roughness leads to fewer contact points between the biomimetic adhesive protrusions and the substrate, resulting in variations in the total adhesion force due to differences in the number of protrusions in contact with surfaces of varying roughness. Therefore, the performance of the biomimetic adhesive shoes can be estimated based on different levels of frictional forces.

In conclusion, increasing the contact area and reducing the compliance of the elastomer in the direction of the frictional force can significantly enhance the tangential adhesion force. The variable-modulus biomimetic adhesive unit optimizes this mechanism for effective performance across various surface conditions.

Three variants were considered to analyze the multi-level, cross-scale performance of the biomimetic adhesive shoes. The biomimetic adhesive shoe with a buffering layer and a biomimetic adhesive layer is designated as BA-S (buffer and adhesion shoes). The control group, comprising a shoe with an adhesive layer without a buffering layer, is labeled as N-B-S (no-buffer and adhesion shoes). The N-B-A (no-adhesion and buffer shoes) group refers to the biomimetic adhesive shoe with a buffering layer but without an adhesive layer (Figure 4a). The experiment was conducted with a constant impact speed of 50 cm/s, a preload of 70 N, and an impact applied to the aluminum plate substrate connected to sensors. The buffering time for the N-B-S group is approximately 2 s, compared to the BA-S and N-B-A groups (~3 s) (Figure 4b). This reduced buffering time in the N-B-S configuration results from the absence of a buffering layer, which shortens the compression phase during impact and accelerates step completion.

Comparing the BA-S and N-B-S groups, evidently, the lack of biomimetic microstructures in the N-B-S group results in lower adhesion force and shorter adhesion time than in the BA-S group. Force analysis during compression and tension of the biomimetic adhesive shoe (Figure 4b inset) shows that during compression, pressure forces are the primary forces at play, concentrating the compression in the cushioning layer through force transmission.

In the microgravity environment, comparison of the compression behavior of BAS and N-B-S under the same stepping pressure (Figure 4c) shows that the maximum compression of BAS during a single step can reach 9.9 mm, which is significantly greater than that of N-B-S (2.2 mm). This indicates that the cushioning layer of the biomimetic adhesive shoes can effectively reduce shocks and impacts during stepping and walking. As illustrated in the inset of Figure 4c, during rapid descent and ground contact, the cushioning layer of BAS first undergoes compression, thereby prolonging the force application time, attenuating the impact, and subsequently recovering.

The maximum adhesion force for BA-S (75 N) is higher than that for N-B-A (50 N). This difference can be explained by the schematic in Figure 4d, which illustrates the detachment process from the edge of the biomimetic adhesive shoe. As the BA-S shoe features micropillars on its contact surface, cracks that form during detachment must overcome additional obstacles during their expansion. When encountering protrusions, the cracks are forced to bypass these protrusions, increasing the crack propagation path length and energy consumption. This phenomenon is described by Equation (4) [29]:

$$\frac{F_s}{F_c} = \frac{E_s I_s w_s}{E_c I_c w_c} \tag{4}$$

where $\frac{F_s}{F_c}$ denotes the adhesion enhancement factor, E_s is the elastic modulus of the stiff region, I_s is the area moment of inertia of the stiff region, and w_s is the actual contact width of the stiff region. Similarly, E_c, I_c, and w_c represent the elastic modulus, area moment of inertia, and actual contact width of the compliant region, respectively.

The relationship reveals that increasing the contact width of the compliant region (w_c) effectively enhances the overall adhesion performance. Conversely, the N-B-S group lacks surface microstructures and presents a relatively smooth contact interface, enabling cracks to propagate more easily, thereby reducing adhesion strength.

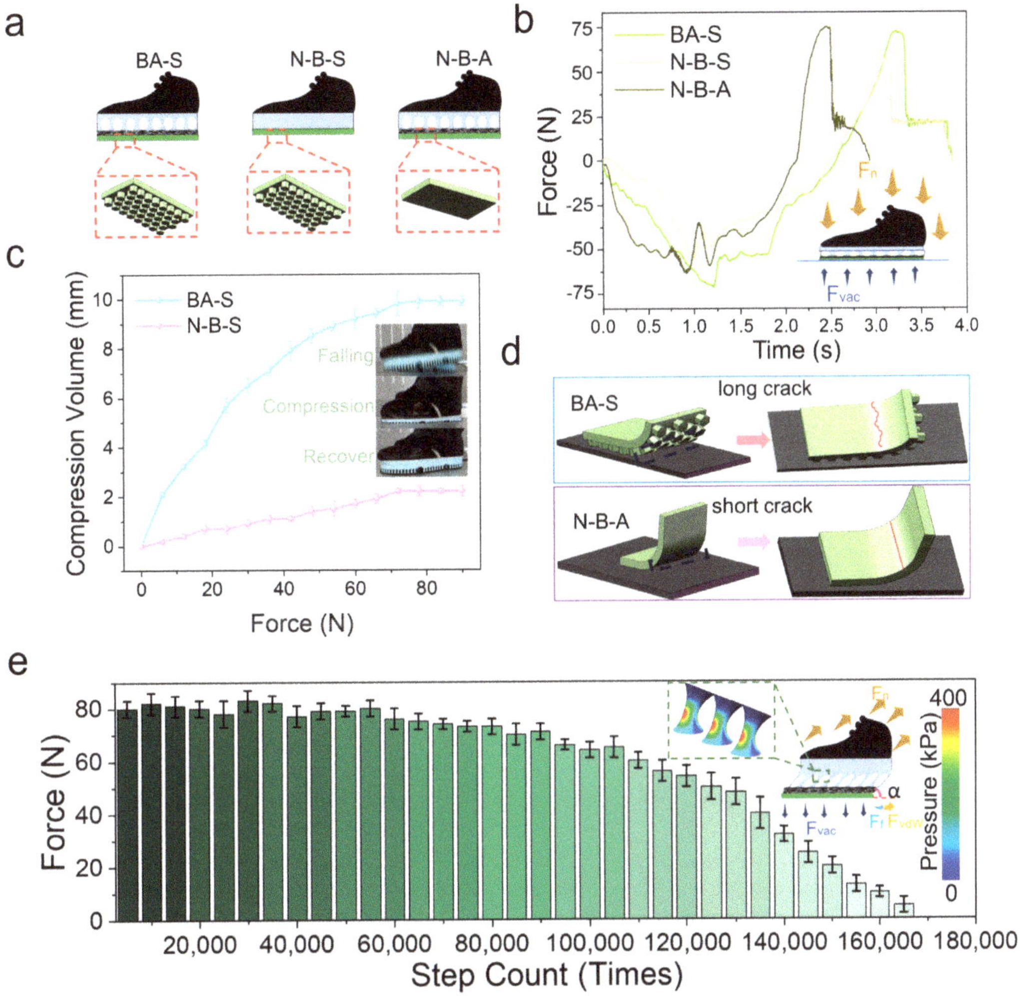

Figure 4. (**a**) Shoes with buffer layer and microstructure layer (BA-S), shoes with no buffer layer (N-B-S), and shoes with buffer layer and no microstructure layer (N-B-A); (**b**) mechanical cushioning responses of the three shoe types under a single impact, with the inset presenting the cushioning mechanics of the biomimetic adhesive shoes upon impact; (**c**) compression cushioning performance of BAS and N-B-S under varying impact conditions, with the inset showing the morphological evolution of the cushioning layer in BAS under a single impact; (**d**) crack propagation behavior of BAS and N-B-A during edge tearing; (**e**) variation in the adhesion force of BAS with increasing step cycles, with the inset illustrating the mechanical response of BAS under tensile loading and the corresponding stress distribution within the cushioning layer; contact pressure distribution (measurement).

Furthermore, protrusions on the contact surface induce stress concentration under loading. These stress concentration zones hinder crack propagation because greater energy is required for a crack to continue growing through these localized high-stress regions. This effect is particularly pronounced on surfaces with microstructural protrusions, improving overall adhesion strength due to increased crack resistance.

Figure 4e illustrates the relationship between the adhesion force of biomimetic adhesive shoes and the number of steps under microgravity conditions. The experiment was conducted at a stepping frequency of 30 steps/min on an aluminum alloy substrate. As shown in Figure 4e, the adhesion force of BAS decreased by approximately 12.5% within the first 90,000 steps, followed by a rapid decline thereafter. This degradation is attributed to the repeated action of normal preload, tensile force, and tangential friction, which damaged the micropillar structures of the adhesive layer, thereby reducing the intermolecular contact area between the adhesive layer and the substrate and ultimately lowering the adhesion force. The inset of Figure 4e illustrates the forces acting on BAS when stretched at an angle α, including the tensile force (F_n), ground reaction force (F_{vac}), adhesive resistance force (F_{vdW}), and tangential frictional force (F_f). Mechanical analysis of the cushioning layer indicates that, during tensile detachment at angle α, tangential frictional and normal adhesion forces are mitigated by the buffering effect of the micropillars, which also facilitates uniform stress distribution and concentrates the load within the micropillars. Within a two-dimensional plane-strain framework consisting of a rigid loading plate, an elastic buffer layer, an effective adhesion/friction interface, and a rigid substrate, a normal preload was applied to establish contact, followed by a small rotational perturbation to realize angled peeling. The computed interfacial normal contact pressure p_n and the von Mises stress field within the buffer layer conform to the colormap (0–400 kPa) in Figure 4e. With increasing peeling angle α, pressure magnitude and shear–tension coupling at the trailing edge are markedly amplified; the peak p_n increases and migrates toward the trailing edge, indicating an edge-dominated detachment process. Concurrently, the buffer layer exhibits a through-thickness "stress-diffusion cone," characterized by smoother iso-stress contours and a reduced peak-to-mean ratio, evidencing redistribution and diffusion of interfacial loads and the suppression of local stress concentrations. These trends suggest that a moderate buffer thickness promotes pressure homogenization and peak attenuation without compromising controllability, whereas an overly thin layer is insufficient for diffusion, and an overly thick layer introduces excessive compliance. The numerical results are consistent with the spatial distribution of the pressure scale in Figure 4e and provide a mechanistic rationale for the enhanced stability and comfort of the biomimetic adhesive footwear under angled loading α.

Under microgravity conditions, the biomimetic adhesive shoe demonstrates considerable application potential owing to its strong adhesive capabilities. One such application is in astronaut exercise, aiming to mitigate the risk of muscle atrophy. In this study, volunteers wore the biomimetic adhesive shoes while suspended by a constant-force suspension system to simulate microgravity, with EMG sensors attached to relevant lower limb muscle groups.

During leg-raising movements, the adhesion force provided by the biomimetic adhesive shoes simulated ground reaction forces to stimulate the muscular system. A control experiment was conducted with participants wearing shoes that lacked the biomimetic adhesive layer. An aluminum alloy plate with a surface roughness of Sa = 0.2 μm was used as the contact substrate. The stepping frequency was set to 30 steps/min, corresponding to a detachment speed of 75 cm/s for the adhesive shoe. The average detachment force recorded was approximately 80 N.

Figure 5a–e show EMG signals and corresponding RMS values from the main force-generating muscle groups during stepping, both with and without the biomimetic adhesive shoes. The multiple peaks in each EMG trace indicate repeated activation of the respective muscle groups during the stepping cycles.

EMG signals (RMS) were consistently weaker when participants did not wear the adhesive shoes. The biceps femoris exhibited the highest RMS values among the recorded muscle groups, indicating its dominant role in leg lifting regardless of shoe condition. However, for muscles such as the tibialis anterior, peroneus longus, medial gastrocnemius, and vastus lateralis, EMG values were significantly higher when participants wore the biomimetic adhesive shoes, suggesting that detachment motions relied more heavily on calf muscle contractions. *An* engineering threshold was defined from the $\Delta EMG\%$–F_n curve as the knee point of a piecewise linear fit with two lines; this knee point represents the minimum F_n that yields a clear increase in $\Delta EMG\%$. As a stability cross-check, a $\geq$95% gait success rate (absence of premature detachment or major slip) was additionally required. Group-level recommendations derived from these knee points are provided in Dataset S2; per-subject thresholds with fit diagnostics and 95% confidence intervals are listed in Dataset S3; the paired raw cells used for estimation ($\Delta EMG\%$ and F_n) are provided in Dataset S4.

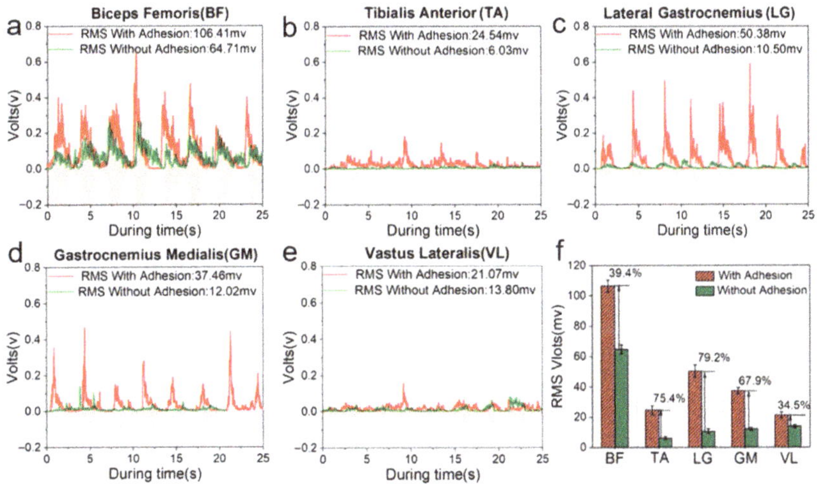

Figure 5. (**a–e**) Root mean squared results of the biceps femoris muscle, tibialis anterior, peroneal lateralis muscle, peroneal medialis, and vastus lateralis muscle in a microgravity environment, with the raw EMG data showing shades; (**f**) root mean squared average value of EMG leg lifting movements in subjects with (red) and without (green) biomimetic adhesive shoes in a microgravity environment.

Figure 5f directly compares EMG amplitudes for each muscle group with and without the adhesive shoes. The results show significant increases in EMG signal amplitudes when using the biomimetic adhesive shoes, with improvements ranging from 34.5% to 79.2%. Statistical details (normality checks; paired t/Wilcoxon; one-/two-way ANOVA for shoe $\times$ speed/roughness; p-values, effect sizes, and 95% CIs) are summarized in Dataset S5. This demonstrates the effectiveness of the biomimetic adhesive shoes in stimulating lower limb muscle activity.

Compared to traditional exercise devices, these shoes stimulate key flexor groups such as the biceps femoris and gastrocnemius during microgravity stepping tasks, supporting their application in astronaut musculoskeletal conditioning.

4. Conclusions

Inspired by the omnidirectional mobility of climbing organisms, this study proposes a multi-level, variable-modulus, detachable biomimetic adhesive shoe. Experimental investigations under various detachment speeds, angles, and preload conditions demonstrate that the biomimetic adhesive shoe possesses low preload, strong adhesion, and detachable capabilities. During stepping exercises, the shoe generates plantar adhesion forces ranging from 50 to 105 N. The shoe comprises an upper layer, a cushioning layer, and an adhesive layer. The adhesive layer adapts to substrates with different curvatures, roughness, and materials, while the cushioning layer effectively mitigates impacts during the detachment process, extending the duration of the force and ensuring stable detachment.

Additionally, the EMG and heart rate data collected from participants wearing the biomimetic adhesive shoes show significant muscle stimulation, particularly in the biceps femoris and other flexor muscles. This research highlights the potential for developing a biomimetic adhesive shoe designed explicitly for flexor muscle training, offering new exercise methods and operational modes for astronauts.

Supplementary Materials: The following supporting information can be downloaded at: https://www.mdpi.com/article/10.3390/biomimetics10120800/s1, Figure S1: Contact surface roughness maps and parameters measured by laser confocal microscopy across the tested Sa range (0.02–5 μm); Figure S2: Constant-tension suspension system for the simulated-microgravity trials—device photo and schematic; the mechanical-performance explanation is provided in Supplementary Methods S3.1, with statics and geometric derivations (Equations (S1)–(S4)); Figure S3: Electrode placement for lower-limb EMG (BF, TA, LG, GM, VL); Video S1: Demonstration of suspension-based microgravity stepping (BA/N-A/No-shoe conditions); Dataset S1 (CSV): Per-subject, per-condition paired observations for the EMG–contact-mechanics study (metadata, loads, EMG_RMS, $\Delta EMG\%$, trial indices); Dataset S2 (CSV): Group-level threshold recommendations from two-segment fits of ΔEMG% versus F_n with the ≥95% gait-success engineering check; Dataset S3 (CSV): Per-subject thresholds (e.g., F_{knee} and stability-checked values) with fit diagnostics and 95% CIs; Dataset S4 (CSV): Raw paired cells used for threshold estimation ($\Delta EMG\%$ and F_n by subject/condition/level) including trial identifiers and covariates; Dataset S5 (CSV): Statistical summary for EMG and adhesion (normality tests; paired *t*/Wilcoxon; one-/two-way ANOVA for shoe × speed/roughness) with *p*-values, effect sizes (Cohen's *d*, η^2/partial-η^2), and 95% confidence intervals.

Author Contributions: Conceptualization, Y.J., K.J. and Z.D.; Methodology, Y.J., Z.L., P.Z., C.L., X.W., X.Y., Z.D. and K.J.; Software, Y.J. and P.Z.; Validation, Y.J., Z.L. and X.W.; Formal Analysis, Y.J., Z.L., P.Z., X.W., Z.D. and K.J.; Investigation, Y.J., Z.L., P.Z. and X.W.; Resources, Y.J. and X.Y.; Data Curation, Y.J. and K.J.; Writing—Original Draft Preparation, Y.J.; Writing—Review & Editing, Y.J. and C.L.; Visualization, Y.J. and P.Z.; Supervision, Y.J., P.Z., X.Y. and K.J.; Project Administration, K.J.; Funding Acquisition, Z.D. and K.J. All authors have read and agreed to the published version of the manuscript.

Funding: This work was supported by the Space Medical Experiment Project of CMSP (No. HYZHXMH01015 and No. HYZHXM01009), by the National Natural Science Foundation of China (No. 52075249 and No. 62233008), and the National Defense Basic Scientific Research program of China (No. JCKY2024205C012), the SCNU Training Program of Innovation for Undergraduates (No. 202510287041Y).

Institutional Review Board Statement: The data collection for this experiment has been for-mally endorsed by Scientific Research Ethics Committee of Jiangsu Institute of Sports Sci-ence (Approval ID: JISS20220009, date of approval: 14 February 2022).

Informed Consent Statement: Informed consent was obtained from all subjects involved in the study.

Data Availability Statement: The main data that support the results in this study are available within the paper and the Supporting Information or are available for research purposes from the corresponding authors upon reasonable request.

Acknowledgments: The authors thank the astronauts of the Shenzhou series for their support of the experiment, and thank Nanjing Adhesion Technology Co., Ltd. for bionic adhesive material processing support.

Conflicts of Interest: The authors declare no conflicts of interest.

References

1. Juhl IV, O.J.; Buettmann, E.G.; Friedman, M.A.; DeNapoli, R.C.; Hoppock, G.A.; Donahue, H.J. Update on the effects of microgravity on the muscular system. *NPJ Microgravity* **2021**, *7*, 28. [CrossRef] [PubMed]
2. Meenakshi, P.; Hargens, A.R. Ground-Based Analogs for Human Spaceflight. *Front. Physiol.* **2020**, *11*, 716. [CrossRef] [PubMed]
3. Lang, T.; Van Loon, J.J.W.A.; Bloomfield, S.; Vico, L.; Chopard, A.; Rittweger, J.; Kyparos, A.; Blottner, D.; Vuori, I.; Gerzer, R.; et al. Towards human exploration of space: The THESEUS review series on muscle and bone research priorities. *NPJ Microgravity* **2017**, *3*, 8. [CrossRef] [PubMed]
4. Yang, F.; Chen, C.S.; Yang, C.H.; Chen, R.J.; Liu, Z.R.; Wen, L.; Xiao, H.F.; Zhou, L.B.; Geng, B.; Xia, Y.Y. Exosome-mediated perturbation of the immune-bone metabolism axis: A mechanistic investigation into bone loss in a simulated microgravity environment. *Artif. Cell Nanomed. Biotechnol.* **2025**, *53*, 494–513. [CrossRef]
5. Vico, L.; Collet, P.; Guignandon, A.; Lafage-Proust, M.-H.; Thomas, T.; Rehailia, M.; Alexandre, C. Effects of long-term microgravity exposure on cancellous and cortical weight-bearing bones of cosmonauts. *Lancet* **2000**, *355*, 1607–1611. [CrossRef]
6. Hagio, S.; Ishihara, A.; Terada, M.; Tanabe, H.; Kibushi, B.; Higashibata, A.; Yamada, S.; Furukawa, S.; Mukai, C.; Ishioka, N. Muscle synergies of multidirectional postural control in astronauts on Earth after a long-term stay in space. *J. Neurophysiol.* **2022**, *127*, 1230–1239. [CrossRef]
7. Araki, K.; Tsujiuchi, N.; Ito, A.; Ohira, Y.; Kamibayashi, K. Analysis of walking characteristics in microgravity environment using anti-gravity treadmill. In Proceedings of the Dynamics Design Conference, Dubrovnik, Croatia, 21–24 May 2018; p. 522.
8. Zhang, J.; Zhou, B.; Nie, J.; Dong, X.; Ding, L. Advances in the study of the musculoskeletal multi-rigid-body dynamic modeling in manned space flight. *Chin. J. Space Sci.* **2025**, *45*, 783–794. [CrossRef]
9. Wu, B.; Gao, X.; Qin, B.; Baldoni, M.; Zhou, L.; Qian, Z.; Zhu, Q. Effect of microgravity on mechanical loadings in lumbar spine at various postures: A numerical study. *NPJ Microgravity* **2023**, *9*, 16. [CrossRef]
10. Zuobing, X.; Qixuan, Z.; Yunwei, N.; Di, Z. Adhesion advances: From nanomaterials to biomimetic adhesion and applications. *Soft Matter* **2022**, *18*, 3447–3464. [CrossRef]
11. Singh, A.; Srivastav, S.; Yadav, K.; Chandran, D.S.; Deepak, K.K. Neurohormonal responses to oscillatory lower body negative pressure in healthy subjects. *Acta Astronaut.* **2021**, *186*, 396–402. [CrossRef]
12. Sonawane, R.; Patil, S.; Rahaman, J.; Mukherjee, D. Effect of microgravity on bone Tissue: Mechanisms of osteodegeneration and advanced treatment modalities. *Biochem. Biophys. Res. Commun.* **2025**, *771*, 152055. [CrossRef]
13. Childress, S.D.; Williams, T.C.; Francisco, D.R. NASA Space Flight Human-System Standard: Enabling human spaceflight missions by supporting astronaut health, safety, and performance. *NPJ Microgravity* **2023**, *9*, 31. [CrossRef] [PubMed]
14. Tait, K.; Lewis, J. Improving Working Conditions for Astronauts: An Electronic Personal Restraint System for Use in Microgravity Environments. *J. Purdue Undergrad. Res.* **2012**, *2*, 70–77. [CrossRef]
15. Xiaoli, W.; Longqing, Y.; LvZhou, Y.; Xiao, Z.; YiJing, Y. Precise Control of Local Spin States in an Adsorbed Magnetic Molecule with an STM Tip: Theoretical Insights from First-Principles-Based Simulation. *J. Phys. Chem. Lett.* **2018**, *9*, 2418–2425.
16. Poerschke, B.; Gorb, S.; Schaber, C. Adhesion of Individual Attachment Setae of the Spider Cupiennius salei to Substrates With Different Roughness and Surface Energy. *Front. Mech. Eng.* **2021**, *7*, 702297. [CrossRef]
17. Li, J.; Wei, Q.Q.; Zhao, X.W.; Sui, T.H.; Ji, J.X.; Liu, J.L. The gradient enhanced adhesion mechanisms of biomimetic mushroom-shaped microcolumn. *Colloids Surf. A-Physicochem. Eng. Asp.* **2025**, *727*, 138454. [CrossRef]
18. Flenner, S.; Schaber, C.F.; Krasnov, I.; Stieglitz, H.; Müller, M. Multiple Mechanical Gradients are Responsible for the Strong Adhesion of Spider Attachment Hair. *Adv. Mater.* **2020**, *32*, 2002758. [CrossRef]
19. Sato, J.; Shimizu, T.; Takahashi, T.; Suzuki, M.; Aoyagi, S. Production adhesion mechanism imitating gecko by optical shaping and measurement of adhesion force. In Proceedings of the JSME annual Conference on Robotics Mechatronics, Kitakyushu, Japan, 2–5 June 2018; pp. 2P1–E11.
20. Alexandra, P.; Victoria, G.; Anastasia, K.; Yuliya, K.; Valeriy, B.; Sergey, S. Cytoskeleton Markers in the Spinal Cord and Mechanoreceptors of Thick-Toed Geckos after Prolonged Space Flights. *Life* **2022**, *12*, 100. [CrossRef]
21. Gulimova, V.; Proshchina, A.; Kharlamova, A.; Krivova, Y.; Barabanov, V.; Berdiev, R.; Asadchikov, V.; Buzmakov, A.; Zolotov, D.; Saveliev, S. Reptiles in Space Missions: Results and Perspectives. *Int. J. Mol. Sci.* **2019**, *20*, 3019. [CrossRef]
22. Autumn, K.; Niewiarowski, P.H.; Puthoff, J.B. Gecko Adhesion as a Model System for Integrative Biology, Interdisciplinary Science, and Bioinspired Engineering. *Annu. Rev. Ecol. Evol. Syst.* **2014**, *45*, 445–470. [CrossRef]
23. Huber, G.; Mantz, H.; Spolenak, R.; Mecke, K.; Jacobs, K.; Gorb, S.N.; Arzt, E. Evidence for capillarity contributions to gecko adhesion from single spatula nanomechanical measurements. *Proc. Natl. Acad. Sci. USA* **2005**, *102*, 16293–16296. [CrossRef]
24. Suthisomboon, T.; Rukpanich, T.; Asawalertsak, N.; Borijindakul, P.; Manoonpong, P. VENOM: Versatile, Adhesive, and Soft Material for Various Surface Adhesion. In Proceedings of the 2021 IEEE 4th International Conference on Soft Robotics (RoboSoft), New Haven, CT, USA, 12–16 April 2021.
25. Benvidi, F.H.; Bacca, M. Theoretical limits in detachment strength for radially-symmetric bi-material adhesives. *J. Appl. Mech.* **2021**, *88*, 121007. [CrossRef]

26. Sey, M. Probing the mechanical properties of soft and hard biomaterials: Polydimethylsiloxane (PDMS) micropillars and alligator teeth. In Proceedings of the MME Undergraduate Research Symposium, Washington, MO, USA, 10 November 2022.
27. Wang, W.; Liu, Y.; Xie, Z. Gecko-Like Dry Adhesive Surfaces and Their Applications: A Review. *J. Bionic Eng.* **2021**, *18*, 34. [CrossRef]
28. Shuai, L.; Hongmiao, T.; Jinyou, S.; Haoran, L.; Duorui, W.; Weitian, Z. Switchable Adhesion for Nonflat Surfaces Mimicking Geckos' Adhesive Structures and Toe Muscles. *ACS Appl. Mater. Interfaces* **2020**, *12*, 39745–39755. [CrossRef]
29. Hwang, D.G.; Trent, K.; Bartlett, M.D. Kirigami-Inspired Structures for Smart Adhesion. *ACS Appl. Mater. Interfaces* **2018**, *10*, 6747–6754. [CrossRef]

biomimetics

Article

Development of Three Alternative Strategies for the Binding of Cells to Functionalized DeepTip™ AFM Probes

Raquel Tabraue-Rubio [1,2,3], Laura Yuste Muñoz [1,2], Marcos Vázquez [4,5], Rafael Daza [1,2], Luis Colchero [1,2,3], María Eugenia Fernández-Santos [4,5,6], Manuel Elices [2], Fivos Panetsos [3,7,8], Gustavo V. Guinea [1,2,8,9] and José Pérez-Rigueiro [1,2,8,9,*]

1 Center for Biomedical Technology, Universidad Politécnica de Madrid, 28223 Pozuelo de Alarcón, Madrid, Spain; luis.colchero@ctb.upm.es (L.C.)
2 Departamento de Ciencia de Materiales, ETSI Caminos, Canales y Puertos, Universidad Politécnica de Madrid, 28040 Madrid, Spain
3 Bioactive Surfaces S.L., C/Puerto de Navacerrada 18, 28260 Galapagar, Madrid, Spain
4 Department of Cardiology, Gregorio Marañón General University Hospital and Instituto de Investigación Sanitaria Gregorio Marañón (IiSGM), 28007 Madrid, Spain; mariuge@fibhgm.org (M.E.F.-S.)
5 Centro de Investigación Biomédica en Red–Enfermedades Cardiovasculares (CIVERCV), RICORS TERAV, ISCIII, 28029 Madrid, Spain
6 ATMPs Production Unit, IiS-Gregorio Marañón Health (IisSGM), Gregorio Marañón General University Hospital, 28007 Madrid, Spain
7 Neurocomputing and Neurorobotics Research Group, Faculty of Biology and Faculty of Optics, Universidad Complutense de Madrid, 28040 Madrid, Spain
8 Biomaterials and Regenerative Medicine Group, Instituto de Investigación Sanitaria del Hospital Clínico San Carlos (IdISSC), Calle Prof. Martín Lagos s/n., 28040 Madrid, Spain
9 Biomedical Research Networking Center in Bioengineering Biomaterials and Nanomedicine (CIBER-BBN), 28029 Madrid, Spain
* Correspondence: jose.perez@ctb.upm.es

Abstract

The efficient design of biohybrid materials requires controlling the interaction between the cell and the material for a wide range of possible combinations. Single cell force spectroscopy (SCFS), an atomic force microscopy (AFM) experimental procedure based on the binding of an individual cell to an AFM cantilever and the assessment of the adhesion force between the cell and a target substrate, represents one of the most promising alternatives to characterize the interaction between cell and material. However, SCFS relies on the efficient binding of the cell to the AFM in order to avoid drawbacks, such as the detachment of the cell. In this work, three different versatile and robust procedures are presented that allow for the binding of either non-adherent ($CD4^+$ T-lymphocytes) or adherent (mesenchymal stem cells, MSC) cells to the AFM probe. The three crosslinking strategies comprise (1) the streptavidin/biotin system, (2) sulfhydryl group-based crosslinkers, and (3) "click" (bioorthogonal) chemistry. Additionally, three decoration schemes of the functionalized AFM probes are explored: a specific antibody, concanavalin A, and direct binding of the cell through azide-derivatized membrane proteins. Differences are observed between these alternatives and it is found that the strength of the interaction (in decreasing order) is as follows: specific antibody, concanavalin A, and binding through azide-derivatized proteins.

Keywords: biohybrid material; single cell force spectroscopy; atomic force microscopy; crosslinking chemistry

 https://doi.org/10.3390/biomimetics11020095

1. Introduction

Biohybrid materials constitute a thriving research field that arises from the combination of living cells and inert materials [1,2]. This combination, on the one hand, may serve to create a protective environment for the cells [3], while the material may take advantage of the biological constituent in order to respond to external stimuli [4]. The opportunities opened by biohybrid materials are reflected in the large number of intended applications, including, for instance, Materials Science [5] and robotics [6]. In this context, it is worth highlighting the attempts to produce novel biomaterials and tissue engineering scaffolds from this type of composite material [2,7,8].

The expected extension of the field is clearly perceived by considering the enormous amount of alternatives that arise from combining various cell lineages [9,10] and different materials [11,12]. Consequently, it is necessary to rely on experimental techniques that allow for an efficient characterization of the interaction established between cells and materials in order to have reliable guidance for the design and implementation of new biohybrid materials [13].

Among the increasing number of applications of atomic force microscopy to biological systems [14–16], its usage for the characterization of the interaction between cells and material surfaces allows us to exploit the excellent performance of AFM, both in terms of spatial and force resolution [17]. These experiments usually require the binding of a cell that must be adhered to the AFM cantilever. Subsequently, the attached cell is taken to contact the substrate while controlling both the force and the time of contact [18,19]. This experimental procedure leads to the so-called single cell force spectroscopy (SCFS) experiments [20].

However, the whole experimental scheme outlined above depends critically on the robustness and reliability of the cell-AFM cantilever adhesion. Several binding approaches have been implemented, including the usage of glutaraldehyde [17] and of biological molecules, such as concanavalin A, fibronectin, and poly-L-lysine, that are adsorbed on the surface of the AFM cantilever [21,22]. However, all these approaches tend to be affected by problems such as the uncontrolled detachment of the cell from the cantilever.

In this context, we present here three crosslinking strategies that allow for the efficient and reliable binding of cells to a functionalized AFM probe. Amine-functionalized DeepTip^{TM} AFM probes are used and decorated with each one of the crosslinking strategies, which comprise (i) the biotin–streptavidin system, (ii) thiol crosslinking chemistry, and (iii) bioorthogonal ("click") chemistry. When required, antibodies or concanavalin A were covalently bound to the AFM cantilever through these chemistries, and the successful decoration of the probe was initially assessed through specific fluorophores. The binding of cells to the decorated cantilevers was verified with two different cell lineages: non-adherent CD4$^+$ lymphocytes and adherent mesenchymal stem cells (MSCs). Although all strategies were observed to lead to the binding of the cell to the probe, it was found that the strongest adhesion corresponded to the antibody-decorated probes.

2. Materials and Methods

This section provides a detailed description of the experimental procedures followed for the following: (i) the extraction and maintenance in culture of CD4$^+$ T cells and of the cell culture of mesenchymal stem cells (MSCs); (ii) the functionalization of MSCs with azide groups; (iii) the decoration of AFM probes with the biomolecules of interest for subsequent cell attachment, using three different crosslinking chemistries (biotin/streptavidin, thiol chemistry, and bioorthogonal chemistry), including the corresponding assessment of each probe decoration through fluorescence microscopy; and (iv) verification of the cell binding to the functionalized cantilevers.

https://doi.org/10.3390/biomimetics11020095

2.1. T Cells Extraction and Culture

T lymphocytes were obtained from *Mus musculus* CD1 mice. The mice were housed with ad libitum access to food and water, under controlled temperature and humidity conditions, as well as regulated light–dark cycles. All animals used were healthy mice of both sexes, aged between 2 and 4 months. No animals were sacrificed for the isolation of this cell type, and all procedures described below were conducted in accordance with Spanish legislation on animal experimentation and welfare.

T cells were collected from the peritoneal cavity, following a standardized laboratory protocol. Briefly, 1 to 3 mL of sterile 1X phosphate-buffered saline (PBS) were injected into the peritoneal cavity, the abdomen of the animal was gently massaged, and approximately 80% of the injected volume was recovered. The recovered volume contained various types of immune cell, including T lymphocytes.

The peritoneal cavity cells suspended in PBS were centrifuged at 396× *g* for 10 min at 4 °C (Rotina 380R, Hettich, Kirnlengern, Germany). The supernatant was then discarded, and the cell pellet was resuspended in complete RPMI medium, consisting of RPMI 1640 medium (Roswell Park Memorial Institute 1640, Gibco) supplemented with 10% fetal bovine serum (FBS, Cytiva), 200 mM L-glutamine (Gibco), and 1% penicillin–streptomycin (P/S, Sigma-Aldrich). Subsequently, to eliminate any erythrocytes that were potentially present due to the extraction procedure, red blood cell lysis was performed using BS PharmLyse™ solution (BD Biosciences), following the manufacturer's instructions.

To separate the non-adherent T lymphocyte population from other adherent cells in the sample, the cell suspension was placed into wells containing a migration inhibitory factor (Kartell) for 30 min at room temperature. After this incubation, the supernatant—expected to contain only $CD4^+$ and $CD8^+$ T lymphocytes—was carefully collected, using a pipette. T lymphocytes were visually identified by light microscopy and counted using a Neubauer chamber prior(Sigma Aldrich, Inc. Missouri, USA) to culture.

For T cell culture, TexMACS™ medium (Miltenyi Biotec) supplemented with 10% FBS, 1% L-glutamine, 1% P/S, and 0.01 mM 2-mercaptoethanol (Sigma-Aldrich) was used, and cells were seeded at a minimum density of 14,000 cells/cm^2. Cultures were maintained in an incubator under controlled conditions at 37.5 °C and 5% CO_2.

2.2. MSCs Culture

Bone-marrow-derived mesenchymal cells from CD1 mice were used (hereafter MSCs). MSCs were isolated from the spinal bone marrow of mice and expanded, following established protocols [23]. For their culture, DMEM medium supplemented with 10% FBS, 1% L-glutamine, and 1% P/S was used. Cultures were maintained in an incubator under controlled conditions at 37.5 °C and 5% CO_2; expansions were made once the cells reached approximately 80% confluence. Expansions were performed according to a standard protocol under sterile conditions. Briefly, the medium was removed, and the Petri dish was washed with 1X PBS. After discarding the PBS buffer, 0.05% trypsin-EDTA (Gibco) was added and allowed to act in the incubator for 3 to 5 min. Subsequently, trypsin was inactivated by adding DMEM, and the cell suspension was collected into a centrifuge tube and centrifuged at 423× *g* for 5 min. Supernatant was discarded and the pellet was resuspended in complete DMEM. Cells were counted using a Neubauer chamber and seeded at an approximate density of 5000 cells/cm^2.

2.3. MSCs Functionalization via Bioorthogonal Chemistry

MSCs were functionalized with an azide-labeled sugar (mannose), 1,3,4,6-tetra-O-acetyl-N-azidoacetylmannosamine ($Ac_4ManNAz$, hereafter) (Sigma-Aldrich). For this purpose, MSCs obtained following the previous protocol were expanded at a seeding

https://doi.org/10.3390/biomimetics11020095

density of 10,000 cells/cm^2 and incubated for 24 h. Afterwards, $Ac_4ManNAz$ was added at 40 μM and cells were incubated for an additional 48 h.

2.4. Assessment of MSCs Azide Functionalization

To assess the functionalization of the MSCs, cells were labeled with a fluorescent molecule coupled to a bioorthogonal chemistry reagent, BP fluor 488–dibenzocyclooctyne-N-hydroxysuccinimidyl ester, isomer 5 (BP fluor 488–DBCO, BroadPharm). The culture medium was removed, and the cells were washed twice with 1X PBS and fixed with 4% paraformaldehyde in PBS for 15 min at room temperature. Two additional washes with PBS were performed for 15 min each. BP fluor 488–DBCO was then added at 25 μM for 20 min in the dark at room temperature. Samples were washed three times with PBS for 5 min each. Cells were subsequently stained with DAPI at a 1:1000 dilution for 5 min in the dark at room temperature, followed by three washes with PBS for 5 min each. Negative controls included cells not treated with $Ac_4ManNAz$, but incubated with BP fluor 488–DBCO, and cells not treated with either $Ac_4ManNAz$ or the fluorescent DBCO (Table 1). Fluorescence microscopy was used to capture at least three images in random areas of each culture well, to obtain a representative sample. All fluorescence images were acquired at 10× magnification with identical imaging settings (DAPI: exposure 300 ms, gain 2×, gamma 1.4; BP fluor 488–DBCO: exposure 446.8 s, gain 2×, gamma 1).

Table 1. Conditions for the negative controls used for the characterization of MSCs functionalized with $Ac_4ManNAz$.

	Sample	Control 1	Control 2
Cells	MSCs + $Ac_4ManNAz$	MSCs	MSCs
BP fluor 488–DBCO	✓	✓	✗

2.5. Functionalization of AFM Probes

All functionalizations were performed using AFM DeepTip™ SiN R67 AFM probes (chips), kindly provided by, Bioactive Surfaces, Galapagar, Madrid, Spain.

2.6. Biotin-NHS Ester/Streptavidin Functionalization

AFM probe biotinylation began by briefly immersing the chips in 2-propanol (Scharlab) to remove impurities, followed by incubation in a solution of Sulfo-NHS-LC-biotin EZ-Link (biotin-NHS, Thermo Fisher Scientific) at 5 mg/mL in 1X PBS for 30 min at room temperature. The chips were washed with PBS and then incubated in a 0.5 mg/mL streptavidin solution in PBS for 30 min at room temperature. After incubation, the chips were washed with PBS and stored at 4 °C for up to 7 days prior to use

2.7. Assessment of Biotin-NHS/Streptavidin Functionalization

Biotin-NHS functionalization was verified using streptavidin conjugated to fluorescein isothiocyanate (FITC) (excitation: 498 nm, emission: 517 nm) (streptavidin-FITC, Molecular Probes). After biotin-NHS incubation, chips were incubated in a 0.5 mg/mL streptavidin-FITC solution in PBS for 30 min at room temperature in the dark. Samples were washed, and fluorescence images were acquired at 10x magnification (exposure: 1 s, gain: 2.1×, gamma: 0.83) and analyzed using ImageJ. Negative controls were prepared by incubating samples with PBS only, instead of a biotin-NHS solution.

https://doi.org/10.3390/biomimetics11020095

2.8. Decoration with Concanavalin A

Concanavalin A (ConA) is a lectin extracted from *Canavalia ensiformis* that is known for binding specifically to glycoprotein receptors that are present in cell membranes. It was used for MSC assays due to its ease of storage, low cost compared to antibodies, and efficient MSC binding. Chips were functionalized with biotin-NHS and streptavidin, as described above, and subsequently incubated with biotinylated ConA (Sigma-Aldrich) at 0.4 mg/mL in PBS for 30 min at room temperature. Chips were washed three times with PBS and were ready for use.

2.9. Thiol Modification and Verification

To functionalize the AFM chips via thiol chemistry, both molecule and substrate must contain either thiol (–SH) or pyridyl disulfide groups. Modifications were applied to both of them to introduce these groups for covalent attachment.

2.10. Derivation of the Biomolecule with Thiol Groups

Streptavidin, lacking native thiol groups, was chemically modified to enable surface attachment through thiol groups, as shown in Figure 1. Streptavidin from *Streptomyces avidinii* (Sigma-Aldrich) was dissolved at 2 mg/mL in PBS and reacted with 20 mM sulfosuccinimidyl 6-(3′-(2-pyridyldithio)propionamide)hexanoate (sulfo-LC-SPDP, Thermo Scientific) for 50 min. The resulting modified streptavidin with a pyridyldisulfide group (streptavidin-PDP from now on) was purified using size-exclusion chromatography columns (MWCO 7.000, Thermo Scientific).

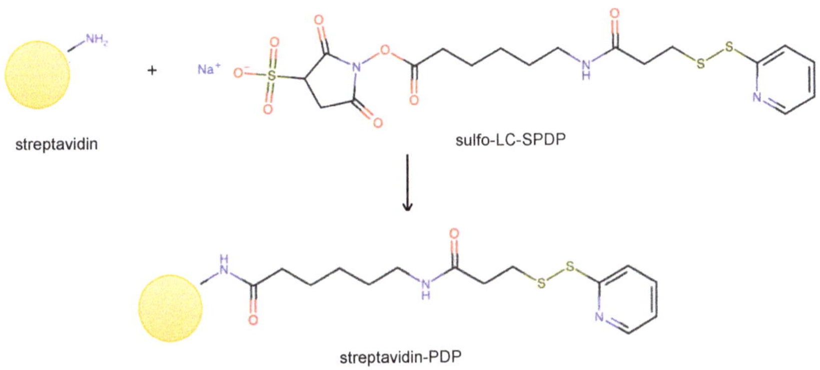

Figure 1. Streptavidin modification using sulfo-LC-SPDP.

2.11. Functionalization of the AFM Probes with Thiol Groups

The AFM chips were cleaned with 2-propanol and incubated with 2.5 mg/mL sulfo-LC-SPDP for 50 min at room temperature. Afterwards, they were washed five times with 1 mM ethylenediaminetetraacetic acid (EDTA, Sigma-Aldrich) in PBS 1X (PBS-EDTA) and then reduced with 3 mg/mL tris(2-carboxyethyl)phosphine (TCEP, Sigma-Aldrich) in PBS-EDTA for 15 min at room temperature, followed by three washes with PBS-EDTA. Then, 40 µg/mL of streptavidin-PDP in PBS-EDTA was incubated with the functionalized chips for at least 18 h at 4 °C. The chips were washed with PBS 1X containing 0.2% TWEEN® 20 (Sigma-Aldrich) and 0.1% sodium dodecyl sulfate (SDS, Fisher Bioreagents) for 30 min under gentle stirring at room temperature. Samples were finally rinsed with PBS 1X. The whole process is summarized in Figure 2.

https://doi.org/10.3390/biomimetics11020095

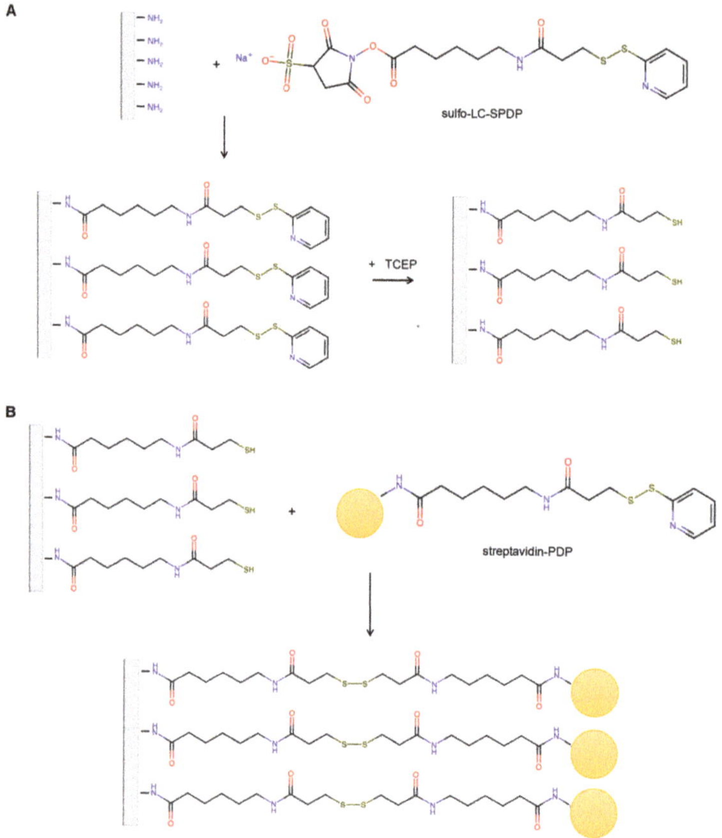

Figure 2. AFM probe decoration with streptavidin via sulfo-LC-SPDP. (**A**) Obtaining sulfhydryl groups on the surface by adding sulfo-LC-SPDP and subsequent reduction with a reducing agent (TCEP). (**B**) Decoration of the AFM probe with streptavidin.

2.12. Assessment of the Thiol Streptavidin Decoration

The decoration of the chips with streptavidin was assessed by using biotinylated horseradish peroxidase (biotin-HRP, Invitrogen). The chips were incubated with a solution of biotin-HRP at 100 ng/mL in PBS-EDTA for 30 min under gentle stirring at room temperature. The chips were then washed with a solution of PBS 1X containing 0.2% TWEEN® 20 (Sigma-Aldrich) and 0.1% sodium dodecyl sulfate (SDS, Fisher Bioreagents) for 30 min under stirring. Subsequently, chips were incubated with 1-Step™ Turbo TMB-Elisa (TMB, Thermo Scientific) for 5 min under stirring, and the enzymatic reaction was stopped by adding 2 M sulfuric acid. TMB (3,3′,5,5′-tetramethylbenzidine) undergoes a reaction catalyzed by HRP that leads to an increase in the absorbance of the solution at a wavelength of 450 nm.

Solutions were collected and absorbance was measured at 450 nm. Controls included AFM probes without sulfo-LC-SPDP and without streptavidin-PDP, adding PBS-EDTA instead (Table 2). A calibration curve was prepared by using serial dilutions of biotin-HRP (102.4 ng/mL to 0.2 ng/mL), adding to each solution TMB for 5 min and stopping the reaction with 2 M sulfuric acid. Subsequently, absorbance was measured at 450 nm.

https://doi.org/10.3390/biomimetics11020095

Table 2. Conditions for the negative controls used for verification of streptavidin-PDP immobilization on the AFM probe surface.

	Chips	Control 1	Control 2
Sulfo-LC-SPDP	✓	✗	✓
Streptavidin-PDP	✓	✓	✗

2.13. Decoration with Anti-CD4 Antibodies

Streptavidin-PDP-decorated chips were blocked with 2% BSA for 1 h at room temperature, incubated with biotinylated anti-CD4 antibody at 10 µg/mL for 1 h, and rinsed twice with PBS 1X. Correct decoration was confirmed using Alexa Fluor® 488-conjugated secondary antibody (1:800) for 1 h at room temperature in the dark. Negative controls lacked the primary antibody. As a final step, every sample was rinsed with PBS 1X. Fluorescence images were captured at 20× magnification (exposure: 1 s, gain: 2.1×, gamma: 0.83) and analyzed using ImageJ.

2.14. Functionalization Through Bioorthogonal Chemistry

Chips were washed in 2-propanol and incubated overnight with dibenzocyclooctyne-N-hydroxysuccinimidyl ester (DBCO ester, Sigma-Aldrich) at a concentration of 250 µg/mL in dimethyl sulfoxide (DMSO, PanReac Applichem) with 0.5% diisopropylethylamine (DIPEA, Thermo Scientific) at room temperature, in the dark, under gentle agitation. Chips were washed with PBS 1X containing 0.2% TWEEN® 20 and 0.1% SDS for 15 min and subsequently rinsed with PBS 1X. Samples could be stored at 4 °C for up to one week, although they were preferably used the same day as prepared. A schematic drawing of this process is shown in Figure 3.

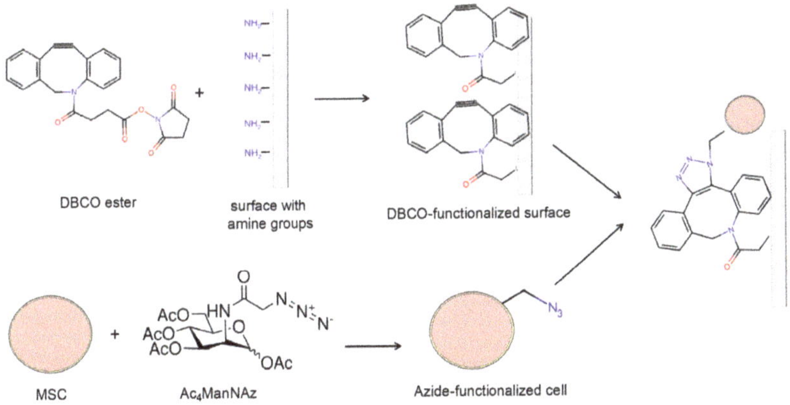

Figure 3. Cell binding to AFM probe via bioorthogonal chemistry. Above: AFM probe decoration with DBCO. Below: MSC decoration with azide groups via metabolic glycoengineering.

2.15. Procedure fo the Assessment of the Functionalization Through Bioorthogonal Chemistry

DBCO ester functionalized chips were incubated with 200 µg/mL FAM-azide, isomer 6 (excitation: 492 nm, emission: 517 nm) (6-FAM-azide, Lumiprobe) in DMSO for 2 h at room temperature in the dark with gentle agitation. Chips were washed with 10% SDS for 5 min under gentle agitation and three times with PBS 1X for 5 min. Negative controls were incubated in the DMSO/DIPEA 0.05% solution. Fluorescence imaging was performed

https://doi.org/10.3390/biomimetics11020095

at 20× magnification (exposure: 1 s, gain: 2.1×, gamma: 0.96). Images were analyzed using ImageJ.

2.16. AFM Probe Fluorescence Image Analysis

All fluorescence images were analyzed using ImageJ (v1.52n, NIH, Maryland, USA). Each image was analyzed individually. RGB color channels were separated to isolate the specific fluorophore, and fluorescence intensity was specifically measured in areas corresponding to the cantilevers. Data obtained in arbitrary units was adjusted to a scale of 10 for each experiment independently, with 10 being the highest value among all replicates.

2.17. Cell Binding to the AFM Probes

Measurements were performed using a Nanolife AFM (Nanotec Electrónica, Spain) and WSxM 5.0 software [24]. The Nanolife AFM is equipped with an inverted optical microscope to monitor cell positioning, relative to the cantilever on transparent sample holders. Experiments were conducted nominally at 25 °C. DeepTip™ SiN R67 chips were functionalized and decorated according to the protocols described above.

Cells were expanded according to their respective protocols, counted with a Neubauer chamber, and separated for each assay. For T lymphocytes, 30,000 cells from the day of extraction or the following day were used. For MSCs, 10,000 cells from passages 10–16 were used. For Ac_4ManNAz-functionalized MSCs, 10,000 cells from passage 16 after 48 h of incubation with Ac_4ManNAz were used. At least three independent samples were prepared with 0.45 μm-filtered culture medium.

One sample was added to the AFM sample holder, ensuring cells were sufficiently separated to bind just one cell to the AFM probe at a time. Using the inverted microscope, an intact, spherical cell was selected. The cantilever tip was carefully placed over the cell without direct pressure.

Attachment was performed via a slow, controlled approach until contact, maintaining 100 pN force during the contact time. Contact times depended on the chemistry used: $CD4^+$ T cells required 5–15 s to bind to anti-CD4-decorated chips, pristine MSCs required 15–30 s to bind to ConA-decorated chips, and Ac_4ManNAz-functionalized MSCs required up to 2 min to bind to DBCO-decorated chips.

After contact, the cantilever-cell probe was retracted from the surface and allowed to rest for about 5 min to stabilize binding. Successful attachment was confirmed visually through the horizontal movement of the AFM cantilever, while verifying that the cell moved together with the cantilever.

3. Results

3.1. Assessment of Biotin-NHS Ester Functionalization

Figure 4 shows the results obtained from the fluorescence analysis of AFM probes functionalized with biotin–NHS and decorated with streptavidin–FITC (n = 3), as well as the corresponding controls (n = 2). In the control image (Figure 4A, bottom), the cantilevers are silhouetted to allow for visualization. Significant differences are observed between the control group and the streptavidin-FITC-decorated samples, as indicated by Student's *t*-test.

https://doi.org/10.3390/biomimetics11020095

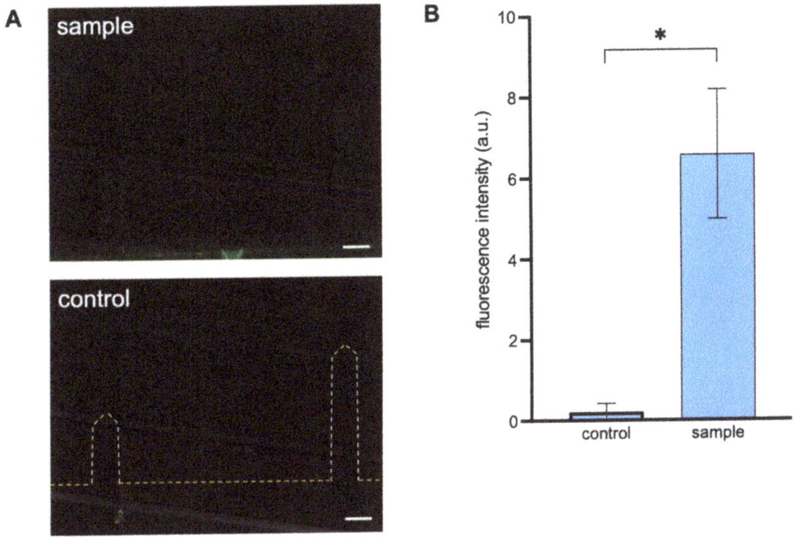

Figure 4. (**A**) Fluorescence microscopy images of streptavidin-FITC-decorated cantilevers (top) and of a control sample (bottom). The control cantilevers are silhouetted to allow for visualization. Scale bar: 40 µm. (**B**) Quantitative analysis of cantilever fluorescence intensity; control n = 2, sample n = 3. Data represent group means ± standard error of the mean. The asterisk indicates statistically significant differences between groups ($p = 0.02$), as established through Student's t-test.

3.2. Assessment of Streptavidin-PDP Immobilization Using Biotinylated Peroxidase

Absorbance measurements of the solutions obtained after the enzymatic reaction of biotin–HRP are shown in Figure 5. The controls indicate the presence of possible residual nonspecific interactions between sulfo-LC-SPDP molecules on the substrate and biotin–HRP (control without streptavidin, n = 2), as well as possible nonspecific interactions between streptavidin–PDP and the nonfunctionalized surface (control without sulfo-LC-SPDP, n = 2). However, as confirmed with Student's t-test, significant differences are observed between the samples (n = 3) and the corresponding controls, indicating that this type of sample is properly functionalized and decorated.

3.3. Assessment of Anti-CD4 Decoration of the AFM Probes

Fluorescence images of anti-CD4-decorated chips are shown in Figure 6, as well as the fluorescence intensity analysis of the cantilevers. Due to the sample size (n = 2 for every group) and since the assumption of normal data distribution was not met, a Mann–Whitney test was performed instead of Student's t-test, as it is considered to be more robust under these conditions. As shown in Figure 6B, significant differences were found between samples and controls, indicating that the chips were successfully decorated with anti-CD4 antibodies. After fluorescence intensity analysis, the brightness and contrast of the images were slightly adjusted to facilitate visualization.

https://doi.org/10.3390/biomimetics11020095

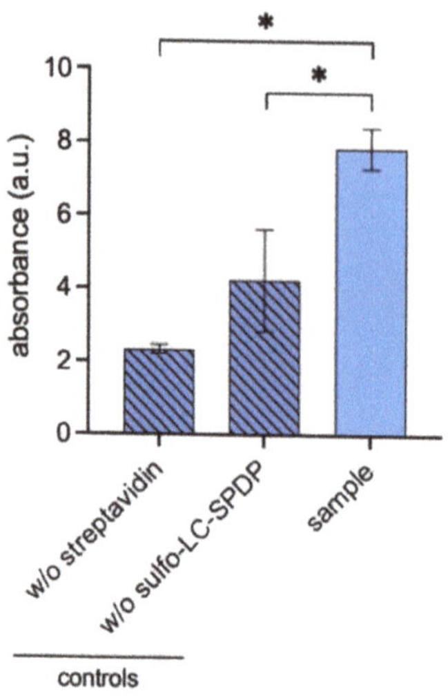

Figure 5. Surface functionalization through thiol groups. Absorbance measurements at 450 nm of TMB solution resulting from the enzymatic activity of biotin–HRP present on the surface of the chip. Data represent group means ± standard error of the mean. Asterisks indicate significant differences between groups ($p < 0.05$), established through Student's t-test. Control w/o streptavidin (n = 2) vs. sample (n = 3): $p = 0.014$; control w/o sulfo-LC-SPDP (n = 2) vs. sample (n = 3): $p = 0.019$.

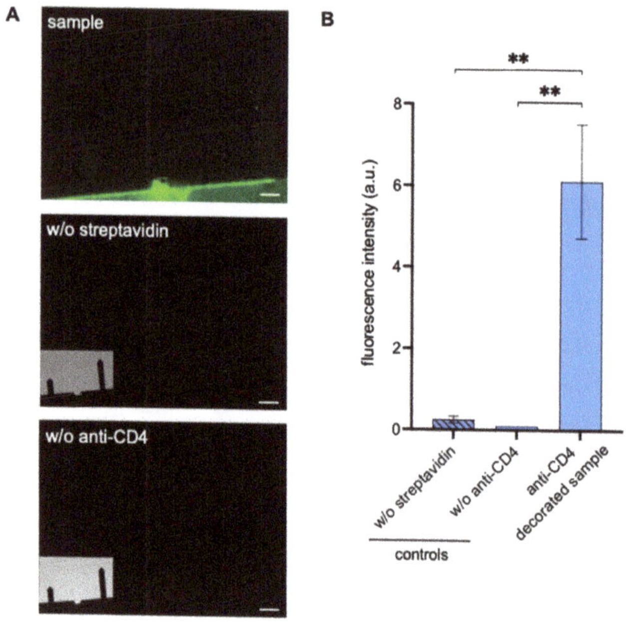

Figure 6. Surface decoration of chips with anti-CD4 antibodies. (**A**) Representative fluorescence microscopy images of a functionalized sample (top), a control sample without streptavidin (middle), and a control sample without anti-CD4 antibody (bottom). Scale bar: 40 µm. (**B**) Analysis of fluorescence intensity across groups (n = 2 for every group). Data represent group means ± standard error of the mean. Asterisks indicate significant differences between groups ($p = 0.0079$), as determined from the Mann–Whitney test.

https://doi.org/10.3390/biomimetics11020095

3.4. Assessment of MSCs Functionalization Through Bioorthogonal Chemistry

In Figure 7A, representative images of $Ac_4ManNAz$-functionalized MSCs are shown. The images correspond to both the (negative) control group (cells not cultured with $Ac_4ManNAz$) and to the functionalized cells obtained after incubation with DBCO–BP Fluor 488. As observed, control cells display a certain residual level of fluorescence, as quantified in Figure 7B. However, the fluorescence observed in the functionalized cells is significantly higher than the residual fluorescence found in the control, as established quantitatively through the comparison of the mean value through Student's *t*-test (Figure 7B). Means correspond to three biological replicates for each group.

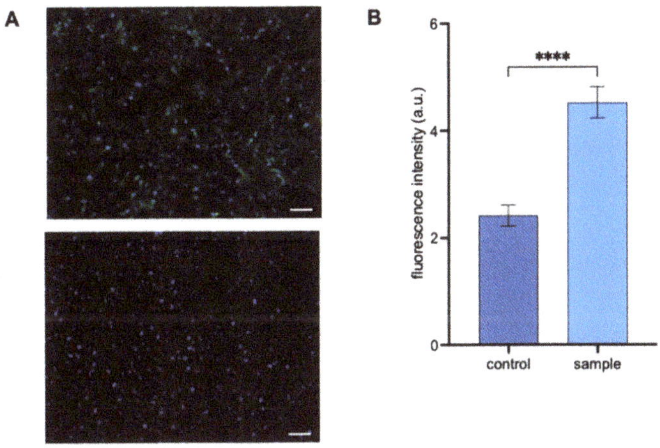

Figure 7. (**A**) Fluorescence microscopy images of MSCs incubated with $Ac_4ManNAz$ (top) and control (bottom). Green: BP Fluor 488–DBCO; blue: DAPI. Scale bars: 80 µm. (**B**) Quantitative analysis of BP Fluor 488–DBCO fluorescence intensity in cells without $Ac_4ManNAz$ (control) and cells incubated with $Ac_4ManNAz$ (sample). Data represent the mean value for each condition ± standard error of the mean; n = 3. Asterisks indicate significant differences ($p < 0.0001$), as determined from Student's *t*-test.

3.5. Assessment of the Functionalization Through Bioorthogonal Chemistry

Figure 8A shows representative images of chips functionalized using the DBCO ester chemistry, and the measured intensities are quantified in Figure 8B. After performing Student's *t*-test, significant differences were observed between the control group (n = 2) and the sample (n = 3), indicating that the surface of the chips were successfully functionalized using this chemistry.

3.6. Assessment of the Cell Binding to the AFM Chips

Optical microscopy images obtained after confirming cell binding to the cantilever through each corresponding crosslinking chemistry are shown in Figure 9 for each of the cell types and crosslinking chemistries. The saturation and brightness of the images have been modified and cells are silhouetted to facilitate visualization. The number of cells tested were four $CD4^+$ T cells, six pristine MSCs, and three azide-functionalized MSCs. Non-silhouetted images are shown in the Supplementary Materials.

https://doi.org/10.3390/biomimetics11020095

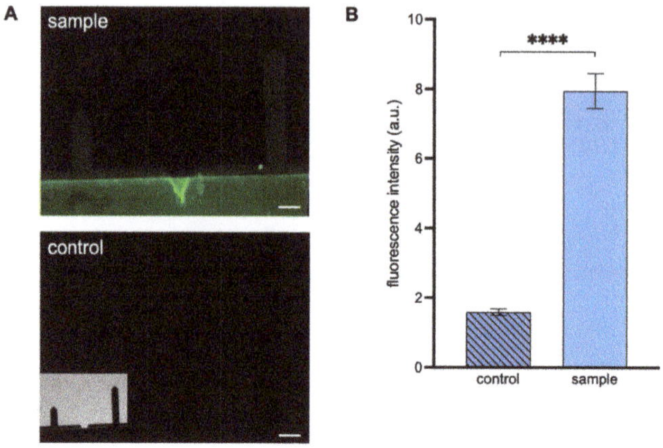

Figure 8. Bioorthogonal chemical functionalization of chip surfaces. (**A**) Fluorescence microscopy images of a DBCO-functionalized sample (top) and of a control sample (bottom). Scale bar: 40 μm. (**B**) Fluorescence intensity analysis of the samples. Data represent group means ± standard error of the mean; control n = 2, sample n = 3. Asterisks indicate significant differences between groups ($p < 0.0001$), as established through Student's t-test.

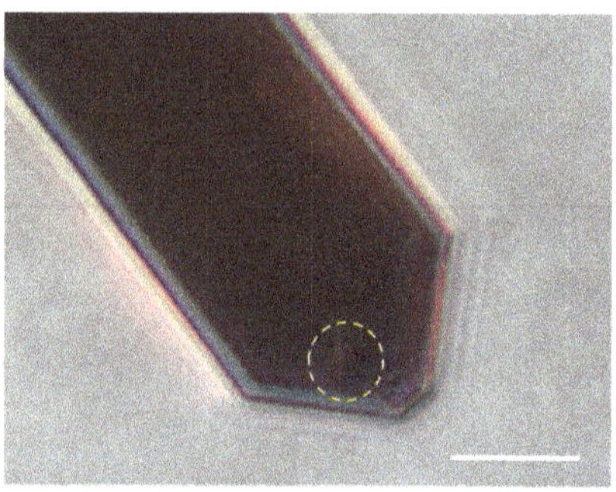

CD4+ T cell.

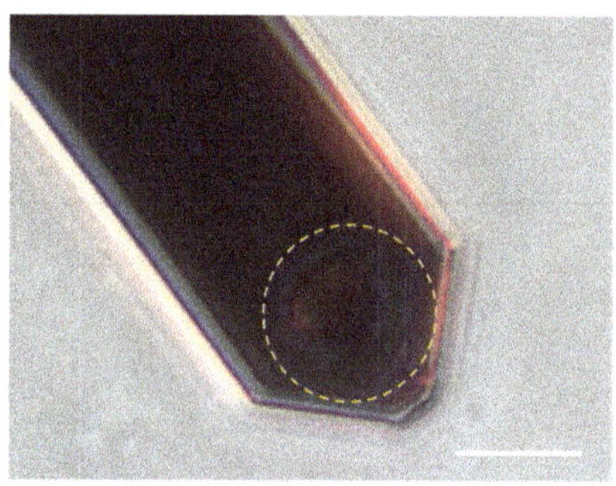

Pristine MSCs cell.

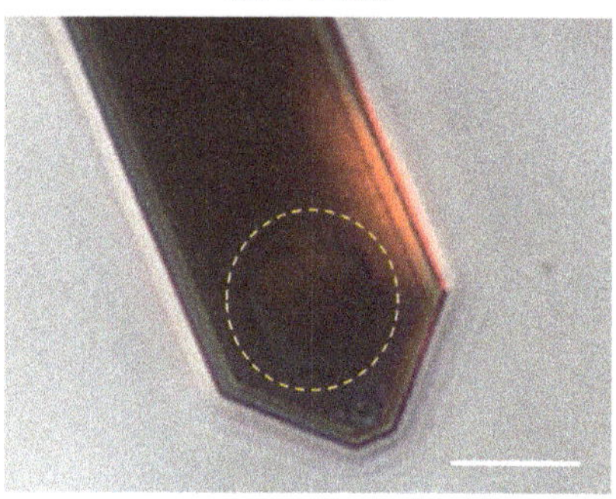

Azide-functionalized MSCs cell.

Figure 9. Optical microscopy images for each cell type bound to the cantilever. (**Top-left**): $CD4^+$ T cell attached to anti-CD4-decorated cantilever, using thiol crosslinking chemistry. (**Top-right**): Pristine MSCs cell attached to concanavalin A-decorated cantilever, using biotin-NHS/streptavidin. (**Bottom**): Azide-modified MSCs cell attached to a DBCO-decorated cantilever via bioorthogonal chemistry. Scale bar: 20 μm.

https://doi.org/10.3390/biomimetics11020095

In order to further assess the efficiency of cell binding to the functionalized cantilevers, as well as to verify the stability of the resulting attachment, a procedure based on the variation in the resonance frequency after cell binding was developed and applied. After waiting 5 min to stabilize the binding of the cell to the cantilever, as indicated above, the cantilever resonance frequency was measured five times. Each resonance frequency measurement is performed by oscillating the cantilever and a typical experiment consists of five independent frequency screenings. Consequently, the cell must remain bound to the cantilever for a duration of at least five minutes, during which the cantilever oscillates.

An example of the amplitude vs. frequency plot obtained from the frequency screening is illustrated in Figure 10A. The black line corresponds to the initial frequency screening and shows a clear peak at the resonance frequency, f_0, of the cantilever. When a cell is bound to the cantilever, the added mass leads to a decrease in the resonance frequency (blue line) and the displacement of the resonance frequency is calculated as $\Delta f = (f_0 - f_1)/f_0$.

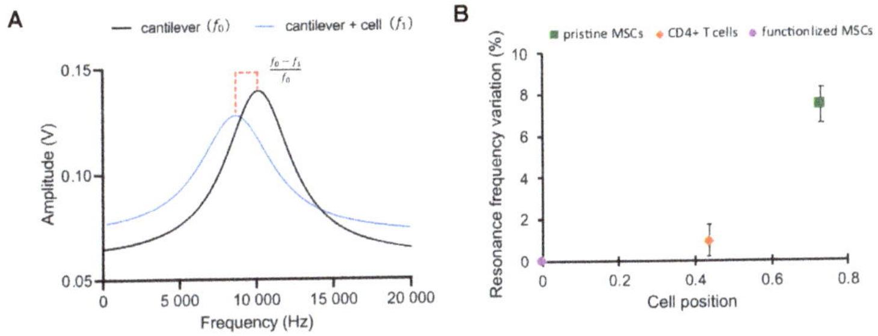

Figure 10. Frequency screening of the cantilever. (**A**) Example of a resonance frequency curve before (black) and after (blue) attaching a cell, including the formula used to calculate the variation in the resonance frequency. (**B**) Mean variation values obtained for pristine MSCs (n = 7) and CD4$^+$ T cells (n = 4). No data are shown on the Azide-functionalized MSCs bound to DBCO-decorated cantilevers since cells were detached during the frequency screening in all cases. Error bars correspond to standard error.

The result of applying this methodology to the various decoration strategies and cell lineages is shown in Figure 10B. Figure 10B shows the variation in the resonance frequency resulting from the binding of pristine MSCs (n = 7) to concanavalin A-decorated cantilevers, and of the binding of CD4$^+$ T lymphocytes (n = 4) to anti-CD4+decorated cantilevers. Each experiment was performed using an independent cell–cantilever pair, since once the cell was attached, it was not possible to detach it from the cantilever. No data are included on the azide-functionalized MSCs binding to DBCO-decorated cantilevers because the binding was not sufficiently stable to maintain the cell attached during the measurements, resulting in cell detachment during frequency screening. This experiment was repeated three times with three different cell-cantilever pairs.

4. Discussion

The use of AFM to investigate the interaction between different cell types and materials through SCFS enables the exploration of these interactions from a perspective that exploits the potential of AFM in terms of both spatial and force resolution. The diversity of approaches available for attaching a cell to the cantilever requires a systematic procedure that not only allows for assessing the advantages and limitations of each binding strategy, but also its optimization. In this work, different strategies that allow for the binding of a cell to the AFM probe are amply explored. Since the main objective of this study is to provide a useful toolkit that can be broadly applied, the proposed crosslinking chemistries are combined with different decorating molecules and cell lineages. The three proposed

https://doi.org/10.3390/biomimetics11020095

crosslinking chemistries comprise the biotin/streptavidin system, a thiol-based crosslinking chemistry, and a "click" bioorthogonal crosslinking chemistry.

It may be worth indicating the characteristic features of each chemistry. To begin with, the biotin/streptavidin is a very versatile approach, as it enables the usage of any biotinylated biomolecule, such as commercially available biotinylated antibodies. In case of necessity, there exist protocols that allow biotinylating of practically any required antibody.

The thiol-based crosslinking chemistry offers the unique possibility of reusing decorated probes by reducing the disulfide bond that maintains the binding molecule that is attached to the surface. This reduction can be achieved through reducing agents, such as TCEP or dithiothreitol (DTT). Although no attempt was made within the context of this work on developing a detailed protocol for the reuse of the probes, its feasibility is demonstrated by the preparation of the SH-functionalized probes that result from the reduction in the original pyridyldisulfide group with TCEP. With respect to the thiol-based crosslinking strategy, we also show how it can be applied even if the decorating molecule lacks a free sulfhydryl group, as shown by the decoration of the SH-functionalized surface with the SPDP-derivatized streptavidin. In this latter case, however, some extra time is required for the derivation of the molecule, although, as illustrated with the derivation of streptavidin, this additional time may be in the range of one hour.

Finally, the last crosslinking chemistry, the biorthogonal ("click") chemistry is of particular interest in the biomedical field because it involves relatively small, highly specific molecules that do not occur naturally in living organisms. This allows for its use in biological environments without causing cellular damage and with minimal risk of interfering with cellular biochemistry or metabolism, as the reagents involved are not present endogenously [25–28]. In order to attach MSCs to the AFM cantilever using this chemistry, cells were successfully functionalized with $Ac_4ManNAz$ through metabolic glycan engineering. The fluorescence images indicate that the cells incorporated azide groups on their surface. However, a significant background signal was detected in the control group, which had not been reported in other studies [27,29,30]. This background signal might result from relatively intense nonspecific interactions between the fluorophore and the non-functionalized cells. In parallel, the cantilevers were decorated with DBCO molecules: a process that was successfully achieved, as shown from the comparison of the fluorescence intensity between controls and functionalized samples. The main advantage of the click chemistry lies in the ability to perform metabolic glycan engineering to incorporate bioorthogonal chemical groups on the cell surface without affecting other cellular functions. This implies that, in principle, the click chemistry might be the most direct technique for attaching cells to the surface of AFM cantilevers of all the three strategies presented in this work. However, as indicated below, this strategy leads to a relatively labile interaction between cell and material.

In those cases in which the cells could not be bound directly to the functionalized probe, it is necessary to use a decorating molecule. Following the rationale of creating a toolkit that is as versatile as possible, two types of biomolecules were employed: a specific antibody and concanavalin A. In addition, the assessment of the decoration of the surface with the SPDP-derivatized streptavidin was performed through the interaction of the streptavidin with a biotinylated enzyme (HRP) which validates not only the presence of functional streptavidin on the surface, but also the preservation of the enzymatic activity after the binding of the molecule to the streptavidin.

Finally, the versatility of the developed protocols was assessed by its application to two different cell lineages, both in terms of size and of its adherent character towards materials. $CD4^+$ lymphocytes are non-adherent cells of a moderate size, while MSC are adherent cells of a relatively large size. It is shown that both cell lineages can be efficiently bound to the

https://doi.org/10.3390/biomimetics11020095

AFM probes. In this regard, optical microscopy images showed that CD4$^+$ T cells, pristine MSCs, and functionalized MSCs were successfully bound to the cantilever through the three different chemistries: biotin-NHS with concanavalin A was used to bind pristine MSCs, thiol chemistry was used to decorate the cantilever with an anti-CD4 antibody in order to bind CD4$^+$ T cells and, lastly, bioorthogonal chemistry was used to bind azide-functionalized MSCs to alkyne-functionalized cantilevers. It was observed, however, that despite applying the same AFM observation procedure and the same force ($\approx$100 pN) to attach the cell to the cantilever in all cases, significant differences were observed in the contact time required for the successful attachment of the cell to the probe. Thus, whereas initially it was assumed that the bioorthogonal chemistry would provide the fastest and most effective attachment [31,32], it was found that longer contact time was required for the cell to adhere (around a minute). Even though the Ac_4ManNAz concentrations used have been shown to be optimal for covering the cell surface, avoiding cell viability issues [29,33] (which means azide density on the cell surface should be sufficient), this crosslinking chemistry also produced the least stable binding, as the cell would detach from the cantilever after approximately 10–15 min. In contrast, both decoration of the cantilevers with anti-CD4 antibodies through the thiol chemistry and with concanavalin A through the biotin/streptavidin system were found to lead to a stable binding of the cell to the AFM probe. It was established, however, that the required contact time to bind the cell to the antibody-decorated cantilever was significantly shorter ($\approx$10 s) than that required for the binding of pristine MSCs to the concanavalin A-decorated AFM probe ($\approx$30 s). In both cases, cells remained stably adhered to the cantilever for a period of at least two hours. It should also be stressed that the stability of the binding of pristine MSCs and CD4$^+$ T lymphocytes to concanavalin A- and antibody-decorated cantilevers, respectively, allowed for the performance of frequency screening measurements. It was not possible to perform those experiments on azide-functionalized MSCs, since cells become detached from the DBCO cantilevers before completing the frequency screening.

Since the objective of this work was the development of a toolkit for the binding of cells to AFM probes that includes a broad range of crosslinking chemistries and decorating biomolecules, and that can be applied to the largest possible number of cell lineages, the interesting question of the interaction between the cells and the substrate is not pursued further here. An example of the power of this type of analysis is offered, for instance, by the comparison of the different adhesion forces of MSCs on titanium, depending on the peptides used for the decoration of the material [20]. In this case, it was possible to correlate the strength of the adhesion with the proliferation of the cells on the material, which was found to be dependent on the particular peptide used for the decoration.

5. Conclusions

In this work, a preparatory experimental study of the performance of three different crosslinking strategies, biotin/streptavidin, thiol-based crosslinking chemistry, and bioorthogonal chemistry, for attaching cells to AFM probes, is presented. Two types of cells, adherent MSCs and non-adherent CD4$^+$ T lymphocytes, were used as model systems to prove the versatility of the procedures. Successful functionalization of AFM probes was demonstrated for all three chemistries through fluorescence microscopy. In turn, MSCs were also successfully functionalized with Ac_4ManNAz, as was also confirmed by fluorescence microscopy. Both MSCs and CD4$^+$ T lymphocytes were successfully attached to cantilevers functionalized with the corresponding chemistries: pristine MSCs were attached to concanavalin A-decorated cantilevers; CD4$^+$ T lymphocytes were bound to anti-CD4-decorated cantilevers, and Ac_4ManNAz-functionalized MSCs were attached via bioorthogonal chemistry to DBCO-decorated chips. It was found, however, that the

https://doi.org/10.3390/biomimetics11020095

binding of the functionalized MSCs to the DBCO-decorated cantilever required a longer contact time and led to a more labile binding compared with the other two attachment strategies. Although within the framework of biohybrid materials, it will be necessary to perform specific studies for each cell lineage–substrate pair of interest, the proposed protocols are shown to constitute a practical toolkit for this type of study. In this context, these protocols may be taken as a convenient starting point for assessing the interaction between materials and cells obtained from different species including, most conspicuously, those of human origin.

Supplementary Materials: The following supporting information can be downloaded at: https://www.mdpi.com/article/10.3390/biomimetics11020095/s1.

Author Contributions: Investigation: R.T.-R., M.V., L.Y.M., R.D. and L.C.; Methodology: R.T.-R. and R.D.; Writing: R.T.-R. and J.P.-R.; Validation: M.E.F.-S., F.P. and M.E.; Supervision: G.V.G. and J.P.-R. All authors have read and agreed to the published version of the manuscript.

Funding: This study was funded by the Ministerio de Ciencia e Innovación (Spain) (grant PID2023-152058OB-I00), by Comunidad de Madrid (Spain) (grants MINA-CM P2022-BMD-7236 and Tec-NanoBio TEC-2024/TEC-158) and by the European Union's EIC-Pathfinder Programme, under the projects THOR (Grant Agreement number 101099719) and ISOS (Grant Agreement number 101130454).

Institutional Review Board Statement: The study was conducted in accordance with the Declaration of Helsinki, and was approved by the Universidad Politécnica Ethic Committee and Comunidad de Madrid (Authorization number: PROEX 010.3/22; 18 February 2022) and the study was developed by following the Spanish animal experimentation legislation (Real Decreto 53/2013 and Orden ECC/566/2015).

Informed Consent Statement: Not applicable.

Data Availability Statement: The original data presented in the study are included in Figures throughout the article. Further reasonable inquiries regarding data access can be directed to the corresponding author.

Conflicts of Interest: DeepTip^{TM} probes were kindly supplied by Bioactive Surfaces S.L., who partially funded this research. Raquel Tabraue-Rubio and Luis Colchero were former employees of Bioactive Surfaces S.L. Fivos Panetsos is founder and one of the owners of Bioactive Surfaces S.L. The remaining authors declare that the research was conducted in the absence of any commercial or financial relationships that be construed as a potential conflict of interest.

References

1. Raman, R.; Langer, R. Biohybrid Design Gets Personal: New Materials for Patient-Specific Therapy. *Adv. Mater.* **2020**, *32*, 1901969. [CrossRef]
2. Rodrigo-Navarro, A.; Sankaran, S.; Dalby, M.J.; del Campo, A.; Salmeron-Sanchez, M. Engineered Living Biomaterials. *Nat. Rev. Mater.* **2021**, *6*, 1175–1190. [CrossRef]
3. Nguyen, P.Q.; Courchesne, N.D.; Duraj-Thatte, A.; Praveschotinunt, P.; Joshi, N.S. Engineered Living Materials: Prospects and Challenges for Using Biological Systems to Direct the Assembly of Smart Materials. *Adv. Mater.* **2018**, *30*, e1704847. [CrossRef] [PubMed]
4. Rivera-Tarazona, L.K.; Campbell, Z.T.; Ware, T.H. Stimuli-Responsive Engineered Living Materials. *Soft Matter* **2021**, *17*, 785–809. [CrossRef]
5. Wangpraseurt, D.; You, S.; Sun, Y.; Chen, S. Biomimetic 3D Living Materials Powered by Microorganisms. *Trends Biotechnol.* **2022**, *40*, 843–857. [CrossRef] [PubMed]
6. Appiah, C.; Arndt, C.; Siemsen, K.; Heitmann, A.; Staubitz, A.; Selhuber-Unkel, C. Living Materials Herald a New Era in Soft Robotics. *Adv. Mater.* **2019**, *31*, e1807747. [CrossRef]
7. Yu, Y.; Wang, Q.; Wang, C.; Shang, L. Living Materials for Regenerative Medicine. *Eng. Regen.* **2021**, *2*, 96–104. [CrossRef]
8. Ganguly, S.; Margel, S. Magnetic Polymeric Conduits in Biomedical Applications. *Micromachines* **2025**, *16*, 174. [CrossRef]

https://doi.org/10.3390/biomimetics11020095

9. Ouyang, L.; Armstrong, J.P.K.; Salmeron-Sanchez, M.; Stevens, M.M. Assembling Living Building Blocks to Engineer Complex Tissues. *Adv. Funct. Mater.* **2020**, *30*, 1909009. [CrossRef]
10. Huang, J.; Liu, S.; Zhang, C.; Wang, X.; Pu, J.; Ba, F.; Xue, S.; Ye, H.; Zhao, T.; Li, K.; et al. Programmable and Printable Bacillus Subtilis Biofilms as Engineered Living Materials. *Nat. Chem. Biol.* **2019**, *15*, 34–41. [CrossRef]
11. Kamperman, T.; Henke, S.; Visser, C.W.; Karperien, M.; Leijten, J. Centering Single Cells in Microgels via Delayed Crosslinking Supports Long-Term 3D Culture by Preventing Cell Escape. *Small* **2017**, *13*, 1603711. [CrossRef] [PubMed]
12. Yeo, M.; Kim, G.H. Anisotropically Aligned Cell-Laden Nanofibrous Bundle Fabricated via Cell Electrospinning to Regenerate Skeletal Muscle Tissue. *Small* **2018**, *14*, e1803491. [CrossRef] [PubMed]
13. Wang, C.; Zhang, Z.; Wang, J.; Wang, Q.; Shang, L. Biohybrid Materials: Structure Design and Biomedical Applications. *Mater. Today Bio* **2022**, *16*, 100352. [CrossRef]
14. Martín-González, N.; Hernando-Pérez, M.; Condezo, G.N.; Pérez-Illana, M.; Šiber, A.; Reguera, D.; Ostapchuk, P.; Hearing, P.; San Martín, C.; de Pablo, P.J. Adenovirus Major Core Protein Condenses DNA in Clusters and Bundles, Modulating Genome Release and Capsid Internal Pressure. *Nucleic Acids Res.* **2019**, *47*, 9231–9242. [CrossRef]
15. Rigato, A.; Miyagi, A.; Scheuring, S.; Rico, F. High-Frequency Microrheology Reveals Cytoskeleton Dynamics in Living Cells. *Nat. Phys.* **2017**, *13*, 771–775. [CrossRef]
16. Zhang, Q.; Yang, J.; Tillieux, S.; Guo, Z.; Natividade, R.d.S.; Koehler, M.; Petitjean, S.; Cui, Z.; Alsteens, D. Stepwise Enzymatic-Dependent Mechanism of Ebola Virus Binding to Cell Surface Receptors Monitored by AFM. *Nano Lett.* **2022**, *22*, 1641–1648. [CrossRef]
17. Viljoen, A.; Mathelié-Guinlet, M.; Ray, A.; Strohmeyer, N.; Oh, Y.J.; Hinterdorfer, P.; Müller, D.J.; Alsteens, D.; Dufrêne, Y.F. Force Spectroscopy of Single Cells Using Atomic Force Microscopy. *Nat. Rev. Methods Prim.* **2021**, *1*, 63. [CrossRef]
18. Holuigue, H.; Nacci, L.; Di Chiaro, P.; Chighizola, M.; Locatelli, I.; Schulte, C.; Alfano, M.; Diaferia, G.R.; Podestà, A. Native Extracellular Matrix Probes to Target Patient- and Tissue-Specific Cell–Microenvironment Interactions by Force Spectroscopy. *Nanoscale* **2023**, *15*, 15382–15395. [CrossRef]
19. Matias, Z.; Lopes, C.S.; Santos, N.C.; Carvalho, F.A. Nanotechnology Meets Medicine: Applications of Atomic Force Microscopy in Disease. *Biophys. Rev.* **2025**, *17*, 359–384. [CrossRef]
20. Álvarez-López, A.; Tabraue-Rubio, R.; Hernández-Escobar, S.; Daza, R.; Colchero, L.; Rezvanian, P.; Elices, M.; Guinea, G.V.; González-Nieto, D.; Pérez-Rigueiro, J. Application of Single Cell Force Spectroscopy (SCFS) to the Assessment of Cell Adhesion to Peptide-Decorated Surfaces. *Int. J. Biol. Macromol.* **2023**, *244*, 125369. [CrossRef]
21. Puech, P.H.; Poole, K.; Knebel, D.; Muller, D.J. A New Technical Approach to Quantify Cell-Cell Adhesion Forces by AFM. *Ultramicroscopy* **2006**, *106*, 637–644. [CrossRef]
22. Smolyakov, G.; Thiebot, B.; Campillo, C.; Labdi, S.; Severac, C.; Pelta, J.; Dague, É. Elasticity, Adhesion, and Tether Extrusion on Breast Cancer Cells Provide a Signature of Their Invasive Potential. *ACS Appl. Mater. Interfaces* **2016**, *8*, 27426–27431. [CrossRef] [PubMed]
23. Martín-Martín, Y.; Fernández-García, L.; Sanchez-Rebato, M.H.; Marí-Buyé, N.; Rojo, F.J.; Pérez-Rigueiro, J.; Ramos, M.; Guinea, G.V.; Panetsos, F.; González-Nieto, D. Evaluation of Neurosecretome from Mesenchymal Stem Cells Encapsulated in Silk Fibroin Hydrogels. *Sci. Rep.* **2019**, *9*, 8801. [CrossRef] [PubMed]
24. Horcas, I.; Fernández, R.; Gómez-Rodríguez, J.M.; Colchero, J.; Gómez-Herrero, J.; Baro, A.M. WSXM: A Software for Scanning Probe Microscopy and a Tool for Nanotechnology. *Rev. Sci. Instrum.* **2007**, *78*, 13705. [CrossRef]
25. Alberti, D.; Grange, C.; Porta, S.; Aime, S.; Tei, L.; Geninatti Crich, S. Efficient Route to Label Mesenchymal Stromal Cell-Derived Extracellular Vesicles. *ACS Omega* **2018**, *3*, 8097–8103. [CrossRef]
26. Chang, P.V.; Prescher, J.A.; Sletten, E.M.; Baskin, J.M.; Miller, I.A.; Agard, N.J.; Lo, A.; Bertozzi, C.R. Copper-Free Click Chemistry in Living Animals. *Proc. Natl. Acad. Sci. USA* **2010**, *107*, 1821–1826. [CrossRef]
27. Yoon, H.I.; Yhee, J.Y.; Na, J.H.; Lee, S.; Lee, H.; Kang, S.W.; Chang, H.; Ryu, J.H.; Lee, S.; Kwon, I.C.; et al. Bioorthogonal Copper Free Click Chemistry for Labeling and Tracking of Chondrocytes In Vivo. *Bioconjug. Chem.* **2016**, *27*, 927–936. [CrossRef] [PubMed]
28. Sojdeh, S.; Panjipour, A.; Yaghmour, A.; Arabpour, Z.; Djalilian, A.R. Click Chemistry-Based Hydrogels for Tissue Engineering. *Gels* **2025**, *11*, 724. [CrossRef]
29. Yun, W.S.; Shim, M.K.; Lim, S.; Song, S.; Kim, J.; Yang, S.; Hwang, H.S.; Kim, M.R.; Yoon, H.Y.; Lim, D.K.; et al. Mesenchymal Stem Cell-Mediated Deep Tumor Delivery of Gold Nanorod for Photothermal Therapy. *Nanomaterials* **2022**, *12*, 3410. [CrossRef]
30. Lim, S.; Kim, W.; Song, S.; Shim, M.K.; Yoon, H.Y.; Kim, B.-S.; Kwon, I.C.; Kim, K. Intracellular Uptake Mechanism of Bioorthogonally Conjugated Nanoparticles on Metabolically Engineered Mesenchymal Stem Cells. *Bioconjug. Chem.* **2021**, *32*, 199–214. [CrossRef]
31. Kolb, H.C.; Finn, M.G.; Sharpless, K.B. Click Chemistry: Diverse Chemical Function from a Few Good Reactions. *Angew. Chemie Int. Ed.* **2001**, *40*, 2004–2021. [CrossRef]

https://doi.org/10.3390/biomimetics11020095

32. Luu, T.; Gristwood, K.; Knight, J.C.; Jörg, M. Click Chemistry: Reaction Rates and Their Suitability for Biomedical Applications. *Bioconjug. Chem.* **2024**, *35*, 715–731. [CrossRef] [PubMed]
33. Altmann, S.; Mut, J.; Wolf, N.; Meißner-Weigl, J.; Rudert, M.; Jakob, F.; Gutmann, M.; Lühmann, T.; Seibel, J.; Ebert, R. Metabolic Glycoengineering in HMSC-TERT as a Model for Skeletal Precursors by Using Modified Azide/Alkyne Monosaccharides. *Int. J. Mol. Sci.* **2021**, *22*, 2820. [CrossRef] [PubMed]

https://doi.org/10.3390/biomimetics11020095

biomimetics

Article

Exploring Adhesive Performance in Horseshoe Bonding Through Advanced Mechanical and Numerical Analysis

C. M. C. Ferreira [1,*], B. D. Simões [1], E. A. S. Marques [2], R. J. C. Carbas [1] and L. F. M. da Silva [2]

1 Institute of Science and Innovation in Mechanical and Industrial Engineering (INEGI), Rua Dr. Roberto Frias, 4200-465 Porto, Portugal; bsimoes@inegi.up.pt (B.D.S.); rcarbas@inegi.up.pt (R.J.C.C.)
2 Department of Mechanical Engineering, Faculty of Engineering, University of Porto, Rua Dr. Roberto Frias, 4200-465 Porto, Portugal; emarques@fe.up.pt (E.A.S.M.); lucas@fe.up.pt (L.F.M.d.S.)
* Correspondence: mferreira@inegi.up.pt

Abstract: Despite technological advancements in various industries, the equine sector still relies on old methods like horseshoeing. Although traditional, the industry is dynamic and well-funded. Therefore, there is a need to modernize these methods with more reliable and less invasive solutions for attaching horseshoes to horse hooves. There are currently several commercial adhesive solutions in the market specifically tailored to this application. In this work, the mechanical properties of two acrylic adhesives were characterized under quasi-static conditions. In the characterization process, tensile, shear, and fracture properties were determined. Subsequently, in-joint behavior was assessed using single-lap joints (SLJ) for both similar and dissimilar adherends. The adherends' materials included steel (St), aluminum (Al), and horse hoof wall (HW), and the following adherend combinations were tested: St–St, Al–Al, and St–HW. A numerical model of similar joints was developed and validated based on experimental results. After its validation, the next steps are the modelling of the real joint and its simulation by considering realistic loading conditions in order to determine the weakest points of the joint. This exploratory study seeks to establish a foundation for investigating alternative adhesive solutions that could address the limitations identified in the solutions studied in this paper.

Keywords: equine industry; horseshoeing; acrylic adhesives; numerical modelling

1. Introduction

Over the past few decades, most industrial sectors have experienced exponential technological advancement, largely driven by the development of new materials and methods. Nevertheless, certain traditional sectors continue to employ longstanding practices. The equestrian industry is one of those sectors, where horseshoe application has remained consistent with methods dating back to medieval times [1].

Despite its traditional methods, the equestrian sector attracts substantial investment and is a dynamic, challenging, and competitive industry. In Europe, it represents a market valued at approximately 100 billion euros a year [2], which, proportionate to the GDP of each member state, translates to an estimated 1700 million euros in Portugal. Consequently, this sector is primed for innovations in its practices, such as the traditional horseshoe application process.

Horses, depending on the demands placed on them, can walk either with or without horseshoes, with the vast majority using them. Unshod horses generally experience less impact damping when their hooves hit the ground compared to shod horses. This can be concerning not only for the hooves but also for the horse's joints, as the shock

https://doi.org/10.3390/biomimetics10010002

transmitted to the limbs can reduce peak performance and overall longevity [3]. Therefore, the horseshoeing process is of the utmost importance to ensure the protection of the hooves as well as the horse's performance [4].

Since medieval times, the horseshoe application process has followed a similar procedure: the horseshoe is heated until incandescent and then placed on the horse's hoof to check its fit. It is then shaped while still hot until it properly fits the hoof. Next, the horse nails are hammered in, and once they emerge on the other side of the hoof, they are cut and bent against the hoof. Finally, the tips are smoothed with sandpaper to refine the hoof. During horseshoe replacement, the horseshoe is first removed by sanding the bent edges of the nails, allowing the horseshoe to be pulled off. These horseshoes are typically made from St or Al alloys [5]. As mentioned earlier, this is an ancient process, and the final quality is highly dependent on the skill of the operator.

Therefore, the process is open to modernization, particularly in the joining method, in which the aim would be to avoid the introduction of horse nails in the hoof while improving reproducibility and reducing the level of skill needed.

Adhesives, when compared to mechanical connections, have much more significant damping and shock-absorbing properties [6–8]. Additionally, they enable the use of other horseshoe materials that would be damaged by horse nails, such as polymers or composites, improving the shock-absorbing properties of the horseshoe even further. As said, even the more traditional metallic horseshoes have better damping characteristics than an unshod horse. However, the horseshoe material can also have a very significant influence on the injuries experienced by horses. According to Sprik et al. [5], who performed drop-weight tests in the same conditions at 8 m/s, determined the probability of limb bones breaking with a steel horseshoe as 75%, and for aluminum as 81%, while when using polyurethane (PU), the bones do not break. Lee and Choi [9] also demonstrated that CFRP horseshoes allow for better walkability and reduced fatigue in horses.

For that reason, adhesive-based solutions have been proposed to attach the horseshoe to the horse's hoof [10–12]. However, these rely on the application of a paste adhesive, which can flow out of the bonded area, causing material waste and also requiring a second step for the removal of the excess adhesive. Furthermore, these adhesives always require a significant time to achieve full strength, which limits the mobility of the horse during this period. Additionally, the dismantling process still consists of simply pulling off the horseshoe by inserting a wedge in an indentation between the adhesive and the hoof [13] which has the potential to damage the hoof if not carefully performed. Although some adhesively bonded solutions have reached the market, their use is still not widespread because they do not instill the necessary confidence in either the farriers, riders, or owners of the horses.

On the other hand, one of the biggest drawbacks, if not the main drawback, associated with the use of adhesive joints to bond horseshoes at this point is still the fact that the currently used process is strongly rooted in the equestrian sector, which is highly averse to change. Therefore, very extensive suitability and durability studies need to be conducted to gain market trust. From research and interviewing horseshoe users, it is also clear that the reason behind these solutions not being widely implemented is that the adhesive interferes with the breathing and elasticity of the hoof [1].

This work aims to lay the groundwork for exploring alternative adhesive solutions to overcome the limitations of current methods. The initial step involved benchmarking and identifying the adhesive solutions currently used in glue-on horseshoeing and then determining their mechanical properties, which will serve as a reference for the development of future solutions. Consequently, two commercial acrylic adhesives, designated as Adhesive A and Adhesive B, were selected. The mechanical properties of both adhesives were assessed through a series of standardized tests, which were then utilized in numerical simulations to evaluate the predictability of in-joint behavior.

2. Experimental Details

2.1. Materials

In the present study, 2 acrylic commercially available adhesives developed for horseshoeing were characterized to establish a reference of mechanical properties useful for the development of future horseshoeing solutions. These two-component acrylic adhesives will be designated as Adhesive A and Adhesive B for confidentiality purposes.

Similar single-lap joints (SLJs) were manufactured with aluminum (Al 6082-T6, σ_y = 255 MPa), high-strength steel (DIN 55 Si 7, σ_y = 1100 MPa), designated as HSS throughout the paper, and mild steel (DIN St 33, σ_y = 120 MPa) adherends.

These materials were selected because both Al and St are commonly used for the production of horseshoes. Usually, Al is used for high-performance horseshoes while mild St is used for standard ones [14].

Dissimilar SLJs were manufactured with mild St and horse hoof wall adherends. These joints are considered representative joints of the real one, which corresponds to the bonding of the horseshoe to the horse hoof. The authors of this paper are aware that the horse hoof wall specimens used in these joints do not fully replicate the real behavior of the surface to which the horseshoes are bonded nor the mechanical properties of the entire structure of the hoof. However, these specimens were chosen because they were the closest available material to realistically mimic the actual surface and behavior of the horse hoof.

A hoof from an adult horse (*Equus caballus* L.) with an unknown physical condition (destroyed for reasons outside the scope of this study) was obtained from the veterinarian hospital of Universidade de Trás-os-Montes e Alto Douro (Vila Real, Portugal), which removed all its living tissues and separated the hoof capsule from the underlying bone. Therefore, the provided sample consisted solely of the keratinized hoof material, without any associated soft tissue or bone.

2.2. Specimen Manufacturing

2.2.1. Bulk Specimens

The bulk specimens were manufactured according to ASTM D412 [15] (injection molding fabrication), which led to 3 mm-thickness specimens [16]. Before use, the mold was degreased with acetone and coated with a release agent. The adhesive was poured into the assembled mold, and the hydrostatic pressure was ensured with 4 screws. The specimens were kept at room temperature (20 °C) for at least 7 days before testing. The dimensions of the bulk specimen are presented in Figure 1. Finally, the specimens were prepared for DIC analysis by coating them with white matte paint followed by the application of black speckling.

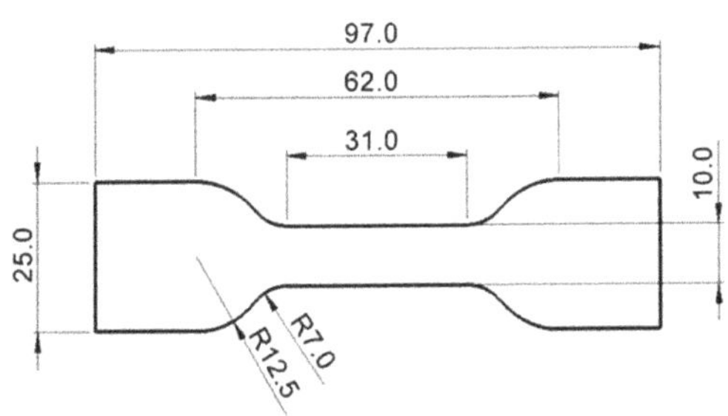

Figure 1. Geometry of the specimens for tensile testing of the adhesives [mm] according to ASTM D412.

2.2.2. TAST Specimens

A dedicated mold was used to manufacture 6 TAST specimens with a 0.5–0.8 mm-thickness adhesive layer and an overlap of 5 mm, which was ensured by 1.5 mm-thick shim tabs.

The adhesive thickness was measured for each specimen using a micrometer. Figure 2 displays the dimensions [mm] of the specimens following ISO 11003-2 standards [17]. The mold base was cleaned with acetone prior to the release agent application.

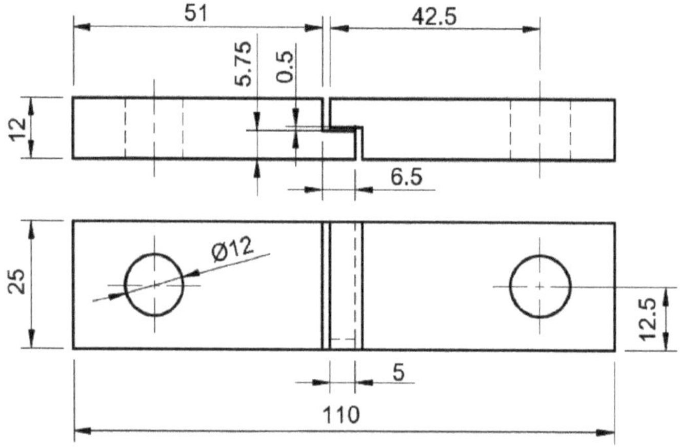

Figure 2. Geometry and dimensions of TAST specimens [mm] according to ISO 11003-2.

Firstly, the adhesive was applied to the lower adherends, followed by the lower shims (1.5 mm thickness) and the top adherends. Next, the upper shims were positioned on each specimen to prevent adhesive overflow onto undesired surfaces, ensuring an almost pure shear condition during testing [16]. Finally, the top plate was placed over the assembly. The specimens were cured at room temperature (20 °C) for 7 days with weight on top of the mold.

After curing, the specimens were carefully removed from the mold, and any excess adhesive was trimmed away using a sharp cutter. The shim tabs were then gently removed from each joint to avoid introducing cracks or damaging the joint.

2.2.3. DCB Specimens

The DCB adherends (DIN 40 CrMnMo 8-6-4 St, σ_y = 800 MPa) were used according to ISO 25217 standard [18]. The DCB specimens' dimensions [mm] are displayed in Figure 3, where *t* represents the thickness of the adhesive layer and *a* the initial crack length.

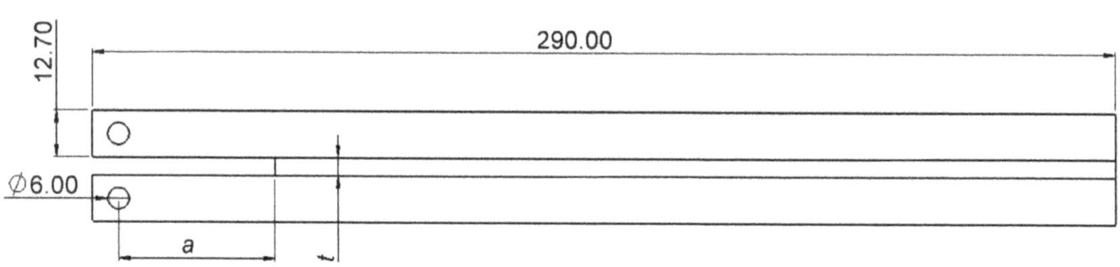

Figure 3. DCB specimen geometry and dimensions [mm].

The specimens were initially cleaned with acetone and then sandblasted. The bottom adherends were placed into a clean mold equipped with guiding pins and pre-coated with a mold release agent. To achieve an adhesive thickness of 0.20 mm, calibrated tape was applied to both sides of each specimen. A 0.10 mm sharp razor blade, with two pieces of 0.05 mm calibrated tape on either side, was used to create the pre-crack. These components were cleaned and coated with a release agent to ensure easy removal after curing. Afterwards, two loading blocks were bonded to the adherends, as suggested by Figure 4, which made possible the introduction of high-precision inclinometers into the test setup. These were responsible for the measurement of the rotations of each substrate at the respective loading point during the test.

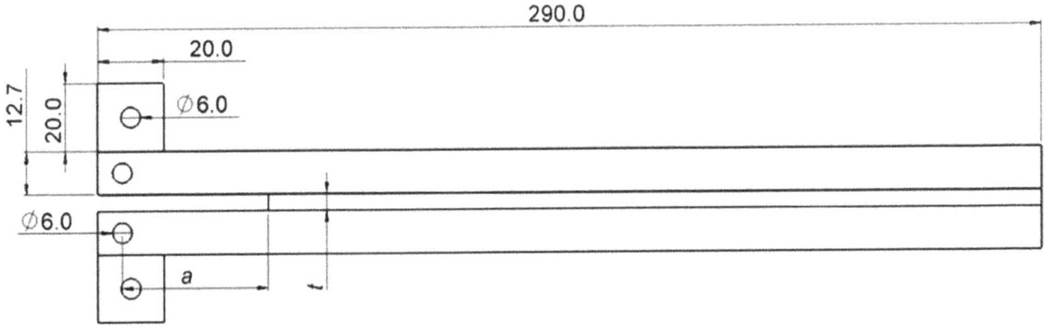

Figure 4. DCB specimen geometry with loading blocks for the introduction of inclinometers in the testing setup.

The adhesive was applied to the bottom and top adherends, which were then assembled, and finally, the mold was closed and loaded with weights to avoid air entrapment. The curing conditions followed the ones from the previous sections.

2.2.4. Horse Hoof Wall Adherends

The supplied hoof provided material for only 3 specimens. Each portion of the hoof wall for each substrate was isolated using an electric band saw (Figure 5c). The length of the substrate was limited by the size of the hoof, which is why dissimilar joints do not have the same dimensions as similar ones. Therefore, a comparison between similar and dissimilar joints will not be addressed. The thickness of the hoof wall adherends was obtained by manual machining (Figure 5e) in order to carefully separate the stratum medium (SM) tissue of the hoof wall from the remaining regions. Therefore, only the SM tissue was used in this study.

Although the hoof wall material in direct contact with the ground is where the horseshoe is bonded in a real scenario, this material was not suitable for specimen preparation due to insufficient material, so it was discarded. The region of the hoof wall facing the coffin bone was also excluded, as it contains tubule-forming dermal papillae [19].

After obtaining the hoof wall adherends, they were cleaned with isopropyl alcohol before the SLJ manufacturing process.

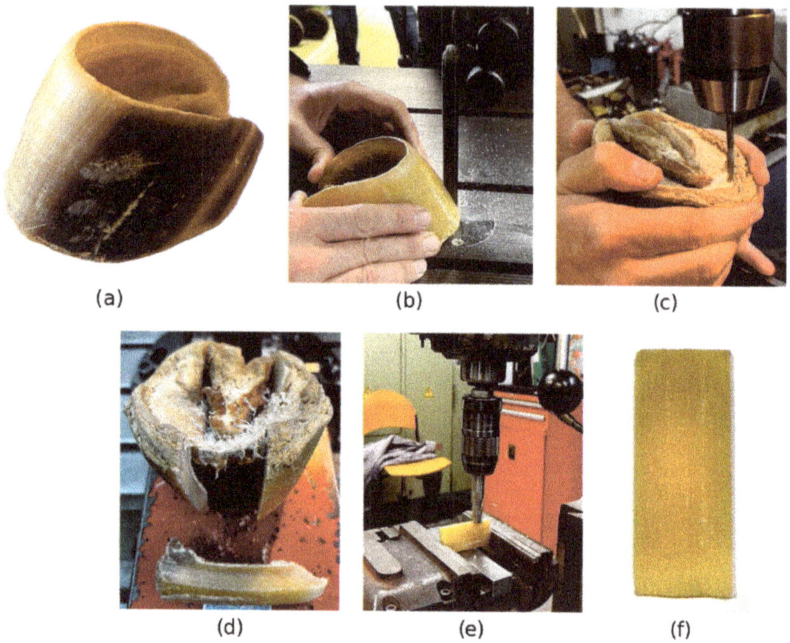

Figure 5. Manufacturing process of the hoof wall adherends from the initial and natural hoof wall (**a**) to the final substrate (**f**). The process involved cutting the hoof wall with a thin electric band saw (**b**) to an intended width, then inducing holes in the sole (**c**) until extraction of the specimen (**d**). The final step is the milling of the specimens (**e**) in order to reach the required dimensions (**f**).

2.2.5. SLJ Specimens

After adherend preparation, the SLJs were manufactured in an Al mold, previously cleaned and coated with release agent.

The Al mold employed for similar SLJs is designed for standard SLJs with overlap lengths of 12.5 mm, 25 mm, and 50 mm. However, given that the manufactured dissimilar SLJs are smaller than these standard dimensions, it was necessary to fabricate a 3D-printed mold. This custom mold was attached to the Al mold using pinned connections and was designed to match the geometry of the dissimilar SLJs, thereby ensuring accurate final joint dimensions.

Following assembly, the lap shear test specimens were kept at room temperature for 7 days with weights placed on top of the Al mold. After curing, the rig was opened, the joints were removed, and the excess adhesive was removed from the overlap area. In similar SLJs, St alignment tabs of 25 × 25 mm were bonded to the ends of each adherend, and for dissimilar SLJs, 3D-printed alignment tabs of 12 × 12 mm were used instead. These alignment tabs were used to minimize load misalignment and peel effects during testing.

The adherend materials used in the manufacturing of similar and dissimilar SLJs included Al, HSS, mild St, and horse hoof wall (HW), and the following adherend combinations were tested: Al–Al, HSS–HSS, Mild St–Mild St, and Mild St–HW. All joints were manufactured with a 12.5 mm overlap and a 0.2 mm adhesive layer thickness ensured by calibrated tape.

- Similar SLJs

Al6082-T6 alloy, HSS, and mild St adherends were used to manufacture the Al–Al, HSS–HSS, and Mild St–Mild St SLJs, respectively. Both St and Al adherends were sandblasted on the overlap region and then cleaned with acetone prior to the manufacturing of SLJs. The adherends' elastic properties are discussed in Section 2.1. The geometry of the similar joints is illustrated in Figure 6.

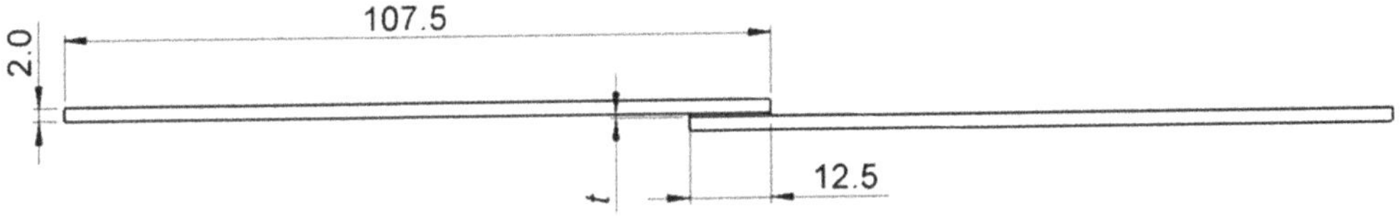

Figure 6. Similar SLJ geometry. The width (out of plane) of each adherend is 25 mm, and t represents the thickness of the adhesive layer, which was defined as 0.2 mm.

- Dissimilar SLJs

Each dissimilar joint was fabricated with a mild St substrate and an HW substrate. Figure 7 displays the geometry used for the manufacturing of these joints. The mild St adherends were also sandblasted on the overlap region and then cleaned with acetone before the SLJ manufacturing process. No surface treatment was applied to the HW substrate to better replicate the real situation, which corresponds to the bonding of the horseshoe to the horse hoof.

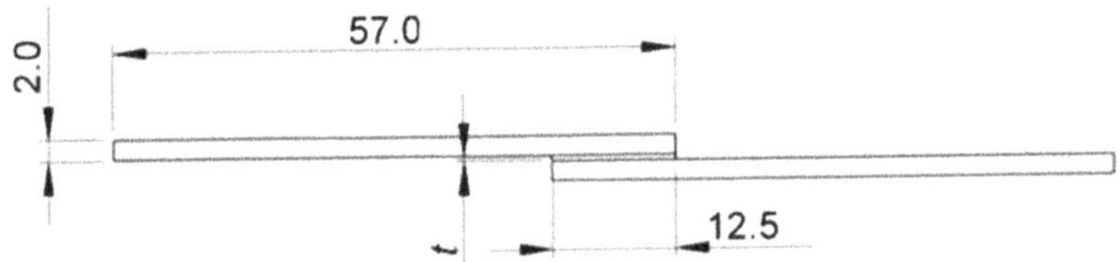

Figure 7. Dissimilar SLJ geometry. The width of each adherend is 12.5 mm, and t represents the thickness of the adhesive layer.

2.3. *Testing Setup*

All tests were performed under quasi-static conditions at room temperature (20 °C) in an INSTRON® 3367 (Illinois Tool Works, Hopkinton, MA, USA) quasi-static machine with a load cell of 30 kN. The adopted quasi-static test speeds were 2.5 mm/min for bulk tensile testing, 0.4 mm/min for TAST, 0.2 mm/min for DCB fracture tests, and 1 mm/min for SLJ testing. For each test, at least three specimens were tested.

2.3.1. Tensile Test

Load–displacement curves were recorded for each test until failure occurred. For each test, it was necessary to synchronize the load applied on the specimen, which was recorded by the quasi-static machine, with the displacement field of the specimen's areas of interest.

For that reason, a digital camera (Canon EOS M5, Canon Inc., Tokyo, Japan) equipped with a Canon EF-M 18–55 mm F/3.5–5.6 lens was placed in front of the specimens, with the lens oriented perpendicular to the observation surface. A laser beam was employed to ensure precise perpendicular alignment between the camera and the specimen. In order to determine the displacement field, a digital image correlation (DIC) approach was implemented using GOM Correlate® (2019) software.

After determining the load–displacement curves, the engineering stress was obtained by dividing the load values by the initial cross-section of the specimen, and the engineering strain was determined considering that the deformation would be the displacement field obtained from DIC divided by the original gauge length of the bulk specimen, L_0.

Consequently, a stress–strain curve was generated for each specimen, allowing the determination of Young's modulus, tensile strength, and tensile strain at failure for each adhesive.

2.3.2. Thick Adherend Shear Test (TAST)

TAST specimens were mounted in the universal testing machine by connecting the end of each adherend to a gripping system, which was then assembled in the machine with stiff steel pins. The St adherends used for each specimen were considerably thicker than the adhesive layer in order to avoid peel stresses commonly seen in SLJ testing due to thin adherends (usually 2–3 mm). Therefore, TAST specimens avoided these mixed-mode load configurations, which ensured a more accurate determination of the shear properties [20]. The recording of the strain field was performed using a clip-on extensometer (2630-100 series, Instron, Norwood, MA, USA).

The engineering stress field was calculated by dividing the load values recorded by the quasi-static machine by the overlap area of the specimens. Finally, a stress–strain curve was obtained for each specimen, followed by the determination of the shear properties of the adhesives: shear modulus and shear strength.

2.3.3. Double Cantilever Beam Test

The DCB specimens were installed in the universal testing machine by connecting the loading blocks to the machine's gripping system with pins.

The J-integral approach requires the synchronization of the load applied on the specimen throughout the test and the rotation angles at the load introduction points. Thus, the load–displacement curve was registered by the quasi-static machine and the rotation angles were recorded with Seika® NA2-10 inclinometers (Wiggensbach, Germany). The DCB specimens were loaded in tension with a small pre-load applied to ensure a stable crack propagation.

- Data-Reduction Scheme

Several data-reduction methods have been used in order to determine the fracture toughness of a material, such as the Simple Beam Theory (SBT), Corrected Beam Theory (CBT), Experimental Compliance Method (ECM), and Compliance-Based Beam Theory (CBBM). All these methods are based on Linear Elastic Fracture Mechanics (LEFM) [21] and can only be applied if the fracture process zone (FPZ) is very small when compared to the length of the component. In order to surpass the limitations of LEFM, other methods have been proposed based on the J-integral approach [22], which is a contour integral that accounts for the energy per unit area needed to create two new surfaces on a cracked body under loading conditions. Rice [23] demonstrated that the J-integral is independent of the contour for any elastic or elastoplastic material and therefore allows the presence of plasticity [24], which means that the calculation of the fracture energy will be more accurate, especially in the case of more ductile adhesives that involve large FPZs.

Preliminary tests indicated that both adhesives examined in this study exhibited high ductility, as will be discussed in Section 4.1.1. Consequently, the assumptions of LEFM were not applicable, and thus the J-integral approach was used to determine the Mode I fracture toughness for each adhesive.

The loads and rotations measured during the tests were applied to the J-integral approach to determine the fracture energy for each test. In a mode I loading configuration, the value of J along the DCB specimen's exterior boundary is given by [25]:

$$J_I = \frac{P}{b}\left(\theta_{up} - \theta_{low}\right) \quad (1)$$

where P is the applied load, b is the width of the specimen, and θ_{up} and θ_{low} are the relative rotations of the upper and lower adherends at the loading points, respectively. This relation is applicable due to the path-independence property of the J-integral in both linear and nonlinear elastic materials while assuming that the region outside the fracture process zone (FPZ) remains elastic during the fracture process. This assumption is commonly accepted in fracture mechanics as it ensures the applicability of the J-integral to evaluate stress intensity factors and fracture toughness in the material [23].

2.3.4. Single Lap Joints

Each joint was assembled on the gripping system of the machine by fixing the end of one adherend to the lower fixture of the machine, which remains static throughout the test, and fixing the end of the opposite adherend to the machine's crosshead resorting to pins. The load–displacement curves were recorded by the machine's software from the beginning of the test up until the specimen's failure.

3. Numerical Modelling

ABAQUS® FE 2017 software was used to perform a numerical analysis where cohesive zone modelling (CZM) was applied to the adhesive layer, resorting to traction-separation laws for mode I and mode II.

The St adherends used in the DCB test procedures were modeled following a linear elastic approach, as were the Al, HSS, and mild St adherends used in similar SLJs. This modelling approach was justified by the observation that the failure loads from the experimental tests indicated that these adherends did not exceed their yielding point, as will be discussed in Section 4.1.4.

All adherends were discretized using ABAQUS® CPE4R plane-strain elements with reduced integration and hourglass control.

The adhesive layer was modelled using COH2D four-node cohesive elements, selected due to their compatibility with the elements chosen for the adherends. For SLJ modelling, a mesh refinement procedure was performed along the adhesive overlap length using a double-bias parameter (0.05–0.2 mm in the x-direction and 0.2–0.3 mm in the y-direction) as suggested by Figure 8, allowing for an increase in mesh density at the overlap edges due to high stress concentrations in these locations.

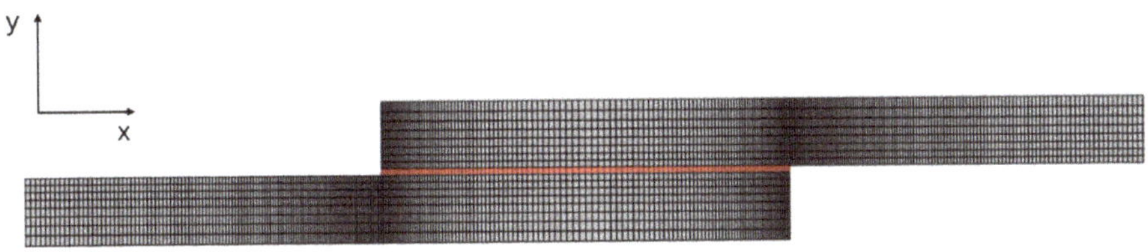

Figure 8. Details of the mesh used for similar SLJs at the overlap region.

The chosen boundary conditions simulated the fixtures used in the experimental tests. The bottom edge of one of the adherends was fixed, and a displacement load was applied to the opposite edge of the other adherend, as displayed in Figure 9.

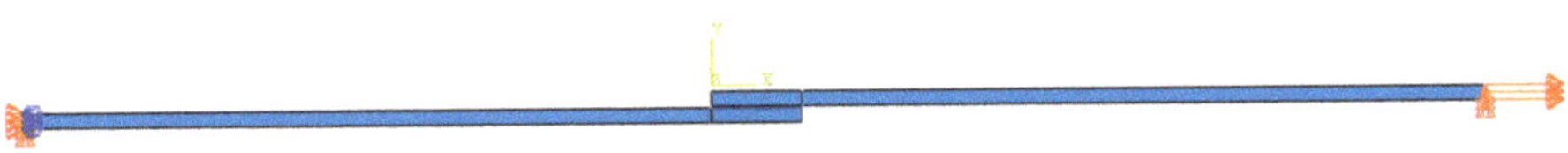

Figure 9. Boundary conditions of the SLJ model.

For the DCB model, mesh refinement was applied to the adherends in the y-direction towards the cohesive layer using a single-bias parameter (Figure 10). This approach increased the mesh density in these regions, which is essential for accurately capturing the stress distribution and modelling crack propagation effectively.

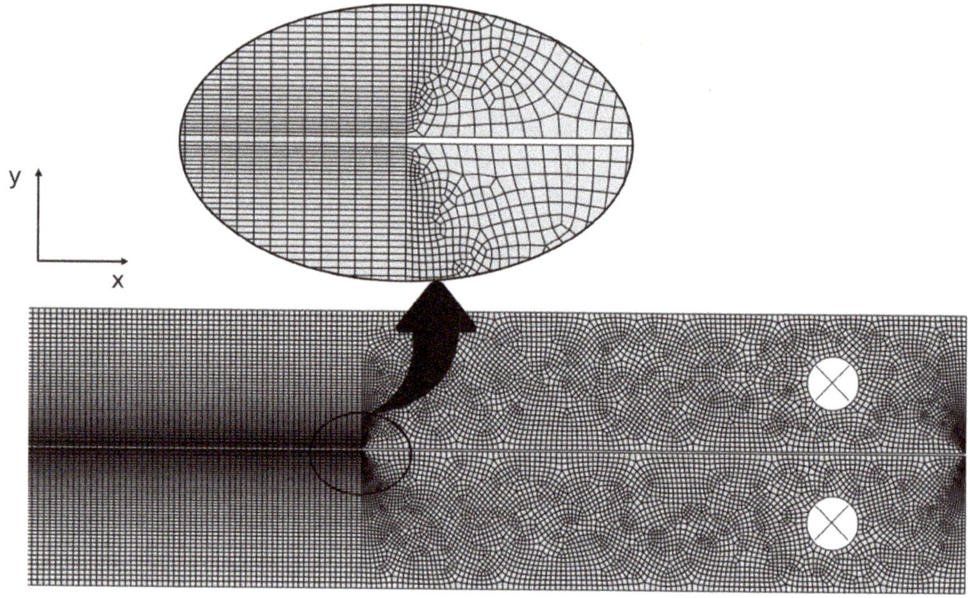

Figure 10. Details of the mesh used for simulating DCB mode I at the cohesive layer.

As for boundary conditions, the hole of the bottom adherend was pinned (Figure 11—RP1), allowing for rotation along the normal axis to the surface, and a displacement along the y-axis was applied to the hole of the upper substrate (Figure 11—RP2).

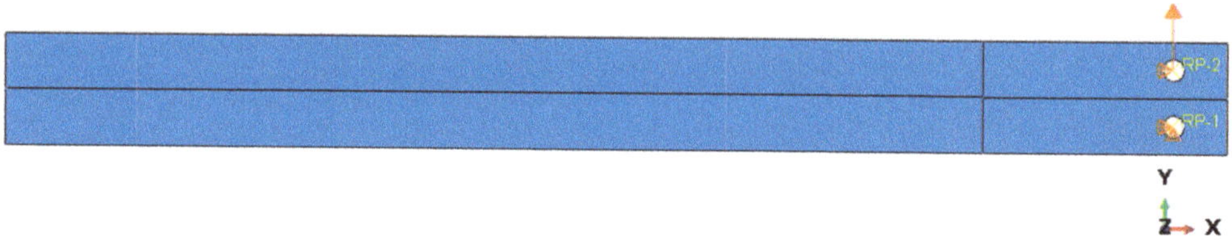

Figure 11. Boundary conditions for the DCB model.

- Cohesive Damage Model

For a pure mode loading (pure Mode I or II), the elastic behavior of the adhesive was described by a linear relationship between traction, σ, and separation, δ, before damage starts to grow. This relation is dependent on the tensile and shear stiffness parameters. These are defined as the ratio between the Young's modulus (Mode I) or the shear modulus (Mode II) and the adhesive thickness (t). After the elastic region, the softening phenomenon begins, leading to a traction separation law defined by:

$$\sigma = (I - D)\delta \tag{2}$$

where I is the identity matrix and D is a diagonal matrix containing the damage parameter, d, in the position corresponding to mode I and mode II.

The objective is to define a damage law that can reproduce the behavior of the adhesive in its respective bonded joints. The triangular shape law is mathematically simple as it assumes a linear elastic response up to the peak stress, followed by progressive damage propagation after the peak stress. Its simplicity enhances computational efficiency, thus reducing the risk of convergence issues. For these reasons, it is a common and simple form for simulating adhesive behavior [26]. However, one of its limitations is the inability to capture the plastic deformation behavior that occurs before damage initiation in ductile adhesives, where stress may remain constant or increase at a reduced rate.

In shear-dominated loading (Mode II), ductile adhesives often show significant plastic deformation before failure. Therefore, the triangular law may not accurately predict the stress distribution or energy dissipation in such scenarios, leading to discrepancies between numerical and experimental results.

As a result, a triangular law was used to describe the adhesive's performance in mode I due to its simplicity and its ability to properly represent the adhesive's behavior in this configuration, as will be discussed in Section 4.2.1. Nevertheless, both adhesives exhibited a nonlinear stress–strain relationship in pure shear conditions due to plastic deformation before the peak stress, as will be discussed in detail in Section 4.1.2. Consequently, a triangular law will not be able to capture this behavior. Pinto et al. [27] proposed a cohesive damage model, characterized by a trapezoidal shape with an increasing stress plateau, to simulate the behavior of a ductile adhesive in a SLJ configuration. Given the similarity in mechanical properties between the adhesives studied in this paper and the adhesive analyzed in [27], the same trapezoidal law was adopted to characterize the adhesive's behavior in mode II.

For mode I, the damage parameter following a triangular shape law is expressed as follows:

$$I = \frac{\delta_I^f\left(\delta_I - \delta_I^0\right)}{\delta_I\left(\delta_I^f - \delta_I^0\right)} \tag{3}$$

where δ_I^0 represents the displacement at damage initiation and δ_I^f represents the displacement at complete failure [27].

For mode II, the damage parameter following the trapezoidal shape of Figure 12 is expressed as follows:

$$d_{II} = \frac{(\delta_{II} - \delta_1)(\delta_2\,\alpha_{II} - \delta_1)}{\delta_{II}\,(\delta_2 - \delta_1)} \qquad \delta_1 \leq \delta_{II} \leq \delta_2 \tag{4}$$

$$d_{II} = 1 + \frac{\alpha_{II}\delta_1(\delta_{II} - \delta_u)}{\delta_{II}(\delta_u - \delta_2)} \qquad \delta_2 \leq \delta_{II} \leq \delta_u \tag{5}$$

where α_{II} stands for the strength ratio

$$\alpha_{II} = \frac{\sigma_u}{\sigma_s} \tag{6}$$

where σ_u corresponds to the ultimate strength and σ_s corresponds to the softening strength in mode II [27].

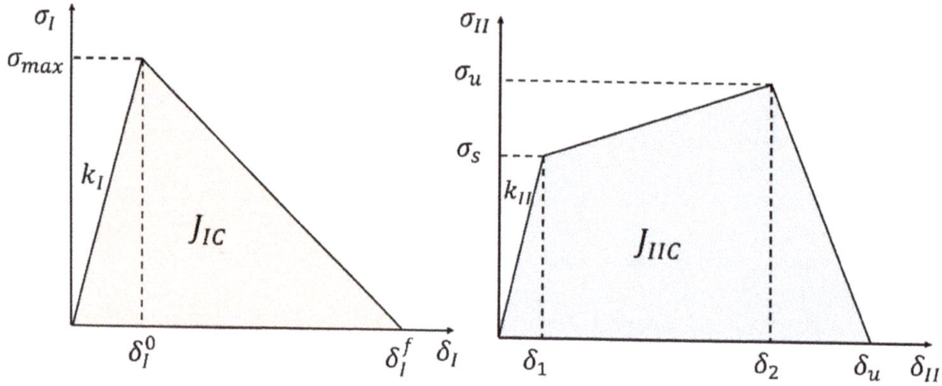

Figure 12. Triangular and trapezoidal shape cohesive laws used for characterizing the behavior of the adhesive in mode I and II, respectively.

The maximum displacement, δ_u, at which complete failure occurs, is determined by equating the area under the softening curve to the critical fracture energy in mode II (J_{IIc}):

$$J_{IIc} = \frac{\sigma_s}{2}(\delta_2 + \alpha_{II}(\delta_u - \delta_1)) \tag{7}$$

Usually, bonded joints are under mixed-mode loading and therefore a mixed-mode damage model had to be implemented. The selected damage criterion was a quadratic nominal stress formulation, which is described by

$$\left(\frac{\sigma_I}{\sigma_{max}}\right)^2 + \left(\frac{\sigma_{II}}{\sigma_s}\right)^2 = 1 \tag{8}$$

where σ_{max} represents the maximum strength in mode I. Therefore, damage begins as soon as this criterion is satisfied.

The definition of the damage criterion for traction–separation laws is based on an energetic approach, where the critical energy release rate values for both mode I and mode II are considered. Thus, it is possible to extrapolate a mixed-mode fracture energy relationship by combining the individual laws into a unified mixed-law formulation. In this case, a linear energetic criterion was used:

$$\frac{J_I}{J_{Ic}} + \frac{J_{II}}{J_{IIc}} = 1 \tag{9}$$

- Cohesive Parameters

The cohesive damage model was implemented in the adhesive layer (t = 0.2 mm) of the adhesive in pure modes I and II. The cohesive parameters in pure mode I to be defined are the maximum strength (σ_{max}), the elastic stiffness (k_I), and the critical fracture energy (J_{Ic}), and in pure mode II are the maximum strength (σ_u), the softening strength (σ_s), the second inflexion point (δ_2), and the critical fracture energy (J_{IIc}). For pure mode I, these parameters were assumed to be equal to the bulk quantities, while J_{Ic} was determined from DCB tests. For pure mode II, the maximum and softening strengths were approximately the same as the TAST quantities as well as the shear modulus. The first inflexion point (δ_1) was calculated from the ratio between the softening strength and the initial shear stiffness of the adhesive (k_I) and the second inflexion point (δ_2) was adjusted to an optimal parameter.

Since fracture tests were only performed for mode I, only the critical fracture energy in mode I was determined, and therefore the critical fracture energy in mode II was estimated according to the literature [27].

4. Results and Discussion

4.1. Experimental Tests

4.1.1. Tensile Test

The surface of the dog-bone specimens was sprayed with matte white paint prior to speckling with dark matte paint. For each adhesive, three specimens were positioned in pinned fixtures for testing. The strains were computed using DIC software, and the stress–strain curve was obtained. The stress–strain curves are illustrated in Figure 13. Table 1 summarizes the average values as well as the standard deviation for Young's modulus, tensile strength, and tensile strain to failure for Adhesive A and Adhesive B.

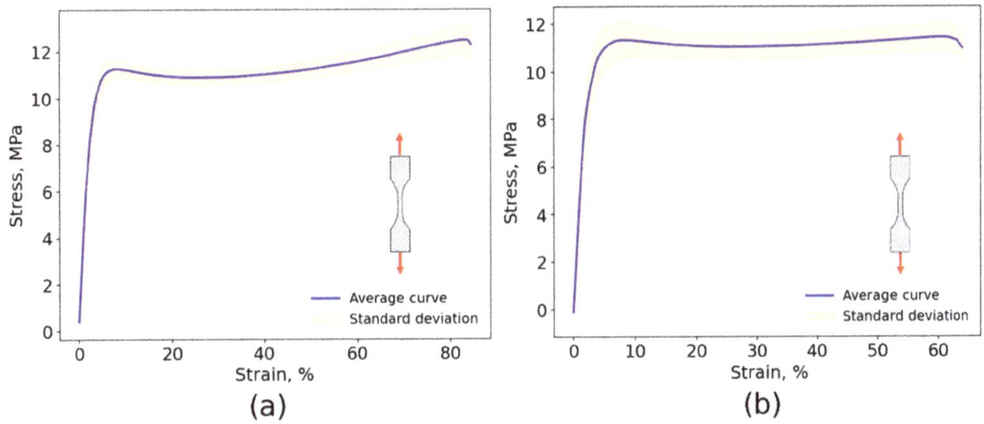

Figure 13. Tensile stress–strain curve of (**a**) Adhesive A and (**b**) Adhesive B.

Table 1. Tensile properties of Adhesives A and B.

Property	Units	Adhesive A	Adhesive B
Young's modulus, E	MPa	572 ± 38	639 ± 68
Poisson's ratio, ν	-	0.30 ± 0.01	0.34 ± 0.03
Tensile strength, σ_f	MPa	12.5 ± 0.7	11.5 ± 0.7
Tensile strain to failure, ε_f	%	85.0 ± 7.9	64.6 ± 6.0

Based on these results, the Poisson's ratio, ν, was estimated using DIC analysis, yielding values of $\nu = 0.30 \pm 0.01$ and $\nu = 0.34 \pm 0.03$ for Adhesive A and Adhesive B, respectively.

Both adhesives display similar tensile properties, suggesting comparable formulations. As shown in Figure 13, the adhesives exhibit highly ductile behavior, with strain-to-failure values reaching approximately 85% for Adhesive A and 65% for Adhesive B.

The area under the stress–strain curve suggests that the material has a high energy absorption capacity, which is beneficial in impact situations such as walking, trotting, and galloping activities. It is also advantageous for accommodating the natural expansion, contraction, and flexing of the horse's hoof.

4.1.2. Thick Adherend Shear Test

The thick adherend shear test strain field was obtained using extensometer-based displacement measurements. All specimens showed a similar failure mode, as illustrated in Figure 14, revealing cohesive failure, which indicated a good adhesion between the adhesive and the adherends. The stress–strain curves are displayed in Figure 15. The shear modulus was calculated using the curve's slope in the elastic region. These results are summarized in Table 2.

Figure 14. Failure mode of TAST specimens: (**a**) Adhesive A, (**b**) Adhesive B.

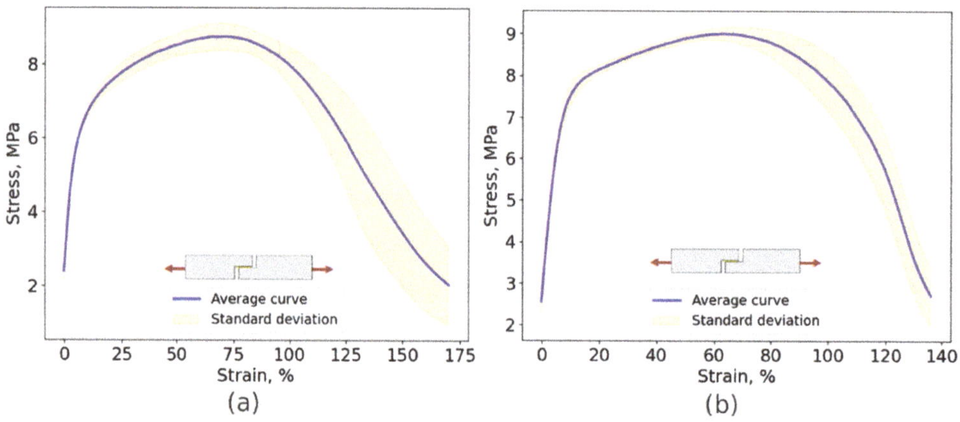

Figure 15. Stress–strain curve for the TAST specimens: (**a**) Adhesive A, (**b**) Adhesive B.

Table 2. Shear properties of Adhesives A and B.

Property	Units	Adhesive A	Adhesive B
Shear modulus, G	MPa	211 ± 17	235 ± 63
Shear strength, τ_f	MPa	9.0 ± 0.2	8.8 ± 0.4
Shear strain to failure, γ_f	%	61.8 ± 7.6	68.4 ± 6.9

According to Table 2, the shear modulus obtained from the TAST is 211 ± 17 MPa and 235 ± 63 MPa for Adhesives A and B, respectively. Considering the Poisson's ratio relation between the Young's and shear modulus in isotropic materials:

$$E = 2G(1 + \nu) \tag{10}$$

The calculated Poisson's ratio would be 0.35 and 0.36 for Adhesives A and B, respectively, which are close enough to the values determined using the DIC method (0.30 for Adhesive A and 0.34 for Adhesive B).

The tensile strength of each adhesive is higher than its shear strength, as shown in Tables 1 and 2, and these values are relatively close due to the high ductility of these adhesives.

For example, when the horse moves, there can be forces that try to pull the horseshoe away from the hoof (e.g., due to the horse's weight or impact on the ground). In this case, tensile forces are predominant, and a higher tensile strength will be able to resist this pulling force and hold the horseshoe in place without detaching.

In the case of shear forces, these occur when there is a sliding motion, such as when the horse walks or runs and the hoof slides along the ground or when the foot is in motion. Therefore, good shear strength is also needed in order to prevent the horseshoe from sliding or detaching from the hoof during the horse's movement.

Consequently, this balance between shear and tensile strength might be beneficial, depending on the specific forces the horse will experience during activities.

4.1.3. DCB Test

For each adhesive, three DCB specimens were used to obtain the critical energy release rate, G_{Ic}. The average curve and its corresponding standard deviation of the DCB test results for both adhesives can be observed in Figure 16. The load peak at which the crack propagation initiates is similar to the studied adhesives, with Adhesive B exhibiting a 10% higher peak load compared to Adhesive A. After the peak load is reached, the crack propagation behaves similarly for both adhesives, as depicted in Figure 16. Figure 17 displays the typical fracture surface for the quasi-static tests for Adhesive A and Adhesive B. The two fracture surfaces show a cohesive fracture behavior, which corroborates the good adhesion between the adhesive and the steel adherends as noted in the TAST results, ensuring that the fracture energy obtained belongs to the adhesive and not to the interface.

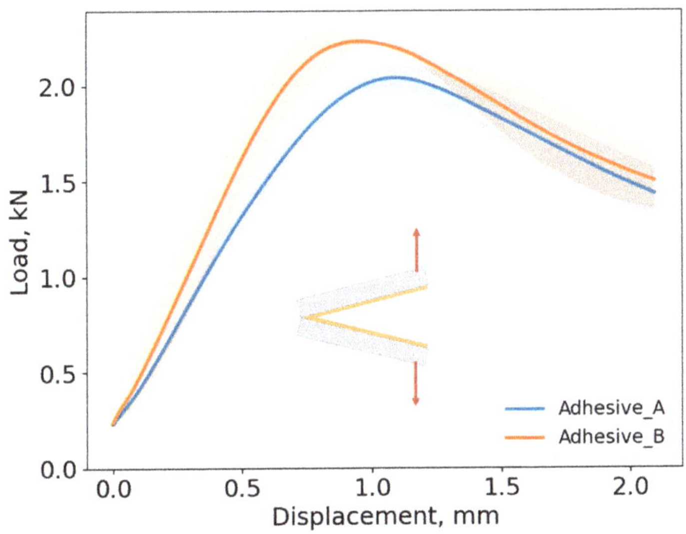

Figure 16. Load–displacement curves of the studied adhesives, where the full line represents the average curve of the shaded area represented the standard deviation.

Figure 17. Failure mode of the DCB specimens for (**a**) Adhesive A and (**b**) Adhesive B, showing cohesive failure for both adhesives.

After testing the DCB specimens, the J-integral data-reduction method was used to assess the fracture energy of each adhesive in quasi-static conditions, resorting to Equation (1). The average curve with the corresponding standard deviation is presented in Figure 18.

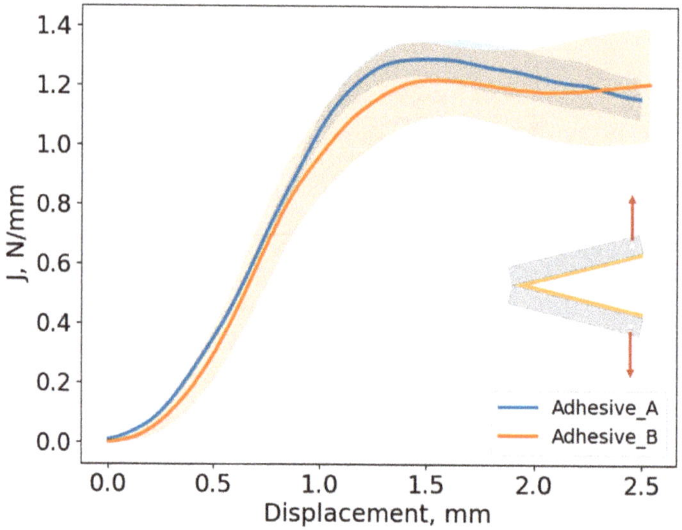

Figure 18. J vs. loading point displacement curves for Adhesives A and B, where full line represents the average curve and the shaded area represents the standard deviation.

The J vs. displacement curves present identical behavior, with a parabolic part at the beginning of the tests followed by an almost linear part that corresponds to the crack propagation. At this stage, it is safe to consider that there is damage in the adhesive layer, and with the damage increase, the adhesives are gradually losing their cohesive capacity. Fracture toughness of G_{Ic} = 1.3 ± 0.07 N/mm and G_{Ic} = 1.24 ± 0.15 N/mm was found for Adhesives A and B, respectively, which once again supports the similar formulations of the adhesives.

These fracture toughness values are characteristic of more ductile adhesives, allowing them to absorb greater amounts of energy before failure and providing enhanced resilience against stress or impact compared to brittle adhesives.

As previously mentioned, the fracture toughness in mode II was estimated according to literature values [27] that consider an adhesive with mechanical properties identical to those of the adhesives studied in this work. Thus, a fracture toughness in mode II of G_{IIc} = 6.0 N/mm was considered as a preliminary value for the numerical simulation.

Table 3 provides an overview of the tensile, shear, and fracture properties determined for each adhesive.

Table 3. Summary of the mechanical properties of Adhesive A and Adhesive B.

Property	Units	Adhesive A	Adhesive B
Young's modulus, E	MPa	572 ± 38	639 ± 68
Poisson's ratio, ν	-	0.30 ± 0.01	0.34 ± 0.03
Tensile strength, σ_f	MPa	12.5 ± 0.7	11.5 ± 0.7
Tensile strain to failure, ε_f	%	85.0 ± 7.9	64.6 ± 6.0
Shear modulus, G	MPa	211 ± 17	235 ± 63
Shear strength, τ_f	MPa	9.0 ± 0.2	8.8 ± 0.4
Shear strain to failure, γ_f	%	61.8 ± 7.6	68.4 ± 6.9
Toughness Mode I, G_{IC}	N/mm	1.3 ± 0.07	1.24 ± 0.15
Toughness Mode II, G_{IIC}	N/mm	6.0	6.0

4.1.4. Similar SLJ

As shown in Table 3, the mechanical properties of each adhesive are almost identical, and therefore it was decided to only present the results of the similar SLJs of Adhesive A. Preliminary SLJ tests were also conducted using Adhesive B to verify that the lap shear strength of both adhesives was comparable, and it was concluded that the in-joint behavior of both adhesives was also similar.

Before testing, St alignment tabs of 25 $\times$ 25 mm were bonded to the end of each adherend to reduce load misalignment and peel effect when testing.

Figure 19 shows the failure mode of Al–Al SLJs, HSS–HSS SLJs, and Mild St–Mild St SLJs for Adhesive A, revealing cohesive failure. This indicates strong adhesion between the sandblasted Al adherends and the adhesive, as well as between the sandblasted St adherends and the adhesive. This enabled the assessment of the adhesive's lap shear strength rather than that of the interface.

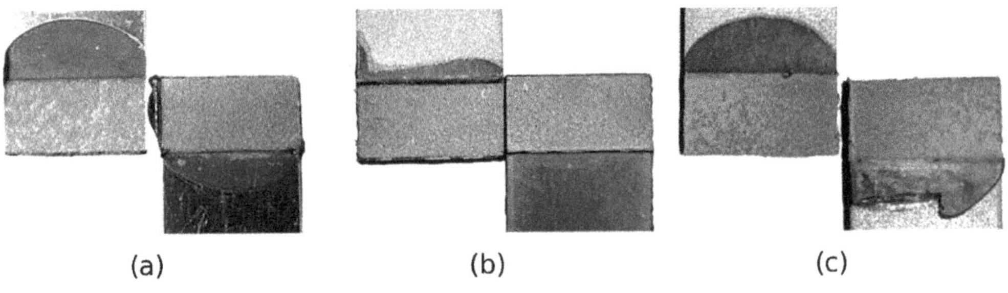

Figure 19. Failure mode of the similar SLJ: (**a**) Al–Al SLJs, (**b**) HSS–HSS SLJs, and (**c**) Mild St–Mild St SLJs, showing cohesive failure in every adherend combination.

In Figure 20, the average curve and its standard deviation of the load–displacement curves and lap shear stress evolution are shown for Adhesive A, considering every studied adherend combination. The shaded area in each curve represents the load standard deviation and the error bar shows the displacement standard deviation.

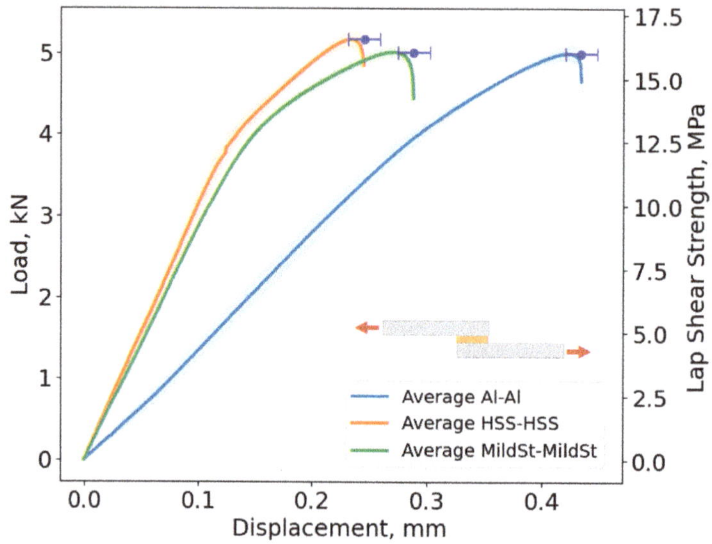

Figure 20. Load–displacement curves of similar SLJs (Adhesive A) for every adherend combination, where the full line represents the average curve and the shaded area the standard deviation.

Based on Figure 20, the maximum lap shear strength for the Al–Al SLJ is 16.0 ± 0.3 MPa, for the HSS–HSS SLJ is 16.5 ± 0.4 MPa, and for the Mild St–Mild St is 16.1 ± 0.3 MPa for Adhesive A. As the adherends remained in the elastic region and did not exhibit plastic deformation prior to adhesive failure, the behavior of the joint is mainly governed by the adhesive's properties. Consequently, the change in slope between 3 and 4 kN (Figure 20) corresponds to the yielding of the adhesive.

However, the change in slope observed in the Al–Al SLJ was less pronounced compared to the St–St combinations. This behavior can be attributed to the significantly lower modulus of elasticity of Al (approximately 1/3 that of St). Therefore, after 4 kN, Al actively contributes to the deformation of the joint together with the adhesive. Since St is much stiffer than Al, at this load level, the deformation of the joint will be governed by the adhesive.

The failure load was similar for every adherend combination, reaching approximately 5 kN. However, as expected, the displacements at the failure of the Al–Al SLJs were higher than those of St–St SLJs due to lower stiffness of the Al adherends when compared to those of St.

Therefore, the lap shear strength values of the studied adhesives should serve as a reference for any future adhesive solutions developed for the horseshoeing process.

This demonstrates that the performance of the adhesive is consistent for Al and St adherends, with respect to the maximum lap shear strength. This is promising, particularly in the context of the horseshoeing process, where both Al and St horseshoes are commonly used. Consequently, it can be ensured that the adhesive will exhibit similar performance for either material as its adhesion properties remain effective regardless of the horseshoe material.

4.1.5. Dissimilar SLJ

As mentioned earlier, the dissimilar SLJs were manufactured with mild St and horse hoof wall adherends. Before SLJ testing, 3D-printed alignment tabs of 12 × 12 mm were bonded to the end of each adherend to avoid load misalignment and peel effects, following the same procedure as for similar joints. Three specimens of dissimilar SLJs were tested, and their load–displacement curves are displayed in Figure 21.

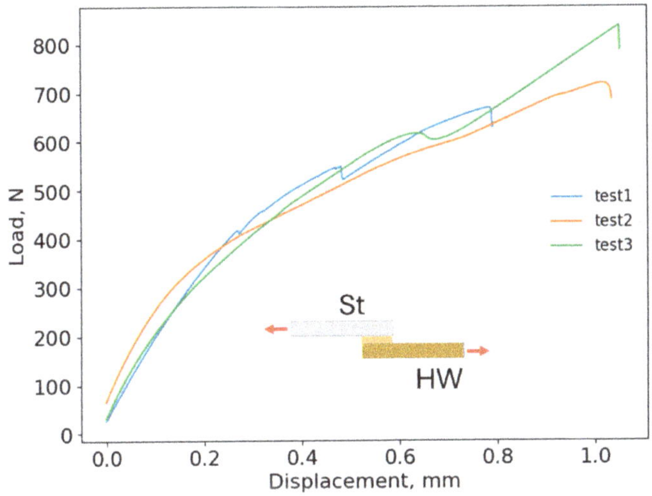

Figure 21. Load–displacement curves of the tested dissimilar joints.

As depicted in Figure 21, all specimens exhibited identical performance, with failure consistently occurring in the hoof substrate, as seen in Figure 22. However, the joints labeled "test1" and "test3" displayed notable sudden load drops followed by immediate recovery. These fluctuations may be attributed to defects introduced in the hoof adherends during the extraction process from the molds and the subsequent cleaning prior to testing.

Figure 22. Failure mode of the tested dissimilar joints, showing a substrate failure in every test.

During the cleaning process of joint "test3", residues of 3D-printed material from the mold remained adhered to the hoof substrate, particularly near the overlap area. To avoid causing additional damage to the same adherend, it was decided to proceed with testing of the joint in this condition (Figure 23). Consequently, the presence of these 3D-printed material traces may have contributed to a higher peak load compared to the other tests.

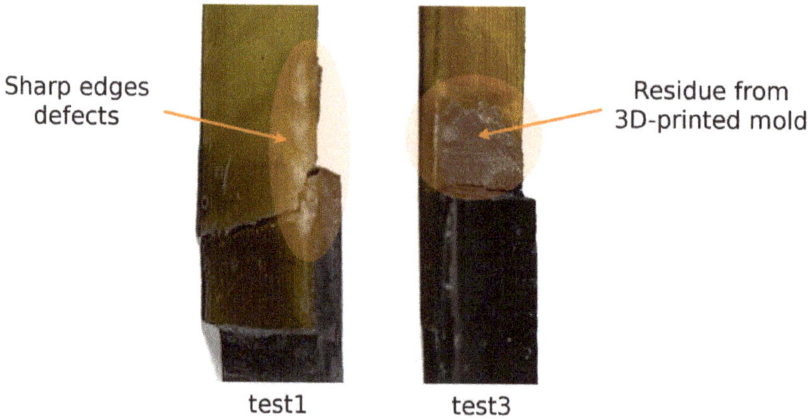

Figure 23. Joint defects created during the cleaning process.

The failure of all dissimilar joints within the hoof substrate indicates that, at least under the current loading conditions, these commercial adhesives possess greater strength than the hoof material. Therefore, if a real joint were subjected to identical loading conditions, it is likely that the horseshoe would detach from the hoof, potentially carrying away hoof material. However, this conclusion cannot be definitively stated, as the hoof substrate used for testing the dissimilar joints does not precisely replicate the mechanical properties or surface roughness of the actual region where the horseshoe is attached, as it originates from the SM region of the hoof wall.

4.2. Numerical Validation

Given the similar mechanical properties of the two adhesives investigated, only Adhesive A will be considered for the development and validation of the numerical model.

4.2.1. Mode I

The cohesive law of the adhesive in mode I was modeled with a triangular shape, characterized by an initial elastic region, followed by damage initiation, which progressively degrades the adhesive's properties. To validate the considered cohesive law, a DCB specimen was numerically simulated during its opening movement, which corresponds to pure mode I loading. The resulting numerical load–displacement curve was then compared to experimental data to assess the agreement between the numerical and experimental results (Figure 24).

Figure 24 demonstrates a strong correlation between the numerical and experimental results, which indicates that a triangular traction–separation law is sufficient to characterize the adhesive's performance in pure mode I. The initial stiffness of the specimen was well captured by the numerical model, and the numerical peak load was also within the standard deviation of the experimental data. The initial crack propagation occurring after 1 mm displacement was also similar between the numerical and experimental curves.

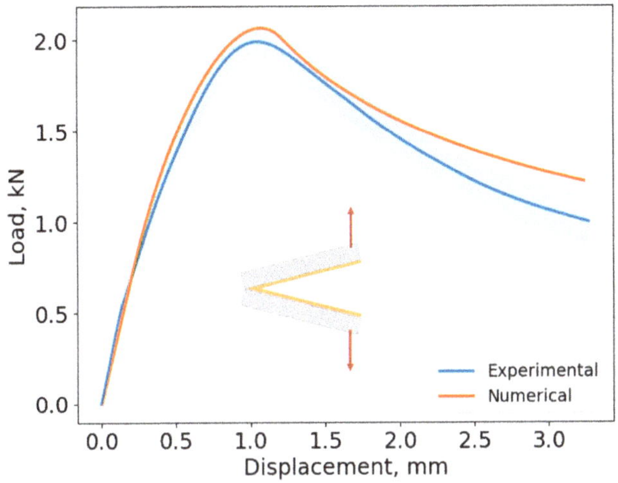

Figure 24. Load vs. displacement curve for both numerical and experimental results of DCB specimens (Mode I). The shaded blue area represents the standard deviation of the experimental results.

4.2.2. Similar SLJ

The cohesive law of the adhesive in mode II was modeled with a trapezoidal shape with increasing stresses, as detailed in Section 3. To validate this cohesive law, a SLJ configuration was modeled using the considered similar adherend combinations, i.e., Al–Al and St–St. The SLJ test does not fully correspond to pure mode II loading but rather to a mixed-mode loading condition. However, the SLJ configuration was selected for mode II validation due to the dominance of in-plane shear stresses throughout the test. To further support this approach, various shapes of cohesive laws in mode I were considered for the SLJ configuration to examine their effect on the load–displacement response. As anticipated, the influence of the mode I cohesive law shape on the P–δ curve was minimal, thereby confirming the suitability of the SLJ configuration for validating the mode II cohesive law.

Figures 25–27 show the comparison between the experimental and numerical load–displacement curves for the Al–Al, HSS–HSS, and Mild St–Mild St SLJs of Adhesive A.

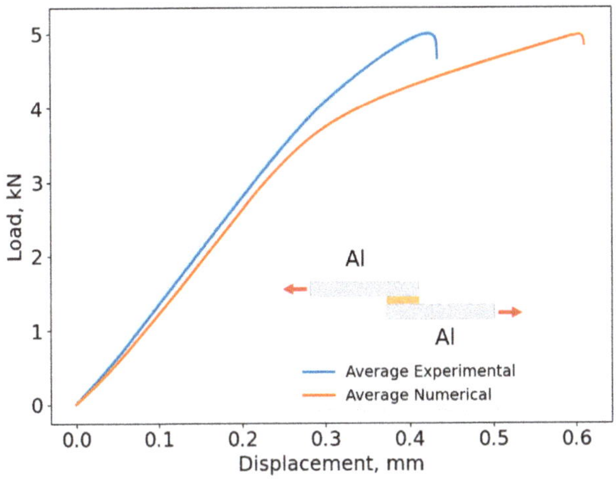

Figure 25. Load–displacement curves for numerical and experimental results of Al–Al SLJs of Adhesive A. The shaded area in the average experimental curve corresponds to standard deviation.

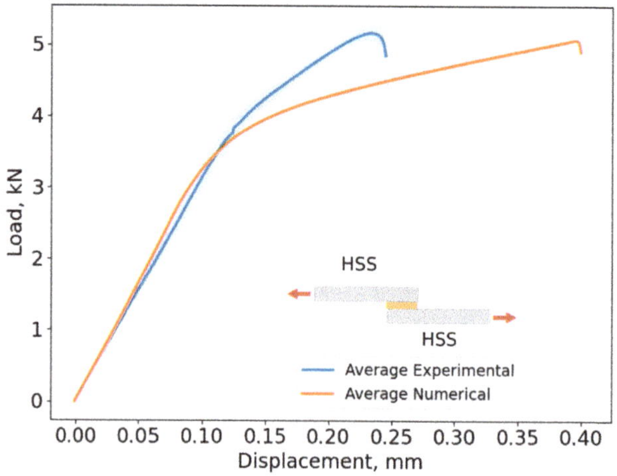

Figure 26. Load–displacement curves for numerical and experimental results of HSS–HSS SLJs of Adhesive A. The shaded area in the average experimental curve corresponds to standard deviation.

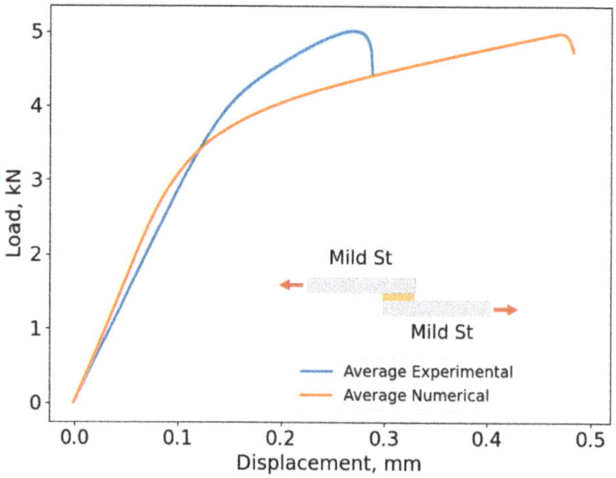

Figure 27. Load–displacement curves for numerical and experimental results of Mild St–Mild St SLJs of Adhesive A.

The cohesive laws described in Section 3 were applied, resulting in the numerical curves shown in Figures 25–27. A comparison between the experimental data and the numerical curves for each adherend combination reveals that the numerical model accurately predicted the failure load of all joints, which was approximately 5 kN. Additionally, the stiffness of each joint was well captured, as the initial elastic region of the numerical curves closely matched the experimental data.

However, discrepancies were observed in the displacement at failure, with the numerical displacements consistently exceeding the experimental values. Specifically, for the Al–Al SLJs, the numerical displacement at failure was 45% higher than the experimental value, while for the St–St SLJs, it was approximately 60% higher.

These discrepancies in failure displacement may be attributed to the strain rate-dependent behavior of ductile adhesives. As noted earlier, the adhesive included in the numerical model has ductile behavior and it is therefore strain rate-dependent. The adhesive's mechanical properties were determined at crosshead speeds recommended by the relevant standards, which do not match the strain rate used in the simulations. This mismatch may explain the observed differences in displacement at failure.

The numerical model also reveals that as soon as the entire adhesive layer begins to experience damage (t = 0.108 s as shown in Figure 28), a change in slope occurs in the pdelta curve (Figure 27), which indicates the onset of plastic deformation in the adhesive.

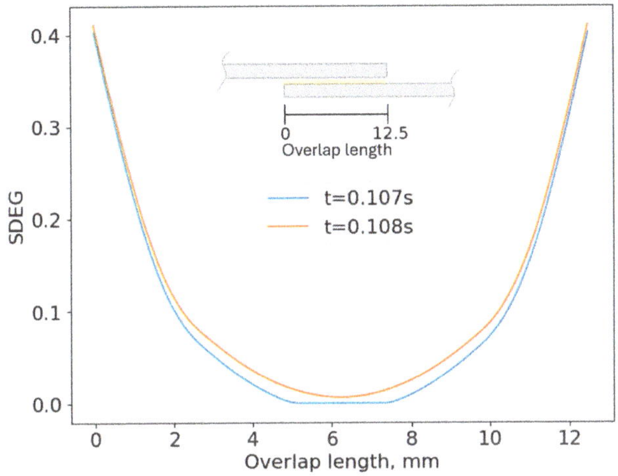

Figure 28. Damage evolution (SDEG) along the overlap length for the Mild St–Mild St SLJs for two consecutive time steps.

As anticipated, the stresses perpendicular to the loading direction are nonzero (Figure 29), confirming the presence of mixed-mode loading during the SLJ test. These stresses contribute to the peel effects experienced by the joint. Nonetheless, it is clear that shear stresses (11) are dominant as they are approximately 12.5 times higher than the normal stresses (22).

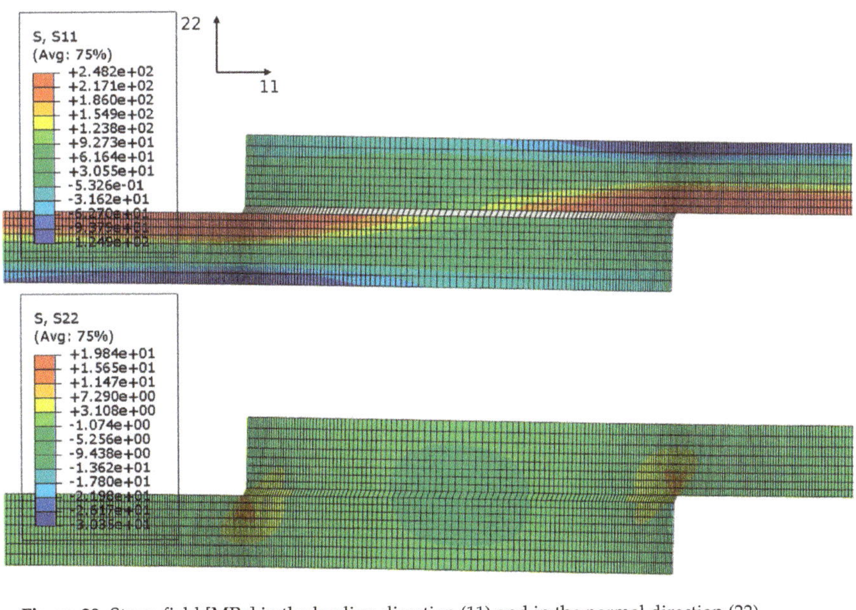

Figure 29. Stress field [MPa] in the loading direction (11) and in the normal direction (22).

Despite the limitations of the numerical model, it provided a reasonable prediction of joint behavior, particularly in terms of failure loads.

With the validation of the cohesive laws in mode I and mode II, it is now possible to simulate more complex loading conditions involving significant mixed-mode loading, as encountered in the real joint configuration, and also more complex geometries.

5. Conclusions

This study examines the mechanical properties of two commercial adhesives currently used in the equine industry, specifically for horseshoeing applications. Both adhesives are acrylic-based, providing an optimal balance of elastic and ductile properties along with the necessary strength for this purpose.

A series of standardized tests revealed that the mechanical properties of both adhesives are quite similar, suggesting a comparable formulation. Therefore, the in-joint behavior in similar SLJs was only analyzed for Adhesive A, resulting in a lap shear strength of approximately 16 MPa for every adherend combination.

This value may serve as a benchmark for developing innovative bonding solutions that could surpass the limitations of the adhesives studied here.

A numerical model was successfully implemented using a cohesive zone modelling (CZM) approach to simulate the performance of the tested similar joints. A triangular traction–separation law was applied for mode I, while a trapezoidal law with increasing stresses was utilized for mode II. These cohesive laws allowed for a moderate prediction of the behavior of similar joints.

The next phase of this research is the development of a model of the real joint by incorporating the geometry of both the hoof and horseshoe into the assembly. This model will then simulate the joint's behavior under realistic geometrical and loading conditions. This approach aims to identify regions requiring adhesive reinforcement, as well as areas where adhesive use may be minimized.

Author Contributions: Conceptualization, C.M.C.F., B.D.S., E.A.S.M., R.J.C.C. and L.F.M.d.S.; data curation, C.M.C.F.; formal analysis, C.M.C.F.; investigation, C.M.C.F.; methodology, C.M.C.F., B.D.S., E.A.S.M., R.J.C.C. and L.F.M.d.S.; supervision, B.D.S., E.A.S.M., R.J.C.C. and L.F.M.d.S.; validation, C.M.C.F., B.D.S., E.A.S.M., R.J.C.C. and L.F.M.d.S.; writing—original draft, C.M.C.F.; writing—review and editing, B.D.S., E.A.S.M., R.J.C.C. and L.F.M.d.S. All authors have read and agreed to the published version of the manuscript.

Funding: The authors gratefully acknowledge the Portuguese Foundation for Science and Technology (FCT) for supporting the work presented here through the individual grant 2024.03586.BD.

Institutional Review Board Statement: Not applicable.

Data Availability Statement: Data is contained within the article.

Conflicts of Interest: The authors declare no conflict of interest.

References

1. Karle, A.S.; Tank, P.H.; Vedpathak, H.S.; Mahida, H.K.; Shah, R.G.; Dhami, M.A. Horseshoeing: An Overview. *Vet. World* **2010**, *3*, 148–151.
2. Williams, J.; Marlin, D. Foreword–Emerging issues in equestrian practice. *Comp. Exerc. Physiol.* **2020**, *16*, 1–4. [CrossRef]
3. Back, W.; Pille, F. The role of the hoof and shoeing. In *Equine Locomotion*; Saunders Limited: Rider Boulevard Rhodes, NSW, Australia, 2013; pp. 147–174.
4. Horan, K.; Kourdache, K.; Coburn, J.; Day, P.; Carnall, H.; Harborne, D.; Brinkley, L.; Hammond, L.; Millard, S.; Lancaster, B.; et al. The effect of horseshoes and surfaces on horse and jockey centre of mass displacements at gallop. *PLoS ONE* **2021**, *16*, e0257820. [CrossRef] [PubMed]

5. Sprick, M.; Fürst, A.; Baschnagel, F.; Michel, S.; Piskoty, G.; Hartnack, S.; Jackson, M.A. The influence of aluminium, steel and polyurethane shoeing systems and of the unshod hoof on the injury risk of a horse kick: An ex vivo experimental study. *Vet. Comp. Orthop. Traumatol.* **2017**, *30*, 339–345. [CrossRef] [PubMed]
6. Silva, L.F.M.; Öchsner, A.; Adams, R. *Handbook of Adhesion Technology*; Springer: Berlin/Heidelberg, German, 2018; Volume 1, ISBN 978-3-642-01168-9.
7. Borges, C.S.P.; Akhavan-Safar, A.; Tsokanas, P.; Carbas, R.J.C.; Marques, E.A.S.; da Silva, L.F.M. From fundamental concepts to recent developments in the adhesive bonding technology: A general view. *Discov Mech. Eng.* **2023**, *2*, 8. [CrossRef]
8. O'Grady, S.E. Basic Farriery for the Performance Horse. *Vet. Clin. Equine Pract.* **2008**, *24*, 203–218. [CrossRef]
9. Lee, J.; Choi, Y. Development of Horseshoe With Carbon Fiber Reinforced Composite. *Int. J. Adv. Res. Eng. Technol.* **2020**, *11*, 53–58.
10. Takahashi, Y.; Yoshihara, E.; Takahashi, T. Comparison of Heel Movement Between two Different Glue-on Type Shoes and Nailed Shoes in Thoroughbreds☆. *J. Equine Vet. Sci.* **2022**, *113*, 103939. [CrossRef]
11. Poupard, D. Glue-on technology and an innovative new technique. In Proceedings of the AAEP Annual Convention, Baltimore, MD, USA, 8 December 2010; pp. 494–498.
12. Yoshihara, E.; Takahashi, T.; Otsuka, N.; Isayama, T.; Tomiyama, T.; Hiraga, A.; Wada, S. Heel movement in horses: Comparison between glued and nailed horse shoes at different speeds. *Equine Vet. J.* **2010**, *42*, 431–435. [CrossRef] [PubMed]
13. How to Remove Glushu Glue on Horse Shoes. Available online: https://www.glushu.com/removal (accessed on 13 November 2024).
14. Huguet, E.E.; Duberstein, K.J. Effects of Steel and Aluminum Shoes on Forelimb Kinematics in Stock-Type Horses as Measured at the Trot. *J. Equine Vet. Sci.* **2012**, *32*, 262–267. [CrossRef]
15. *ASTM D412*; Standard Test Methods for Vulcanized Rubber and Thermoplastic Elastomers-Tension. ASTM International: West Conshohocken, PA, USA, 2016.
16. da Silva, L.F.M.; Dillard, D.A.; Blackman, B.R.K.; Adams, R.D. *Testing Adhesive Joints: Best Practices*; Wiley-VCH Verlag GmbH & Co. KGaA: Weinheim, Germany, 2012; ISBN 978-3-527-32904-5.
17. *ISO 11003-2:2019(E)*; Adhesives—Determination of Shear Behaviour of Structural Bonds, Part 2: Thick-Adherend Tensile-Test Method. International Organization for Standardization: Geneva, Switzerlan, 1993.
18. *ISO 25217:2009(E)*; Adhesives—Determination of the Mode 1 Adhesive Fracture Energy of Structural Adhesive Joints Using Double Cantilever Beam and Tapered Double Cantilever Beam Specimens. International Organization for Standardization: Geneva, Switzerlan, 2009.
19. Kasapi, M.A.; Gosline, J.M. Strain-rate-dependent mechanical properties of the equine hoof wall. *J. Exp. Biol.* **1996**, *199*, 1133–1146. [CrossRef] [PubMed]
20. Yang, Y.; Zhao, J.; Zhang, S.; Marques, E.; Carbas, R.; Machado, J.; da Silva, L. Determination of fracture toughness of an adhesive in civil engineering and interfacial damage analysis of carbon fiber reinforced polymer–steel structure bonded joints. *Proc. Inst. Mech. Eng. Part L J. Mater. Des. Appl.* **2021**, *235*, 2423–2440. [CrossRef]
21. Sun, F.; Blackman, B.R.K. A DIC method to determine the Mode I energy release rate G, the J-integral and the traction-separation law simultaneously for adhesive joints. *Eng. Fract. Mech.* **2020**, *234*, 107097. [CrossRef]
22. Abdel Monsef, S.; Pérez-Galmés, M.; Renart, J.; Turon, A.; Maimí, P. The influence of mode II test configuration on the cohesive law of bonded joints. *Compos. Struct.* **2020**, *234*, 111689. [CrossRef]
23. Rice, J.R. A Path Independent Integral and the Approximate Analysis of Strain Concentration by Notches and Cracks. *J. Appl. Mech.* **1968**, *35*, 379–386. [CrossRef]
24. Bucci, R.J.; Paris, P.C.; Landes, J.D.; Rice, J. J Integral Estimation Procedures. In *Fracture Toughness: Part II*; ASTM International: West Conshohocken, PA, USA, 1972; pp. 40–69. [CrossRef]
25. Paris, A.J.; Paris, P.C. Instantaneous evaluation of J and C. *Int. J. Fract.* **1988**, *38*, R19–R21. [CrossRef]
26. Akhavan-Safar, A.; Marques, E.A.S.; Carbas, R.J.C.; da Silva, L.F.M. Cohesive Zone Modelling-CZM. In *Cohesive Zone Modelling for Fatigue Life Analysis of Adhesive Joints*; Akhavan-Safar, A., Marques, E.A.S., Carbas, R.J.C., da Silva, L.F.M., Eds.; Springer International Publishing: Cham, Switzerland, 2022; pp. 19–42, ISBN 978-3-030-93142-1. [CrossRef]
27. Pinto, A.M.G.; Magalhães, A.G.; Campilho, R.D.S.G.; de Moura, M.F.S.F.; Baptista, A.P.M. Single-Lap Joints of Similar and Dissimilar Adherends Bonded with an Acrylic Adhesive. *J. Adhes.* **2009**, *85*, 351–376. [CrossRef]

biomimetics

MDPI

Article

Numerical Investigation on Drag Reduction Mechanisms of Biomimetic Microstructure Surfaces

Jiangpeng Liu, Jie Xu *, Chaogang Ding, Debin Shan and Bin Guo

National Key Laboratory for Precision Hot Processing of Metals, Harbin Institute of Technology, Harbin 150001, China
* Correspondence: xjhit@hit.edu.cn

Abstract

Biomimetic microstructured surfaces offer a promising passive strategy for drag reduction in marine and aerospace applications. This study employs computational fluid dynamics (CFD) simulations to systematically investigate the drag reduction performance and mechanisms of groove-type microstructures, addressing both geometry selection and dimensional optimization. Three representative geometries (V-groove, blade-groove, and arc-groove) were compared under identical flow conditions (inflow velocity 5 m/s, $Re = 7.5 \times 10^5$) using the shear-stress-transport (SST k-ω) turbulence model, and the third-generation Ω criterion was employed for threshold-independent vortex identification. The results establish a clear performance hierarchy: blade-groove achieves the highest drag reduction rate of 18.2%, followed by the V-groove (16.5%) and arc-groove (14.7%). The analysis reveals that stable near-wall microvortices form dynamic vortex isolation layers that separate the high-speed flow from the groove valleys, with blade grooves generating the strongest and most fully developed vortex structures. A parametric study of blade-groove aspect ratios ($h^+/s^+ = 0.35$–1.0) further demonstrates that maintaining $h^+/s^+ \geq 0.75$ preserves effective vortex-isolation layers, whereas reducing h^+/s^+ below 0.6 causes vortex collapse and performance degradation. These findings establish a comprehensive design framework combining geometry selection (blade-groove > V-groove > arc-groove) with dimensional optimization criteria, providing quantitative guidance for practical biomimetic drag-reducing surfaces.

Keywords: drag reduction; biomimetic microstructure; CFD simulation; SST k-ω model; Ω criterion

1. Introduction

Drag reduction technologies for marine vessels and other fluid transport systems are of great importance because they can significantly decrease fuel consumption, enhance flow rates, and increase operational efficiency, thereby generating substantial economic and environmental benefits [1–4]. Over the past decades, a variety of active and passive drag reduction techniques have been proposed, including polymer additives, microbubble injection, compliant coatings, superhydrophobic surfaces, and bioinspired microstructures [5]. Although these approaches have achieved remarkable reductions in frictional drag, many of them still suffer from inherent drawbacks such as high energy consumption, environmental concerns, or limited durability, which restrict their large-scale application.

Natural organisms, through long-term evolution, have developed a rich repertoire of morphological and interfacial strategies to minimize hydrodynamic resistance under

 https://doi.org/10.3390/biomimetics11010077

harsh environments [6]. Five representative bioinspired concepts have attracted extensive attention in the field of drag reduction: shark skin-inspired microstructure drag reduction, dolphin-skin-inspired compliant walls, penguin-feather-inspired microbubble drag reduction, lotus-leaf-inspired superhydrophobic drag reduction, and pitcher-plant-inspired slippery liquid-infused surfaces [7]. These natural examples demonstrate that appropriately designed surface topography and interfacial properties can passively manipulate the near-wall flow, providing energy-efficient solutions for flow control in marine and fluid engineering.

Among the various bioinspired strategies, interfacial-slip-based approaches such as superhydrophobic surfaces (SHSs), liquid-infused surfaces (LISs), and mucus-inspired slippery coatings have emerged as particularly promising. SHSs combine hydrophobic chemistry with micro/nano-scale roughness to entrap a stable air layer, or plastron, between the roughness elements when submerged [8,9]. The presence of this air–water interface effectively changes the boundary condition from no-slip to partial slip, thereby weakening the near-wall shear and reducing turbulent frictional drag. Extensive reviews and experiments have demonstrated that well-designed SHSs can achieve substantial drag reduction in laminar and turbulent flows, with the drag reduction level strongly depending on texture geometry, gas fraction, and Reynolds number [10]. However, maintaining a stable air layer under high pressure and over long operational periods remains challenging, as plastron depletion and gas dissolution can severely degrade performance.

Inspired by the Nepenthes pitcher plant, slippery liquid-infused porous surfaces (SLIPS) or liquid-infused surfaces (LISs) have been proposed as an alternative to air-based SHSs. In these surfaces, a viscous immiscible lubricant is locked within a textured or porous solid due to capillarity, forming a stable liquid–liquid interface with extremely low contact-angle hysteresis [11,12]. The lubricant layer provides a robust slip boundary, enabling high mobility of aqueous or oily liquids and achieving drag reduction levels on the order of 10–30% for various fluids [13,14]. Compared with SHSs, LISs exhibit better pressure tolerance and self-healing capability; nevertheless, lubricant depletion under continuous shear and mechanical damage is still a critical issue.

Another important class of bioinspired drag-reducing surfaces is based on mucus or mucus-like compliant coatings. Many fish and aquatic organisms secrete a viscoelastic mucus layer on their body surface, which forms a lubricating interface and modifies near-wall turbulence [15]. Experiments have reported drag reduction rates of several tens of percent for fish mucus or mucus-mimicking polymer solutions at appropriate concentrations [16]. Mucus-inspired bionic membranes and slime–groove hybrid surfaces can effectively suppress turbulent fluctuations and attenuate wall-shear stress [17,18]. These mucus or lubricant-based coatings are often combined with underlying microgrooved textures that store and redistribute the liquid phase, again indicating that the geometry of the groove plays a crucial role in determining the overall drag reduction performance. Therefore, a clear understanding of how canonical groove shapes influence the formation of near-wall vortices and local shear stress is essential for the rational design of both hydrophobic and mucus-based drag-reducing coatings.

Despite the rapid development of interfacial-slip technologies, shark-skin-inspired microstructured surfaces remain one of the most attractive strategies for turbulent drag reduction in marine applications, because they are purely passive, require no external energy input, and are compatible with large-scale manufacturing [19,20]. Biological observations reveal that the skin of fast-swimming sharks, such as the shortfin mako shark, is covered with specialized placoid scales whose surfaces exhibit longitudinal riblet-like grooves with characteristic widths below 100 μm [21–24]. These grooves are aligned with the main flow direction and have been hypothesized to lift and pin streamwise vortices

https://doi.org/10.3390/biomimetics11010077

within the viscous sublayer, thereby suppressing lateral momentum transfer and reducing skin-friction drag [19,25,26]. Experimental studies on engineered shark-skin replicas and riblet surfaces—such as triangular, blade-shaped, and semicircular grooves—have demonstrated drag reduction rates of several percent to more than 10% in turbulent flows, confirming the effectiveness of groove-type microstructures [27–31]. "Recent advances in biomimetic riblet design have demonstrated significant potential for practical applications. Mawignon et al. [32] achieved 13.5% drag reduction through optimized cuboidal riblet geometries, while Zani et al. [33] explored multi-scale hierarchical designs for enhanced turbulent boundary layer control. These studies, along with comprehensive reviews by Pakatchian et al. [34], highlight the continued relevance of riblet-based drag reduction research.

Building on these findings, considerable efforts have been devoted to fabricating shark-skin-inspired surfaces using polymers, metals, and composite materials, as well as to exploring their performance in realistic flow conditions. However, most experimental investigations provide limited insight into the internal flow structures inside and above the grooves, because direct measurements of microscale vortices in turbulent boundary layers are extremely challenging. Computational fluid dynamics (CFD) simulations offer a powerful tool to complement experiments by resolving the detailed flow field and elucidating the underlying drag-= reduction mechanisms. Previous numerical studies on riblet and shark-skin-inspired surfaces have revealed that microgrooves can modify the near-wall turbulence structures and reduce wall shear stress [35–37]. Nevertheless, many of these works rely on traditional vortex-identification methods such as the Q and λ_2 criteria, which are threshold-dependent and may miss weak vortices close to the wall. Moreover, systematic quantitative relationships between groove geometry, near-wall vortex characteristics, and drag reduction performance have not yet been fully established.

A critical review of the literature reveals three key gaps that currently limit the practical application of riblet-based drag reduction. First, most studies examine single groove geometries without systematic comparison under identical conditions, preventing clear design preferences. Second, while the relationship between groove dimensions and drag reduction has been explored, the effect of aspect ratio (h^+/s^+) on vortex formation and drag reduction remains poorly understood, lacking quantitative design criteria. Third, previous investigations predominantly rely on traditional vortex identification methods (Q criterion, λ_2 method) that require subjective threshold selection, leading to inconsistent vortex identification and hindering quantitative mechanism analysis. The present work addresses these gaps through systematic CFD simulations. Specifically, three representative groove geometries (V-groove, blade-groove, and arc-groove) are systematically compared under identical flow conditions to establish a quantitative ranking of drag reduction performance correlated with vortex metrics. The effect of aspect ratio (h^+/s^+ ranging from 0.35 to 1.0) on drag reduction performance is parametrically investigated using blade-groove microstructures as a representative case to establish optimal design criteria. Furthermore, the third-generation Ω vortex-identification criterion is employed for threshold-independent vortex analysis, enabling quantitative correlation between vortex characteristics and drag reduction performance. Together, these investigations provide a comprehensive design framework that bridges the gap between fundamental understanding and practical application of biomimetic drag-reducing surfaces. This work extends the current understanding of biomimetic drag-reducing surfaces beyond traditional Q or λ_2 methods.

https://doi.org/10.3390/biomimetics11010077

2. Numerical Methodology

2.1. Geometry and Model Design

In this study, the drag reduction effects of three types of micro-grooves, including V-groove, blade-groove, and arc-groove, were compared under the same Reynolds numbers. Early research by Walsh at NASA demonstrated that streamwise V-shaped riblets effectively reduce frictional drag when their dimensionless height and spacing satisfy $h^+ \leq 25$ and $s^+ \leq 200$ [38]. The relationships for these parameters are given by the following equations:

$$s^+ = 0.172 \times \frac{sURe^{-0.1}}{v} \quad (1)$$

$$h^+ = 0.172 \times \frac{hURe^{-0.1}}{v} \quad (2)$$

where U denotes the inlet velocity, Re the Reynolds number, and ν the kinematic viscosity of the fluid (for water, $\nu = 1 \times 10^{-6}$ m^2/s) [37].

A representative inflow velocity of 5 m/s was selected for simulation. Because drag reduction occurs primarily under turbulent conditions, the simulations were configured to ensure fully developed turbulence. For flow over a flat plate, turbulence forms when $Re > 5 \times 10^5$.

$$Re = \frac{\rho v L}{\mu} \quad (3)$$

In the Reynolds number formula, L represents the reference dimension of the boundary layer, which refers to the length of the plate; μ is the dynamic viscosity; ρ is the fluid density and v is the velocity of the incoming flow.

Based on the calculation results, at a flow velocity of 5 m/s, the fluid can reach a turbulent state when the flow direction distance reaches 130 mm. Thus, the streamwise dimension is set to 150 mm, at which the Reynolds number is approximately 7.5×10^5. The ranges of micro-groove heights and widths that meet the drag reduction requirements at an inflow velocity of 5 m/s are shown in Table 1.

Table 1. The range of microstructure heights and widths.

Velocity (m/s)	Reynolds Numbers	Height (μm)	Width (μm)
5	7.5×10^5	≤112	≤900

Based on the analysis, an aspect ratio of 1 for the microstructures was selected. Thus, the height and width of the three types of microstructures were all set to 100 μm for the drag reduction analysis, as shown in Figure 1. Both the drag reduction effect and the underlying mechanisms were investigated through numerical simulation.

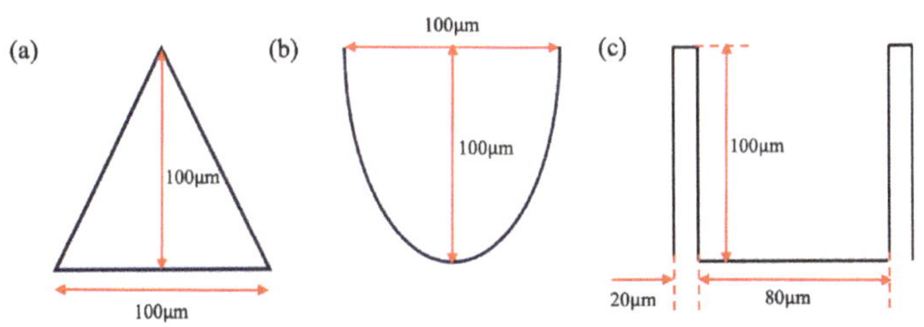

Figure 1. Geometry of microgrooves. (**a**) V-groove, (**b**) arc-groove, and (**c**) blade-groove.

https://doi.org/10.3390/biomimetics11010077

2.2. Establishment of Fluid Drag Reduction Model

Following the above analysis, the streamwise dimension of the model was set to 150 mm. To balance computational efficiency and accuracy, the streamwise dimension of the fluid domain is initially set to 10 mm and then scaled up to 150 mm in Fluent. The analysis focused on the final 10 mm downstream region where turbulence is fully developed. The spanwise dimension was set to ten times the groove width to maintain periodicity.

For turbulent viscous flow, the overall flow can be divided into an outer potential flow region and an inner boundary layer region [39,40]. It is crucial to account for viscous forces inside the boundary layer. The boundary layer thickness is typically much smaller than the characteristic length scale of the flow. However, wall shear stress and other physical quantities near the wall are strongly influenced by viscous effects within the boundary layer [41]. Hence, the boundary-layer thickness δ must be considered when defining the domain height. It is estimated using the following:

$$\delta \approx 0.37 \times \frac{x}{Re_x^{1/5}} \tag{4}$$

where x is the streamwise distance and Re_x the local Reynolds number.

For $Re = 7.5 \times 10^5$, the boundary-layer thickness is approximately 3.7 mm; therefore, the fluid-domain height was set to 6 mm to capture near-wall flow structures. Based on these considerations, the fluid domain model is established as illustrated in Figure 2.

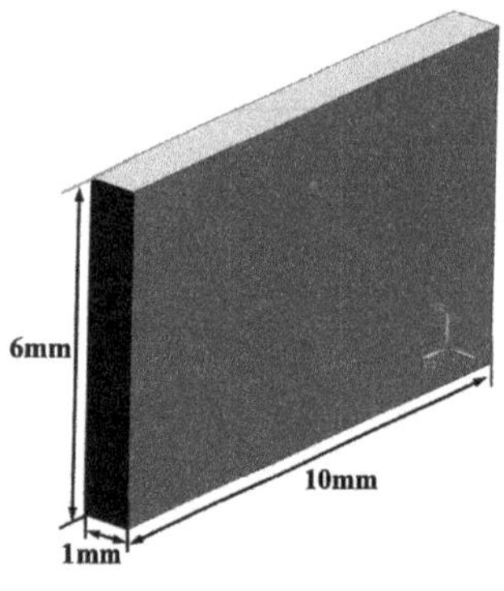

Figure 2. Fluid domain model.

In the mesh generation process, particular attention was paid to the boundary layer formed as the fluid flows along the wall. The fluid velocity gradient is substantial inside the boundary layer, necessitating that the boundary layer mesh maintains good orthogonality with the wall. The first layer height of the mesh is one of the most critical parameters in boundary layer meshing [42]. The dimensionless wall distance y^+ is commonly used to characterize the first mesh layer as follows:

$$y^+ = \frac{u^* y}{v} \tag{5}$$

where u^* represents the friction velocity near the wall surface, y is the distance between the first layer of mesh nodes and the wall, and v is the kinematic viscosity of the fluid.

The formula for calculating the friction velocity near the wall is as follows:

$$u = \sqrt{\frac{\tau_w}{\rho}} \tag{6}$$

https://doi.org/10.3390/biomimetics11010077

τ_w represents the wall shear stress, calculated as $\tau_w = \mu$ ($y = 0$, where μ is the dynamic viscosity, and y is the distance of the first layer of the mesh from the wall).

To further simplify the calculation, the wall friction coefficient, C_f, is introduced, which is defined as follows:

$$C_f = \frac{\tau_w}{\frac{1}{2}\rho U^2} \tag{7}$$

The wall shear stress τ_w can be determined by calculating C_f. The calculation of C_f can be performed using Prandtl's empirical formula when the Reynolds number range is $5 \times 10^5 < Re < 10^7$:

$$C_f = 0.074Re^{-\frac{1}{2}} \tag{8}$$

The SST k-ω model is used for the turbulence model based on previous research. In this case, it is necessary to satisfy $y^+ \leq 1$, with a value close to 1 being ideal [43]. Calculations indicate that the first-layer height of the first layer of the mesh should be at least approximately 1.5 μm. The vertical mesh employed an exponential distribution with a ratio set to 1.1, while the streamwise and spanwise directions have a uniform distribution. Mesh generation was performed using ICEM, using structured grids, as illustrated in Figure 3.

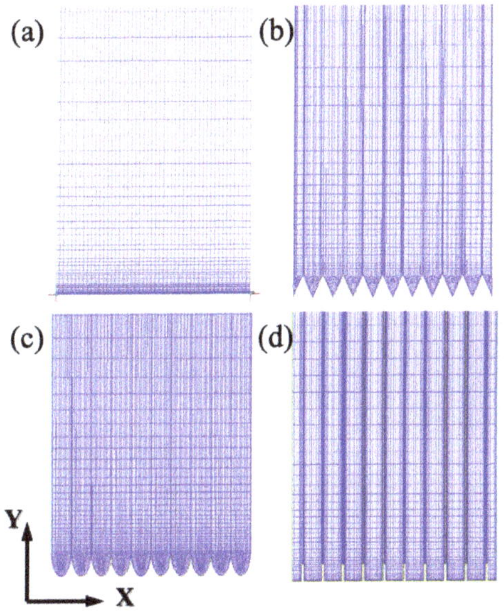

Figure 3. Mesh division of structured grids. (**a**) Smooth surface, (**b**) V-groove, (**c**) arc-groove, and (**d**) blade-groove.

2.3. *Verification of Mesh Independence and Model Accuracy*

Increasing the number of mesh cells generally improves the accuracy of numerical simulation calculations, but an excessive number of cells can reduce computational efficiency and increase unnecessary computational costs [44]. To mitigate the impact of mesh density on calculation accuracy and reasonably reduce the required computational power, it is necessary to verify mesh independence. This verification is primarily conducted by comparing the wall friction coefficients at different mesh numbers. Five mesh configurations were tested: 50 × 50 × 50, 50 × 75 × 75, 50 × 100 × 100, 50 × 125 × 125, and 50 × 150 × 150. These mesh configurations are tested for both smooth and microstructure wall models to establish mesh independence.

https://doi.org/10.3390/biomimetics11010077

Mesh independence was first verified for the smooth wall model using five configurations. The SST k-ω model is selected as the turbulence model, and periodic boundary conditions were applied in the spanwise direction. A velocity inlet boundary condition was specified with a flow speed of 5 m/s, while no-slip conditions were imposed on the walls. A pressure outlet boundary condition was applied. After completing the fluid domain settings, the results are shown in Tables 2 and 3.

Table 2. Mesh independence of the smooth wall verification scheme.

X × Y × Z	Count of Mesh	C_f
50 × 50 × 50	12,500	0.00498
50 × 75 × 75	281,250	0.00503
50 × 100 × 100	500,000	0.00504
50 × 125 × 125	781,250	0.00505
50 × 150 × 150	1,125,000	0.00510

Table 3. Mesh independence of the microstructure wall verification scheme.

X × Y × Z	Count of Mesh	C_f
50 × 50 × 50	12,500	0.00123
50 × 75 × 75	281,250	0.00123
50 × 100 × 100	500,000	0.00153
50 × 125 × 125	781,250	0.00123
50 × 150 × 150	1,125,000	0.00142

For the smooth wall model, the wall friction coefficient stabilizes around 0.005 when the mesh is set as 50 × 75 × 75. The wall friction coefficient paradoxically increases when the mesh size exceeds 50 × 150 × 150. For microstructure wall models, the simulation results are similar to those of smooth walls. The wall friction coefficient is approximately 0.0012. Therefore, the mesh size is approximately determined to be 50 × 75 × 75 with a total mesh count of about 3×10^5 for subsequent calculations.

To verify the accuracy of SST k-ω turbulence models in simulating fluid drag reduction, simulations were conducted using a mesh number of 50 × 75 × 75 [45]. The wall friction coefficients obtained from the simulations were compared with empirical values derived from Prandtl's formula, as shown in Table 4. The results show that the SST k-ω model keeps the error within 2% of the empirical values, indicating the high accuracy of the SST k-ω model. Therefore, the SST k-ω turbulence model will be used in subsequent numerical simulations.

Table 4. Comparison of simulated values and empirical values of C_f incoming.

Flow Velocity (m/s)	Empirical Value of C_f	Simulated Value of C_f (SST k-ω)	Relative Error (%)
5	0.00495	0.00504	1.8

Considering that the first layer height of the boundary layer mesh was estimated before simulation calculations, the mesh must meet the condition $y^+ \leq 1$ to ensure the accuracy of the simulation results in the boundary layer since the SST k-ω turbulence model was used. As shown in Figure 4, the distribution of y^+ along the flow direction near the wall shows that after the calculations converge, y^+ remains below 1, effectively meeting the requirements of the SST k-ω model for the y^+ range.

https://doi.org/10.3390/biomimetics11010077

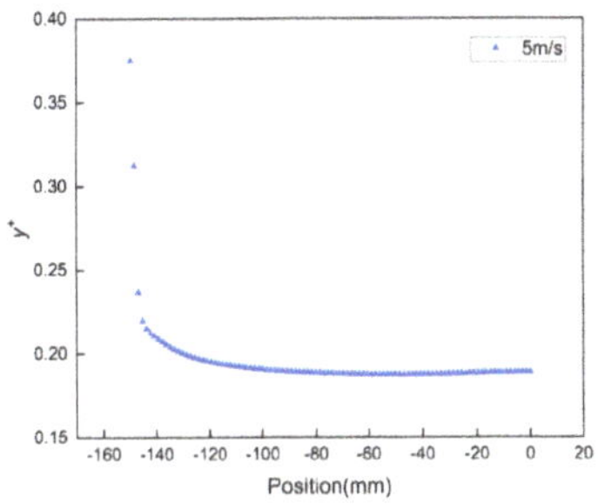

Figure 4. y^+ range of the mesh at different velocities.

3. Results and Discussion

3.1. Drag Reduction Effect of Microstructures

Figure 5 illustrates the wall-shear-stress distributions for the smooth wall and the three microstructured surfaces. The microstructured surfaces significantly affect the distribution and magnitude of the wall shear stress. Excluding the fluid inlet area, the shear stress is uniformly distributed across the smooth surface, with overall high shear stress levels. For V-groove, blade-groove, and arc-groove microstructures, the distribution of shear stress is noticeably uneven, with periodic fluctuations appearing that correspond to the groove geometry. Compared with the smooth wall, the mean wall-shear-stress values on the microstructured surfaces are markedly reduced, confirming the effectiveness of groove-induced drag reduction. The shear stress reaches its peak at the crests of the microstructures and reaches a minimum at the troughs.

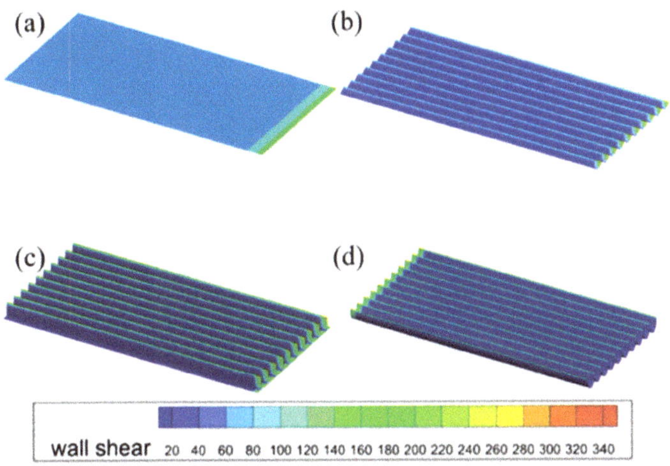

Figure 5. Wall shear stress of different surfaces. (**a**) Smooth surface, (**b**) V-groove, (**c**) blade-groove, and (**d**) arc-groove.

To further investigate the influence of different microstructured surfaces on near-wall flow, a cross-sectional slice, Plane-A, is made at 140 mm from the fluid inlet (where the fluid reaches a full turbulent state). This cross-section intersects with the bottom surface to form a spanwise polyline, Polyline-a, as shown in Figure 6. This approach enables detailed analysis of flow variations across the microstructured region, providing insight into how surface geometry influences near-wall turbulence.

https://doi.org/10.3390/biomimetics11010077

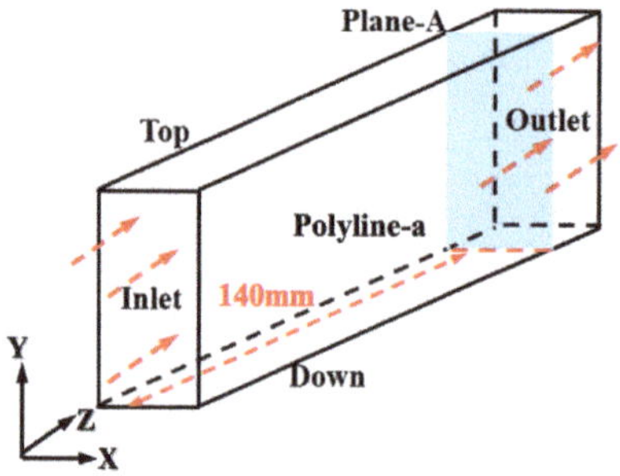

Figure 6. Position of Plane-A and Line-a.

To explore the distribution of wall shear stress affected by different microstructure surfaces further, the shear stress distribution along Polyline-a is extracted using Tecplot. The wall shear stress on microstructured surfaces exhibits periodic variations. The maximum wall shear stress on microstructure surfaces significantly increases compared to that on the smooth surfaces, sometimes by several multiples. However, these stress peaks occur over a very narrow spanwise extent. Consequently, all three microstructure types achieve net reductions in wall shear stress.

The wall shear stress distributions on different microstructured surfaces were analyzed to understand the drag reduction mechanisms. Figure 7a presents the computational domain with the measurement plane indicated, showing the location where wall shear stress data were extracted. The measurement plane is positioned at the mid-span of the microstructured region to capture representative flow characteristics while avoiding edge effects.

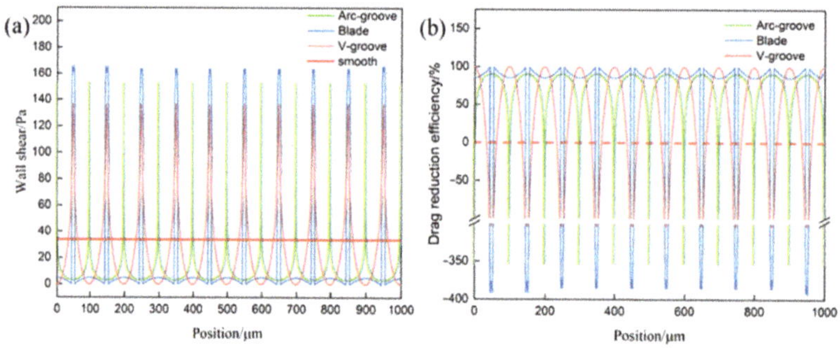

Figure 7. Wall shear stress and drag reduction rate along Polyline-a. (**a**) Distribution of local wall shear stress τ_w along Polyline-a. (**b**) Distribution of the local drag reduction rate η along Polyline-a.

To further quantify the drag reduction effects of different microstructure surfaces, a method is proposed that involves comparing the wall shear stress on smooth and microstructure surfaces to characterize the rate of drag reduction [46]. The formula for calculating the drag reduction rate is as follows:

$$\eta = \frac{\tau_{smooth} - \tau_{groove}}{\tau_{smooth}} \times 100\% \quad (9)$$

The drag reduction rate η is calculated based on the difference between the wall shear stress of a smooth surface (τ_{smooth}) and the microstructure surface (τ_{groove}). When η is greater than 0, it exhibits drag reduction; conversely, when η is less than 0, it exhibits drag increase.

https://doi.org/10.3390/biomimetics11010077

To quantify the effect of these variations, Figure 7b shows the local drag reduction rate η along Polyline-a, calculated from the corresponding wall shear stresses according to Equation (9). Here η represents the local reduction of wall shear stress relative to the smooth wall at the same streamwise position. It can be seen that grooved surfaces achieve positive drag reduction over a large portion of the groove period, with pronounced peaks above the groove valleys where the shear stress is strongly suppressed. The blade-groove surface exhibits the highest local drag reduction rate, followed by the V-groove and then the arc-groove, which is consistent with the ranking inferred from the wall shear stress profiles in Figure 7a. In addition, the average drag reduction rate η is obtained by calculating along the Polyline-a, where the blade microstructure achieves the highest drag reduction rate, reaching up to 18.2%. The V-groove microstructure follows with a drag reduction rate of 16.5%, and the arc-groove microstructure has a drag reduction rate of 14.7%.

This line-averaged quantity represents the overall reduction of wall shear stress along the evaluation line relative to the smooth wall and follows the same ranking (blade-groove > V-groove > arc-groove) as the local distributions in Figure 7b.

3.2. Flow Field and Velocity Distribution

The influence of different microstructures on the distribution of various physical fields near those surfaces is investigated, as can be seen in Figure 8. The velocity contours over different microstructured surfaces at the cross-section of Plane-A indicate that the velocity gradient is relatively small near the smooth surface, and the area of low velocity near the bottom is much smaller than that near the microstructure surfaces. In addition to having relatively larger low-velocity areas near the microstructure surface, there are also significant velocity gradients apparent near these microstructure surfaces. The phenomenon demonstrates that microstructures on surfaces can significantly affect fluid dynamics, creating regions with varied velocities that can influence the overall behavior of the fluid flow.

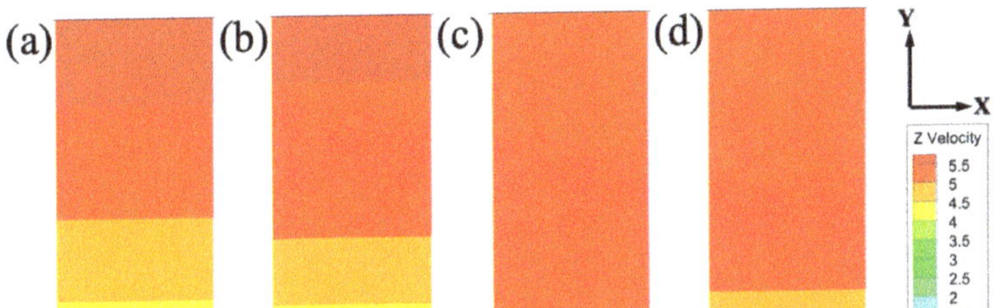

Figure 8. Distribution of fluid velocity in Plane-a of different surfaces. (**a**) Smooth surface, (**b**) V-groove, (**c**) arc-groove, and (**d**) blade-groove.

To investigate the intrinsic mechanisms of drag reduction by microstructured surfaces further, this paper analyzes the vortex structures near the wall. In previous studies on fluid drag reduction through vortex structure analysis, second-generation criteria such as Q, λ_1, Δ, and λ_{ci} are commonly used to identify vortex structures. However, the accuracy of these second-generation criteria depends on threshold values related to specific cases and can be affected by shear, which often influences the determination of vortex structures.

In the present study, the third-generation Ω vortex identification method proposed by Liu et al. [47] was adopted for near-wall vortex structure analysis. The Ω criterion is defined as the ratio of the vorticity squared to the summation of the vorticity squared and

https://doi.org/10.3390/biomimetics11010077

the deformation squared, representing the relative strength of rotation in the flow field. Mathematically, the Ω parameter is expressed as follows:

$$\Omega = \frac{||B||_F^2}{||A||_F^2 + ||B||_F^2 + \varepsilon} \tag{10}$$

where $||A||$ represents the magnitude of the strain rate tensor (symmetric part of the velocity gradient tensor), $||B||$ denotes the magnitude of the vorticity tensor (antisymmetric part), and ε is a small positive number to avoid division by zero [48]. When $\Omega > 0.5$, the rotation dominates over deformation, indicating the presence of a vortex. In practice, $\Omega = 0.52$ has been demonstrated to effectively capture vortex structures across various flow conditions without requiring case-specific threshold adjustment.

Compared with traditional vortex identification methods such as the Q criterion and the λ_2 method, the Ω criterion exhibits several important advantages [48–50]. First, it is effectively threshold independent: while Q and λ_2 require case-dependent and often subjective selection of iso-values, which can lead to inconsistent vortex identification among different flows, the Ω criterion admits a nearly universal threshold ($\Omega = 0.52$) that can be applied in a wide range of configurations. Second, Ω has a clearer physical interpretation, as it directly quantifies the relative dominance of rotation over strain and is bounded between 0 (pure strain) and 1 (pure rotation), in contrast to the less intuitive magnitudes of Q and λ_2 iso-surfaces. Third, the Ω method can capture both strong and weak vortices within a single threshold, whereas inappropriate choices of Q or λ_2 thresholds may highlight only intense vortices while suppressing weaker but dynamically relevant structures. Finally, Ω is a normalized quantity confined to the interval, which facilitates direct and quantitative comparison of vortex structures across different geometries and flow conditions, unlike Q, whose magnitude may vary substantially between cases.

To provide quantitative evidence for the vortex-induced drag reduction mechanism, the area-averaged Ω values and vortex coverage ratios were calculated for each groove type as shown in Table 5. The blade-groove surface exhibits the highest average Ω value of 0.68 in the near-wall region ($y^+ < 10$), compared to 0.61 for V-groove and 0.54 for arc-groove surfaces. Furthermore, the vortex coverage ratio, defined as the percentage of groove area occupied by regions with $\Omega > 0.52$, reaches 78% for blade-grooves, 65% for V-grooves, and 52% for arc-grooves. These quantitative metrics directly correlate with the observed drag reduction rates (18.2%, 16.5%, and 14.7%, respectively), providing strong evidence that higher vorticity concentration and greater vortex coverage lead to more effective drag reduction.

Table 5. Quantitative vortex metrics and drag reduction rates.

Groove Type	Average Ω	Coverage %	Drag Red. %
Blade-groove	0.68	78	18.2
V-groove	0.61	65	16.5
Arc-groove	0.54	52	14.7

Figure 9 illustrates the near-wall vortex structures identified using the Ω criterion across different surface configurations. As expected, the smooth surface (Figure 9a) shows no coherent vortical structures. In contrast, the V-groove surface (Figure 9b) develops symmetric double-vortex patterns within each cavity, with elongated structures extending from the groove tips along both sidewalls; however, the vortex cores do not penetrate to the groove bottom. The blade-groove surface (Figure 9c) displays the most fully developed vortex structures, which occupy the entire cavity depth and form distinct secondary vortex

https://doi.org/10.3390/biomimetics11010077

clusters at the groove valleys, thereby creating effective vortex-isolation layers. The arc-groove surface (Figure 9d) exhibits rounded vortex shapes that conform to its curved profile, with penetration depths intermediate between those of V-groove and blade-groove configurations. Notably, the degree of vortex development follows the same ranking as the drag reduction performance: blade-groove > V-groove > arc-groove, confirming the direct link between vortex characteristics and drag reduction effectiveness.

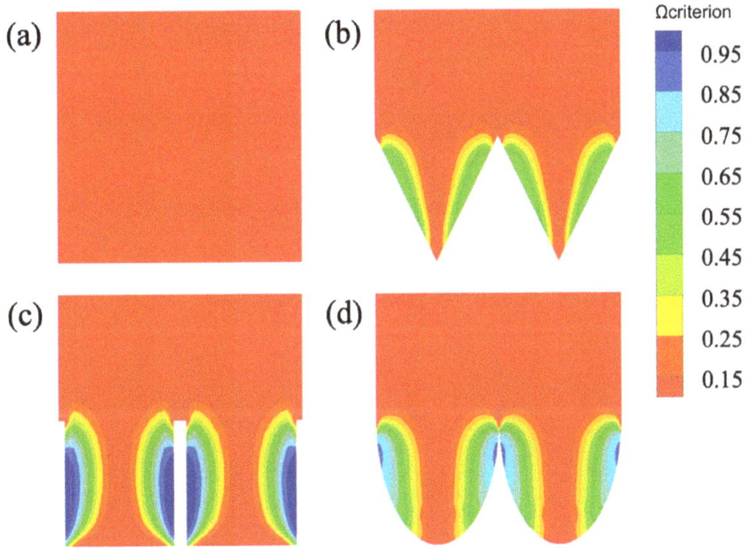

Figure 9. Vortex structures near the wall of different surfaces. (**a**) Smooth surface, (**b**) V-groove, (**c**) blade-groove, and (**d**) arc-groove.

Additionally, combining this with the analysis from the wall shear stress contour maps, it can be observed that for all three types of microstructure surfaces, the areas on either side of the microstructures are regions of low wall shear stress where various sizes of vortex structures are present. These vortex structures effectively separate the microstructure surfaces from the high-velocity fluid regions, preventing direct contact between them and thereby significantly reducing the wall shear stress between the fluid and wall. This mechanism contributes directly to drag reduction. Notably, the blade-groove microstructures exhibit significantly larger vortex structures near the wall compared to the other two types, which aligns with their stronger drag reduction performance as calculated earlier. This correlation confirms the earlier computed drag reduction rate comparisons.

Based on the streamline distribution diagrams of the smooth and microstructure surface, as shown in Figure 10, small, uniformly distributed vortex clusters form at the bottom of the blade-groove microstructures. The presence of these vortex clusters acts as vortex-isolation layers, effectively reducing fluid drag. Within the V-groove microstructures, complete vortex clusters do not form inside the microstructures. Instead, a larger vortex cluster forms at some distance from the top of the microstructures. Although this also contributes to drag reduction, its effect is weaker than that of the blade-groove microstructures. For the arc-groove microstructures, no significant vortex clusters are observed either inside the microstructures or near the microstructure walls.

https://doi.org/10.3390/biomimetics11010077

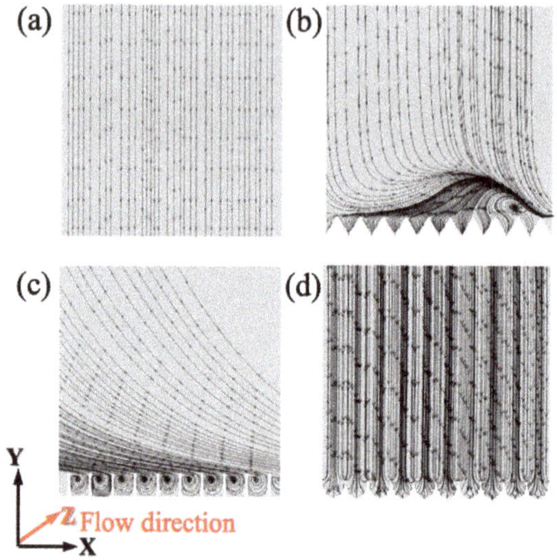

Figure 10. Streamline of different microstructure surfaces in Plane-A. (**a**) Smooth surface, (**b**) V-groove, (**c**) blade-groove, and (**d**) arc-groove.

3.3. *Effect of Aspect Ratio on Drag Reduction Performance*

To establish practical design guidelines, the influence of groove aspect ratio (h^+/s^+) on drag reduction performance was systematically investigated using blade-groove microstructures. Five aspect ratios (h^+/s^+ = 1.0, 0.75, 0.6, 0.45, and 0.35) were examined while maintaining a constant groove width of 100 μm. The analysis focused on wall shear stress distribution, velocity gradient, and vortex structure evolution to elucidate the underlying mechanisms.

Figure 11 presents the wall shear stress distribution for blade-groove surfaces at different aspect ratios. The overall distribution pattern shows high shear stress at the groove tips and low shear stress along the groove sidewalls, which remains consistent across all aspect ratios. However, the aspect ratio significantly affects both the maximum shear stress and the extent of low-stress regions.

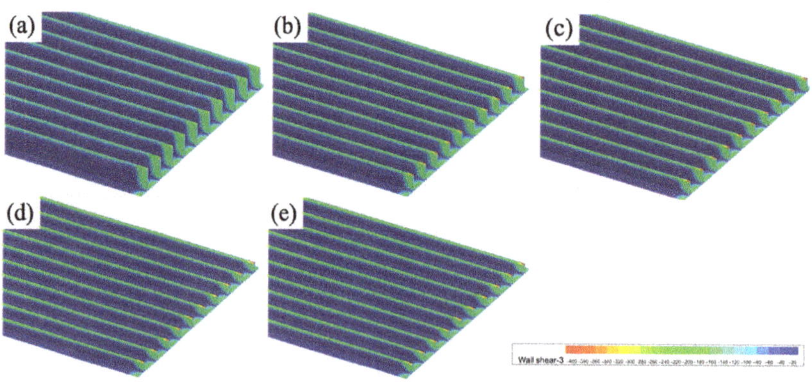

Figure 11. Wall shear stress of different aspect ratios. (**a**) h^+/s^+ = 1.0, (**b**) h^+/s^+ = 0.75, (**c**) h^+/s^+ = 0.6, (**d**) h^+/s^+ = 0.45, and (**e**) h^+/s^+ = 0.35.

https://doi.org/10.3390/biomimetics11010077

Figure 12 presents the spanwise distributions of wall shear stress (Figure 12a) and local drag reduction rate (Figure 12b) along Polyline-a for different aspect ratios. The wall shear stress profiles reveal a non-monotonic dependence on aspect ratio: peak stress at groove tips increases progressively as h^+/s^+ decreases from 1.0 to 0.45. But then drops at $h^+/s^+ = 0.35$; meanwhile, stress in the groove valleys rises sharply once h^+/s^+ falls below 0.6. The drag reduction rate profiles directly reflect these stress variations in terms of practical performance. When $h^+/s^+ \geq 0.75$, the drag reduction rate remains consistently high across most of the groove period, confirming that effective vortex-isolation layers are maintained. As h^+/s^+ decreases to 0.6 and below, the drag reduction rate in groove valley regions deteriorates markedly, and at $h^+/s^+ = 0.35$, certain spanwise locations even exhibit negative values, indicating local drag increase rather than reduction. The highest drag reduction rates are achieved at $h^+/s^+ = 1.0$ and $h^+/s^+ = 0.75$, demonstrating that higher aspect ratios favor drag reduction. These results collectively establish $h^+/s^+ = 0.6$ as a critical design threshold, below which drag reduction performance deteriorates rapidly.

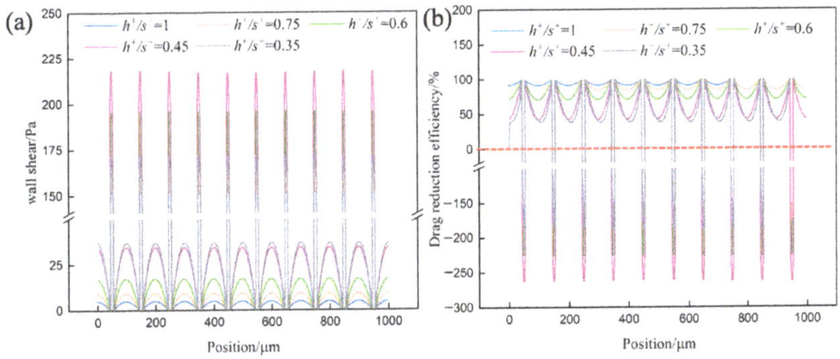

Figure 12. Wall shear stress and drag reduction rate along Polyline-a of different h^+/s^+. (**a**) Distribution of local wall shear stress τ_w along Polyline-a. (**b**) Distribution of the local drag reduction rate η along Polyline-a.

Figure 13 presents the vortex structures identified by the Ω criterion for blade-groove surfaces at different aspect ratios. At $h^+/s^+ = 1.0$~0.75, well-defined elongated vortex structures form along the groove sidewalls, with distinct secondary vortex clusters occupying the groove valleys. These valley vortices serve as effective vortex-isolation layers that separate the high-speed main flow from the groove bottom. As the aspect ratio decreases to 0.6 and below, the vortex structures undergo significant morphological changes: the elongated sidewall vortices become shorter and wider, and more critically, the secondary vortex clusters in the groove valleys progressively weaken at $h^+/s^+ = 0.35$. This disappearance of valley vortices directly explains the sharp increase in wall shear stress observed at low aspect ratios, as the protective isolation layer is lost and high-speed flow penetrates deeper into the grooves. Despite these morphological changes, the maximum vorticity magnitude remains relatively constant across all aspect ratios, indicating that the drag reduction degradation at low h^+/s^+ stems primarily from changes in vortex spatial distribution rather than vortex intensity. These observations establish $h^+/s^+ = 0.6$ as the critical threshold for maintaining effective vortex-isolation layers in blade-groove microstructures.

https://doi.org/10.3390/biomimetics11010077

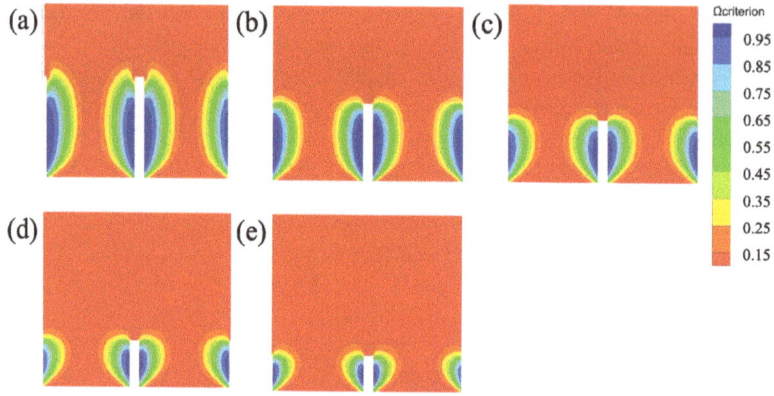

Figure 13. Near-wall vortex structures (identified by Ω criterion) for blade-groove surfaces at different aspect ratios. (**a**) $h^+/s^+ = 1.0$, (**b**) $h^+/s^+ = 0.75$, (**c**) $h^+/s^+ = 0.6$, (**d**) $h^+/s^+ = 0.45$, and (**e**) $h^+/s^+ = 0.35$.

3.4. Engineering Implications

The present study provides quantitative design guidelines for biomimetic drag-reducing surfaces by systematically investigating both groove geometry and dimensional parameters. For the Reynolds number and groove size considered here (groove height and width both 100 μm, aspect ratio $h^+/s^+ = 1.0$), the blade-groove surface achieves the highest drag reduction rate of 18.2%, followed by the V-groove (16.5%) and arc-groove (14.7%). This clear performance hierarchy demonstrates that groove cross-section geometry significantly influences drag reduction effectiveness, even when overall groove dimensions are constrained by manufacturing considerations.

By combining the distributions of local drag reduction rate along Polyline-a with the near-wall vortex structures identified by the Ω-criterion, the analysis reveals that blade-grooves promote a stable and fully developed cavity vortex that effectively isolates the main flow from the groove valleys. For the V-groove surface, the cavity vortex is weaker and high-speed fluid partially penetrates into the grooves, whereas for the arc-groove surface, the flow more easily impinges onto the groove bottom. These differences in vortex topology explain the observed ranking of drag reduction rates among the three geometries.

From an engineering perspective, these findings can be synthesized into practical design guidelines that address both geometry selection and dimensional optimization.

For applications where drag reduction is the primary objective, blade-groove microstructures represent the optimal choice. Their sharp edges generate strong and stable cavity vortices that form effective buffer layers, isolating the high-speed main flow from the wall and substantially reducing wall shear stress. When manufacturing constraints preclude sharp blade edges, V-grooves provide a practical alternative that balances drag reduction performance with geometric robustness and ease of fabrication. Arc-shaped grooves, though less effective for drag reduction, may be preferred in applications where fouling resistance or structural integrity outweighs hydrodynamic performance.

Beyond geometry selection, dimensional optimization is equally critical. The parametric study on blade-groove microstructures demonstrates that maintaining an aspect ratio of $h^+/s^+ \geq 0.75$ ensures well-developed vortex-isolation layers and optimal drag reduction, whereas reducing h^+/s^+ below 0.6 triggers vortex collapse and significant performance degradation. These findings establish a comprehensive design framework: the ranking of blade-groove > V-groove > arc-groove guides geometry selection, while the aspect ratio criterion of $h^+/s^+ \geq 0.75$ ensures optimal dimensional design for practical biomimetic drag-reducing surfaces.

https://doi.org/10.3390/biomimetics11010077

4. Conclusions

This study presents a systematic CFD investigation of biomimetic drag-reducing microstructures, examining groove geometry effects, aspect ratio optimization, and vortex-based drag reduction mechanisms. The principal findings are as follows. A comparative analysis of three representative groove geometries under identical flow conditions ($Re = 7.5 \times 10^5$, groove dimensions 100 μm) reveals a clear performance hierarchy: blade-groove achieves the highest drag reduction rate of 18.2%, followed by V-groove (16.5%) and arc-groove (14.7%). The blade-groove owes its superior performance to sharp edges that induce strong, stable cavity vortices forming effective vortex-isolation layers, which separate the high-speed main flow from the groove valleys and markedly reduce wall shear stress.

A parametric study of blade-groove aspect ratios ($h^+/s^+ = 0.35$–1.0) demonstrates that dimensional optimization is as important as geometry selection. At $h^+/s^+ \geq 0.75$, well-developed vortex-isolation layers persist, and drag reduction remains effective. When h^+/s^+ falls below 0.6, secondary vortex clusters in the groove valleys weaken progressively and eventually vanish, allowing high-speed flow to penetrate deeper into the grooves and causing rapid performance degradation. These results identify $h^+/s^+ = 0.6$ as a critical design threshold for blade-groove microstructures.

Methodologically, the third-generation Ω vortex-identification criterion adopted here provides threshold-independent, quantitative vortex recognition that surpasses traditional Q and λ_2 methods. By correlating vortex metrics (average Ω values and coverage ratios) directly with drag reduction rates, the Ω criterion offers clear mechanistic insight into how groove geometry governs near-wall flow structures.

Collectively, these findings yield a comprehensive design framework for biomimetic drag-reducing surfaces: the geometry ranking (blade-groove > V-groove > arc-groove) guides cross-section selection, while the aspect ratio guideline ($h^+/s^+ \geq 0.75$ recommended; $h^+/s^+ > 0.6$ as the minimum) ensures favorable dimensional design. This framework links fundamental vortex dynamics with practical engineering requirements, offering quantitative guidance for groove-type drag-reducing surfaces in marine and aerospace applications. Future work will pursue experimental validation of optimized blade-groove structures and scalable fabrication methods, such as hot-rolling, to facilitate industrial implementation.

Author Contributions: Conceptualization, J.X.; methodology, J.L. and J.X.; software and simulation, J.L.; validation, J.L., B.G. and D.S.; formal analysis and investigation, J.L.; resources, J.X., B.G. and D.S.; writing—original draft preparation, J.L.; writing—review and editing, J.X. and C.D.; supervision, J.X. All authors have read and agreed to the published version of the manuscript.

Funding: This research received no external funding.

Data Availability Statement: The data that support the findings of this study are available from the corresponding author upon reasonable request.

Acknowledgments: The authors gratefully acknowledge the computational resources provided by the National Key Laboratory for Precision Hot Processing of Metals at Harbin Institute of Technology.

Conflicts of Interest: The authors declare no conflicts of interest.

References

1. Ahmadzadehtalatapeh, M.; Mousavi, M. A Review on the Drag Reduction Methods of the Ship Hulls for Improving the Hydrodynamic Performance. *Int. J. Marit. Technol.* **2015**, *4*, 51–64.
2. Bushnell, D.M.; Moore, K.J. Drag Reduction in Nature. *Annu. Rev. Fluid Mech.* **1991**, *23*, 65–79. [CrossRef]
3. Engineering, R.A.O. *Future Ship Powering Options: Exploring Alternative Methods of Ship Propulsion*; Royal Academy of Engineering: London, UK, 2013.

https://doi.org/10.3390/biomimetics11010077

4. Gu, Y.; Yu, S.; Mou, J.; Wu, D.; Zheng, S. Research Progress on the Collaborative Drag Reduction Effect of Polymers and Surfactants. *Materials* **2020**, *13*, 444. [CrossRef] [PubMed]
5. Grüneberger, R.; Kramer, F.; Wassen, E.; Hage, W.; Meyer, R.; Thiele, F. Influence of Wave-Like Riblets on Turbulent Friction Drag. In *Nature-Inspired Fluid Mechanics: Results of the DFG Priority Programme 1207 "Nature-Inspired Fluid Mechanics" 2006–2012*; Tropea, C., Bleckmann, H., Eds.; Springer: Berlin/Heidelberg, Germany, 2012; pp. 311–329.
6. Dean, B.; Bhushan, B. Shark-skin surfaces for fluid-drag reduction in turbulent flow: A review. *Philos. Trans. R. Soc. A Math. Phys. Eng. Sci.* **2010**, *368*, 4775–4806. [CrossRef] [PubMed]
7. Bixler, G.D.; Bhushan, B. Bioinspired rice leaf and butterfly wing surface structures combining shark skin and lotus effects. *Soft Matter* **2012**, *8*, 11271–11284. [CrossRef]
8. Lee, C.; Kim, C.-J. Underwater Restoration and Retention of Gases on Superhydrophobic Surfaces for Drag Reduction. *Phys. Rev. Lett.* **2011**, *106*, 014502. [CrossRef]
9. Samaha, M.A.; Tafreshi, H.V.; Gad-El-Hak, M. Superhydrophobic surfaces: From the lotus leaf to the submarine. *Comptes Rendus Mécanique* **2012**, *340*, 18–34. [CrossRef]
10. Lang, A.; Motta, P.; Hidalgo, P.; Westcott, M. Bristled shark skin: A microgeometry for boundary layer control? *Bioinspir. Biomim.* **2008**, *3*, 046005. [CrossRef]
11. Wong, T.-S.; Kang, S.H.; Tang, S.K.; Smythe, E.J.; Hatton, B.D.; Grinthal, A.; Aizenberg, J. Bioinspired self-repairing slippery surfaces with pressure-stable omniphobicity. *Nature* **2011**, *477*, 443–447. [CrossRef]
12. Epstein, A.K.; Wong, T.-S.; Belisle, R.A.; Boggs, E.M.; Aizenberg, J. Liquid-infused structured surfaces with exceptional anti-biofouling performance. *Proc. Natl. Acad. Sci. USA* **2012**, *109*, 13182–13187. [CrossRef]
13. Solomon, B.R.; Khalil, K.S.; Varanasi, K.K. Drag reduction using lubricant-impregnated surfaces in viscous laminar flow. *Langmuir* **2014**, *30*, 10970–10976. [CrossRef] [PubMed]
14. Chen, Q.H.; Zhang, C.Q.; Cai, Y.K.; Luo, X.; Wang, B.; Song, Q.; Liu, Z. Periodically oriented superhydrophobic microstructures prepared by laser ablation-chemical etching process for drag reduction. *Appl. Surf. Sci.* **2023**, *615*, 156403. [CrossRef]
15. Daniel, T.L. Fish mucus: In situ measurements of polymer drag reduction. *Biol. Bull.* **1981**, *160*, 376–382. [CrossRef]
16. Hoyt, J. Hydrodynamic drag reduction due to fish slimes. *Swim. Fly. Nat.* **1975**, *2*, 653–672.
17. Oh, K.X.; Nugroho, B.; Hutchins, N.; Monty, J.P. Meandering riblets targeting spanwise spatial oscillation of turbulent boundary layers. In Proceedings of the 18th Australasian Fluid Mechanics Conference, Launceston, Australia, 3–7 December 2012.
18. Okabayashi, K. Direct numerical simulation for modification of sinusoidal riblets. *J. Fluid Sci. Technol.* **2016**, *11*, 0015. [CrossRef]
19. Walsh, M.J. Riblets as a viscous drag reduction technique. *AIAA J.* **1983**, *21*, 485–486. [CrossRef]
20. Bixler, G.D.; Bhushan, B. Shark skin inspired low-drag microstructured surfaces in closed channel flow. *J. Colloid Interface Sci.* **2013**, *393*, 384–396. [CrossRef]
21. Liu, Z.-H.; Dong, W.; Xiong, Y.; Xia, F. Analysis on Factors and Mechanism of Drag Reduction by Grooved Surface. *J. Ship Mech.* **2007**, *11*, 820–831.
22. Lang, A.W.; Bradshaw, M.T.; A Smith, J.; Wheelus, J.N.; Motta, P.J.; Habegger, M.L.; Hueter, R.E. Movable shark scales act as a passive dynamic micro-roughness to control flow separation. *Bioinspir. Biomim.* **2014**, *9*, 036017. [CrossRef]
23. Luo, Y.; Zhang, D.; Liu, Y. Recent Drag Reduction Developments Derived from Different Biological Functional Surfaces: A Review. *J. Mech. Med. Biol.* **2015**, *16*, 1630001. [CrossRef]
24. Fu, Y.F.; Yuan, C.Q.; Bai, X.Q. Marine drag reduction of shark skin inspired riblet surfaces. *Biosurface Biotribol.* **2017**, *3*, 11–24. [CrossRef]
25. Luo, Y.; Liu, Y.; Zhang, D.Y. Hydrodynamic testing of a biological sharkskin replica manufactured using the vacuum casting method. *Surf. Rev. Lett.* **2015**, *22*, 1550030. [CrossRef]
26. Luo, Y. Recent progress in exploring drag reduction mechanism of real sharkskin surface: A review. *J. Mech. Med. Biol.* **2015**, *15*, 1530002. [CrossRef]
27. Han, X.; Zhang, D.; Li, X.; Li, Y. Bio-replicated forming of the biomimetic drag-reducing surfaces in large area based on shark skin. *Chin. Sci. Bull.* **2008**, *53*, 1587–1592. [CrossRef]
28. Chen, H.; Zhang, X.; Zhang, D.; Pan, J.; Hagiwara, I. Large-scale equal-proportional amplification bio-replication of shark skin Based on solvent-swelling PDMS. *J. Appl. Polym. Sci.* **2013**, *130*, 2383–2389. [CrossRef]
29. Graybill, M.T.; Xu, N.W. Experimental studies of bioinspired shark denticles for drag reduction. *Integr. Comp. Biol.* **2024**, *64*, 742–752. [CrossRef]
30. Cui, X.; Chen, D.; Chen, H. Multistage gradient bioinspired riblets for synergistic drag reduction and efficient antifouling. *ACS Omega* **2023**, *8*, 8569–8581. [CrossRef]
31. Mawignon, F.J.; Qin, L.; Kouediatouka, A.N.; Liu, J.; Djandja, O.S.; Turay, M.C.; Winston, D.D.; Yang, H.; Wang, Z.; Wen, J.; et al. Optimized three-dimensional cuboidal shark-inspired riblets for enhanced drag reduction in turbulent flow. *Ocean Eng.* **2025**, *318*, 120199. [CrossRef]

https://doi.org/10.3390/biomimetics11010077

32. Mawignon, F.J.; Liu, J.; Qin, L.; Kouediatouka, A.N.; Ma, Z.; Lv, B.; Dong, G. The optimization of biomimetic sharkskin riblet for the adaptation of drag reduction. *Ocean Eng.* **2023**, *275*, 114135. [CrossRef]
33. Zani, M.R.; Maor, N.S.; Bhamitipadi Suresh, D.; Jin, Y. Turbulent boundary layer control with multi-scale riblet design. *Energies* **2024**, *17*, 3827. [CrossRef]
34. Pakatchian, M.R.; Rocha, J.; Li, L. Advances in riblets design. *Appl. Sci.* **2023**, *13*, 10893. [CrossRef]
35. Zhang, Y.; Ye, Z.; Li, B.; Xie, L.; Zou, J.; Zheng, Y. Numerical analysis of turbulence characteristics in a flat-plate flow with riblets control. *Adv. Aerodyn.* **2022**, *4*, 29. [CrossRef]
36. Cafiero, G.; Iuso, G. Drag reduction in a turbulent boundary layer with sinusoidal riblets. *Exp. Therm. Fluid Sci.* **2022**, *139*, 110723. [CrossRef]
37. Martin, S.; Bhushan, B. Fluid flow analysis of a shark-inspired microstructure. *J. Fluid Mech.* **2014**, *756*, 5–29. [CrossRef]
38. Walsh, M.J. Drag characteristics of V-groove and transverse curvature riblets. In Proceedings of the Symposium on Viscous Drag Reduction, Dallas, TX, USA, 7–8 November 1979.
39. Pang, R.; Sang, W.; Cai, Y. Drag Characteristics of V-Groove and Transverse Curvature Riblets. In *Proceedings of the 2018 Asia-Pacific International Symposium on Aerospace Technology (APISAT 2018), Chengdu, China, 16–18 October 2018*, 9th ed; Springer: Berlin/Heidelberg, Germany, 2019.
40. Devenport, W.J.; Lowe, K.T. Equilibrium and non-equilibrium turbulent boundary layers. *Prog. Aerosp. Sci.* **2022**, *131*, 100807. [CrossRef]
41. Schetz, J.A.; Bowersox, R.D. *Boundary Layer Analysis*; American Institute of Aeronautics and Astronautics: Reston, VA, USA, 2011.
42. Tsinober, A.; Bushnell, D.; Hefner, J. Viscous Drag Reduction in Boundary Layers. *Ser Prog. Astronaut. Aeronaut.* **1990**, *123*, 327–349.
43. Chamorro, L.P.; Arndt, R.; Sotiropoulos, F. Drag reduction of large wind turbine blades through riblets: Evaluation of riblet geometry and application strategies. *Renew. Energy* **2013**, *50*, 1095–1105. [CrossRef]
44. Li, C.; Tang, S.; Li, Y.; Geng, Z. Numerical and experimental investigations on drag-reducing effects of riblets. *Eng. Appl. Comput. Fluid Mech.* **2021**, *15*, 1726–1745. [CrossRef]
45. Yusuf, S.N.A.; Asako, Y.; Sidik, N.A.C.; Mohamed, S.B.; Japar, W.M.A.A. A short review on rans turbulence models. *CFD Lett.* **2020**, *12*, 83–96. [CrossRef]
46. Xu, Y.; Ruan, C.; Zhang, Z. Numerical Study on Drag Reduction of Superhydrophobic Surfaces with Conical Microstructures in Laminar Flow. *J. Appl. Fluid Mech.* **2024**, *17*, 1129–1142. [CrossRef]
47. Liu, C.; Wang, Y.; Yang, Y.; Duan, Z. New omega vortex identification method. *Sci. China Phys. Mech. Astron.* **2016**, *59*, 684711. [CrossRef]
48. Dong, X.-R.; Wang, Y.-Q.; Chen, X.-P.; Dong, Y.; Zhang, Y.-N.; Liu, C. Determination of epsilon for Omega vortex identification method. *J. Hydrodyn.* **2018**, *30*, 541–548. [CrossRef]
49. Wang, D.-D.; Wang, Z.-H.; Fan, Y.-W.; Sun, X.; Gao, Q.-J. Characterization of vortex structures with self-excited oscillations based on Liutex-Omega vortex identification method. *J. Hydrodyn.* **2023**, *35*, 95–111. [CrossRef]
50. Zhang, Y.-N.; Wang, X.-Y.; Liu, C. Comparisons and analyses of vortex identification between Omega method and Q criterion. *J. Hydrodyn.* **2019**, *31*, 224–230. [CrossRef]

https://doi.org/10.3390/biomimetics11010077

biomimetics

MDPI

Article

A New Concept of Sustainable Wind Turbine Blades: Bio-Inspired Design with Engineered Adhesives

Leon Mishnaevsky, Jr. [1,*], Mohsen Jafarpour [2], Johanna Krüger [2] and Stanislav N. Gorb [2]

[1] Department of Wind and Energy Systems, Technical University of Denmark, DK-4000 Roskilde, Denmark
[2] Zoological Institute, Kiel University, 24118 Kiel, Germany; mjafarpour@zoologie.uni-kiel.de (M.J.); johannakrueger96@gmail.com (J.K.); sgorb@zoologie.uni-kiel.de (S.N.G.)
* Correspondence: lemi@dtu.dk

Abstract: In this paper, a new concept of extra-durable and sustainable wind turbine blades is presented. The two critical materials science challenges of the development of wind energy now are the necessity to prevent the degradation of wind turbine blades for several decades, and, on the other side, to provide a solution for the recyclability and sustainability of blades. In preliminary studies by DTU Wind, it was demonstrated that practically all typical wind turbine blade degradation mechanisms (e.g., coating detachment, buckling, spar cap/shell adhesive joint degradation, trailing edge failure, etc.) have their roots in interface degradation. The concept presented in this work includes the development of bio-inspired dual-mechanism-based interface adhesives (combining mechanical interlocking of fibers and chemical adhesion), which ensures, on the one side, extra-strong attachment during the operation time, and on the other side, possible adhesive joint separation for re-use of the blade parts. The general approach and physical mechanisms of adhesive strengthening and separation are described.

Keywords: wind energy; wind turbine blade; sustainability; lifetime

1. Introduction

Large expansion of wind energy is foreseen in Europe in the next years and decades [1]. Europe is expected to install 129 GW of new wind farms over the period 2023–2027, one quarter of which will be offshore. The EU should be building over 30 GW a year on average to meet its 2030 targets. Large and extra-large wind turbines are installed, and the size of blades is increasing. In 2021, the Chinese company MingYang Smart Energy released details of a 264 m tall design with 118 m blades. The Danish firm Vestas developed a 15-megawatt turbine with 115.5 m blades, and Siemens Gamesa Renewable Energy developed a turbine with 108 m blades. Bigger wind turbines allow the capture of more wind and produce more electricity [2].

However, recently, an increased failure rate in wind turbines, largely in newer and bigger models, has been reported [3], leading to large expenses for wind turbine manufactures, due to "extraordinary" repairs and upgrades and multimillion losses on rising warranty provisions for GE, Vestas and Siemen Gamesa in 2022. In average, wind turbine blades require repairs every 2 to 3 years, with costs of the order of 30,000 EUR or more per repair [4,5]. Figure 1a shows a damaged wind turbine. The frequency of wind turbine failure offshore can be up to four times higher than for onshore wind turbines [4], while repair costs for offshore wind turbines are also 2 to 2.5 times higher [5]. The frequent and expensive repairs cause higher energy costs and make wind energy less competitive. Thus, the question of how to prevent wind turbine blade degradation and failure becomes especially urgent, due to more and more larger wind turbines installed offshore.

Figure 1. Illustration of two critical problems of wind energy: (**a**) Wind turbine with broken blades, and (**b**) wind blades accumulated in landfill. Photos from Bigstock © collection.

On the other side, another, to some degree inverse, problem becomes critical now: the sustainability and recyclability of wind turbine blades. Currently, wind turbines are recyclable up to 85%. Roughly two thirds of the remaining 15% come from composite materials [6,7]. At end of their life, due to the current blade material, the amount of waste coming from wind turbine blades is expected to reach 43 million tons globally in 2050 [8] (see Figure 1b, showing wind turbine blades accumulated in landfill). Europe will be the first continent to decommission wind turbines and wind farms [8]. In Europe, predictions on the waste amount are 175 kilotons in 2030 and 325 kilotons in 2050 [9]. The solution for this problem includes the development of new recycling (e.g., pyrolysis, solvolysis) technologies, new wind turbine blade materials (like thermoplastic based, vitrimer based, or natural material-based blades) [6], but also the development of disassemblable blades, with detachable parts, which can be reused.

The concept of this paper is as follows. In the preliminary studies [4], it was demonstrated that practically all common blade degradation mechanisms (e.g., coating detachment, blade buckling and collapse, spar cap/shell adhesive joint degradation, trailing edge failure, etc.) are caused by interface degradation, often related to the micro- and nanoscale defects in interfaces. Therefore, the path to prevent the failure and damage of blades lies in the development of methods to strengthen the interfaces in composite blade structures. Further, the development of disassemblable, sustainable blades also requires controlled interfaces, this time, not extra-strong, but rather detachable, separable. Thus, the main assumption of this work is that the structured, engineered interfaces in wind turbine blades are the key to both high durability and sustainability of composite wind turbine blades. This idea is discussed in this paper.

2. Concept of Interface Controlled Blade: Lifetime and Recyclability

In order to analyze the mechanisms of damage and failure of wind turbine blades, a number of methods can be applied, including post-mortem analysis of failed blades; full scale testing of blades in laboratories with video-observation and structural health monitoring; analysis of databases and collections of incident reports; direct monitoring of blade deformation and degradation during service (for instance, using non-destructive testing and structural health monitoring methods); testing design of subcomponents (e.g., beam), reproducing parts or elements of the blades (e.g., joints or sandwiches); computational modelling of blade deformation and damage [4]. The review of various studies on the root causes of the failure of wind turbine blades, carried out with the use of these methods, is presented in [4]. Table 1 shows the overview of main mechanisms of wind turbine blade, linking the endangered region of wind turbine blade, observed failure mechanisms and their root causes. Figure 2 shows, schematically, the root causes of the main structural failure mechanisms of the blades. It should be noted that while leading edge erosion and other surface effects (like sizing) are most often observed, and also most expensive damage mechanisms, they are not structural failure mechanisms and are not discussed in the following sections.

Table 1. Common mechanisms of damage and failure of wind turbine blades.

Endangered Area	Failure Mechanism	Root Causes
Leading edge	Surface erosion, coating debonding	Multiple rain droplet, hail or sand impacts causing surface damage, and/or stress waves, leading to coating debonding
Adhesive joints/bond lines, between the shell and main spar, or between the spar cap and internal stiffeners	Adhesive layers become damaged or debonded, causing buckling or collapse of blade. Blade collapse occurs if the shear web is detached from the shell and buckles. The most often observed failure mechanisms of blades are transverse cracks in the max chord region (initiated as a skin disbond from the sandwich core) and the disbond of the aft shear web from the blade shell in the root-transition zone (triggered by peeling stresses in the adhesive bond lines).	Damage of adhesive joints in the spar area is one of the main processes leading to blade failure. Shear stress is mainly responsible for debonding. Skin debonding from the adhesive joint is the initial mechanism of failure. Apart from buckling-driven failure, two inside surfaces of the sandwich panels can come into contact during the loading, and this has a strong influence on the failure of the trailing edge sections.
Trailing edge	Debonding in the adhesive joint (e.g., due to peeling stress) and/or by buckling of sandwich panels.	Adhesive joints at the trailing edge are other critically important structural parts depending on the reliability of adhesive bond. Blade failure starts from the crack in the bonding at the trailing edge, and then the crack propagates in the laminate.
Tapered areas, plydrop (thickness transitions), root region	Local stress concentration in tapered areas, with thickness transitions, lead to ply delamination (mid spans in the root region, transition from cylinder to an aerofoil, large laminate panels near the maximum chord)	Local stress concentrations, interlaminar stresses in the vicinity of the ply drop-off which can lead to the delamination of plies and the failure of these regions.
Spar and shear webs; buckling	Structural collapse and buckling of blade. Shell debonding from the adhesive joints is the initial failure mechanism, which lead to unstable debonding propagation.	Sudden progressive collapse of a blade is caused by the coupled phenomenon of delamination (interlaminar failure) and buckling, with compressive fiber failure in the delaminated flange material. Local delamination is initiated at the boundary of the adjoining structural elements and in the center of the compressive flange due to local buckling, leading to a sudden structural collapse.

An important observation is that virtually all the structural mechanisms of blade failure are controlled by weak, damageable interfaces, causing the buckling and collapse, trailing edge detachment or coating debonding.

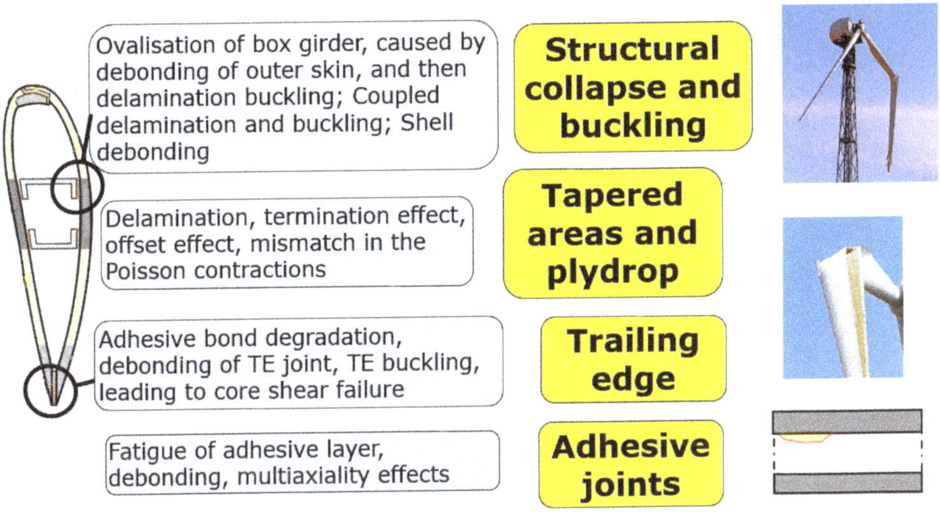

Figure 2. Schema: Damage mechanisms of wind turbine blades. Here, TE = trailing edge.

3. Bio-Inspired Adhesive Layers in Adhesive Joints of Blades

Wind turbines are subjected to long term, permanent randomly varying mechanical and environmental loading, with current usual lifetime of 20 to 30 years. The estimated costs for each repair event are very high, for instance, an average offshore blade repair can cost 30 k EUR, and a new blade can cost 200 k EUR [10].

There are a number of ways to prevent the failure of wind turbine blades, which can be grouped into (a) development of stronger and tougher materials, (b) lighter materials, e.g., carbon composites, which reduce loads on the blades, (c) design solutions, like using transverse cap stiffeners or strengthening trailing edge [4]. However, strengthening of materials is not an infinite process.

In the situation of low and challenging maintenance, extra-long-term service, random varying mechanical and environmental loadings, solutions for the optimization of wind turbine structures can be taken from the biological materials, which are also evolved to bear randomly varying loads over long-term period, without repair option. Looking at strong and tough biological structures (e.g., nacre/shells, skulls/teeth/bones, timber/bamboo) [11], one can observe a quite common feature: these biological composites have extraordinary strength and toughness, which is related with their internal microarchitectures.

These materials at microlevel typically rather often represent layered structures (often, stiff platelets), with nanostructured adhesive layers between stiff platelets). The multiscale design of biomaterials allows the combination of two effects: high stiffness and flexibility, ensured by macroscale structures (e.g., cylindric or cellular wood design, or stiff platelets in nacre) and tough, nanoengineered interfaces, preventing damage even at high deformation.

For instance, nacre, which consists of 95% of $CaCO_3$, has a work of fracture 3000 times higher than that of the monolithic $CaCO_3$ [11]. Its fracture toughness is controlled by nanostructures in nacre interfaces (i.e., mineral bridges between aragonite platelets in biopolymer layers) [12]. Similarly, graded multi-level dentin-enamel interface in teeth, with scalloped structure (convexities toward dentin, and concavities toward enamel) ensures the extraordinarily high toughness of teeth [11]. Figure 3 shows examples of biological extra-tough structures with nanostructured interfaces, microarchitecture of wood as a layered, fibril reinforced material and nacre, mother of pearl, aragonite platelets, biopolymer layer and mineral bridges.

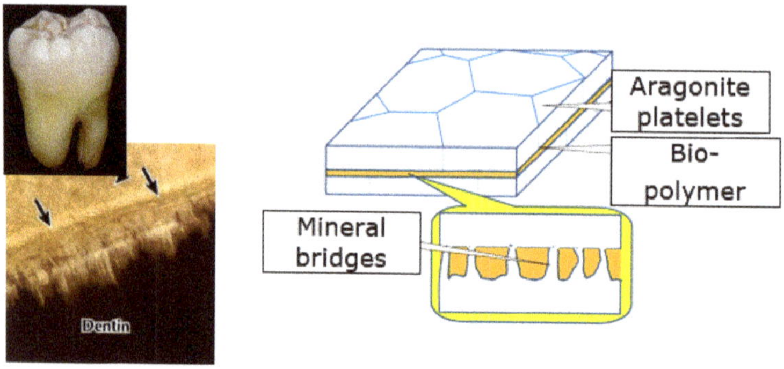

(**a**) (**b**)

Figure 3. Two examples of extra-strong and tough biological composites, with structured interface layers: (**a**) Tooth, with graded, multilevel dentin-enamel interface, and (**b**) nacre, with biopolymer layer with mineral bridges. The arrows show the dentin-enamel interface region in a tooth.

As applied to composite wind turbine blades, this suggests nanostructuring of adhesive joints and interfaces in blades, mimicking the complex structure of biological adhesives and preventing the majority of blade degradation mechanisms.

4. Combined Mechanisms of Adhesion: Mechanical Interlocking and Chemical Adhesion

In a series of works, structural bioinspired adhesives have been developed [11–17], which exhibit smart adhesion capabilities, like controllable adhesion strength, active adhesion control, reversible adhesion to diverse dry and wet surfaces. So, hydrogel-based, reversible adhesives combining the benefits of both liquid and dry adhesives were developed by Cho et al. [18]. Gecko-inspired hexagonal and lamellae patterned adhesives were developed in [19]. In [20], the authors studied so-called probabilistic fasteners, with two surfaces covered with cuticular micro-outgrowths, ensuring quite reliable and reversible attachment ("probabilistic" because the outgrowths on the contacting surfaces do not correspond exactly to each other, and the interlocking takes place without any precise positioning). Such structures are observed in the wing-locking mechanism in beetles and the head arrester in dragon flies. Further, various types of micro- and nano-hierarchical structures of adhesives (e.g., lamellae, setae, branches and spatulae, self-similar pillar structure, etc.) have been investigated [17].

High potential of such structures for the development of new adhesives was demonstrated. However, structured adhesives and also probabilistic fasteners are developed in most cases for relatively low loads and detachable adhesion.

In this section, the possibility of the development of bio-inspired structured adhesives for high load structures is considered.

The idea here is to use a combination of the mechanical and chemical attachment, similar to the adhesive layers in nacre, from polymer and mineral bridges. The adhesion is achieved using the mechanical interlocking of fibrous structures from each side (formed as fibrous nanograss-like, as shown in Figure 4, or brush-like surface structures on sides, or as already mentioned probabilistic fasteners [20]) and is supported via chemical gluing. An example of such mechanically interlocked and also chemically glued biological structure is the red deer skull. Figure 5 shows the frontal skull sutures of a male red deer (*Cervus elaphus*). These sutures must withstand very high stresses during male contests. The insert on Figure 5 shows the mechanically interlocked structures, fibrous joints between the cranial bones and the irregular, but interlocking edges of the bones attached. Sutures provide locomotive shock absorption, to protect the brain [21,22].

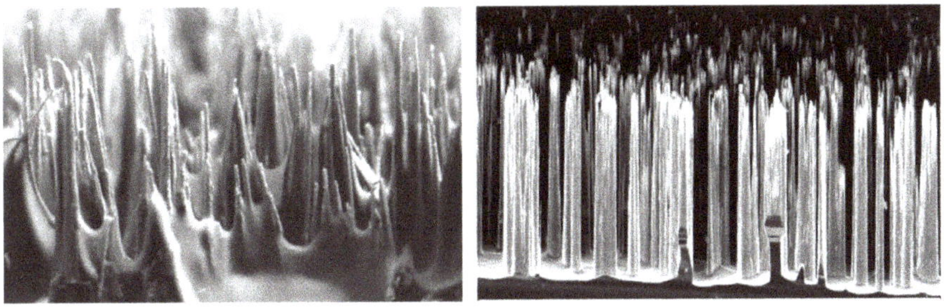

Figure 4. High aspect ratio nanograss, fibrous structures of interfaces (Photo Nicolai Frost-Jensen Johansen, DTU).

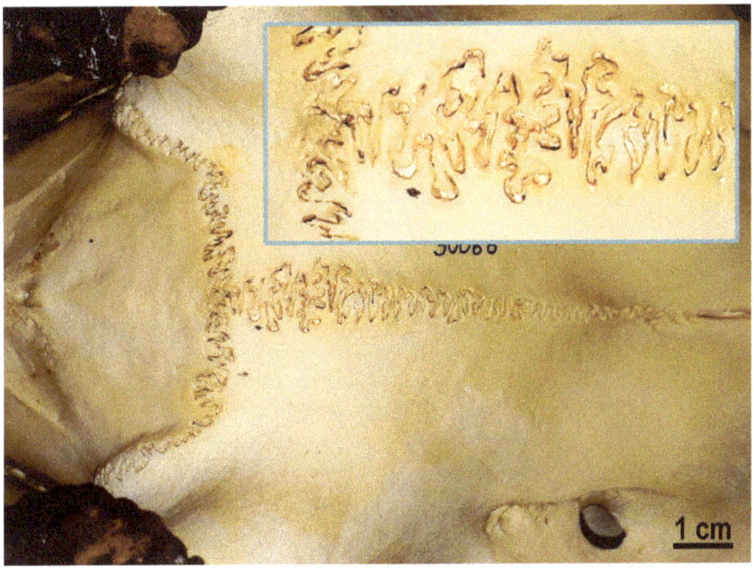

Figure 5. Frontal skull sutures of the male red deer (*Cervus elaphus*). These sutures must withstand very high stresses during male contests. Inset: detailed view.

5. Toughening Potential of Combined Mechanical/Chemical Adhesives: Demonstration

Figure 6 shows the schema of the combination of dual attachment (mechanical and chemical), with mechanically interlocked fibrous structures and glue matrix. While fibrous adhesives are often based on soft attachment mechanisms, like friction between fibers or hydrogen bonds, the strength of combined interlocking-chemical adhesives is based on mechanical effects, in particular, reinforcing effect and also toughening mechanism, fiber bridging of defects.

In order to estimate the effect of the fibrous surfaces on the mechanical properties of the adhesive layer, the simple rule of mixture model was applied. Representing the fibers as vertical cylinders, one can estimate that the shear modulus of the polymer adhesive increases using:

$$\alpha = \frac{G_{reinf}}{G_m} = \frac{v_f G_f + v_m G_m}{G_m} \tag{1}$$

where v—content of fibers or polymer matrix; indices "f" and "m" mean fibers or matrix, respectively; G—shear modulus; "reinf"—index for reinforced material.

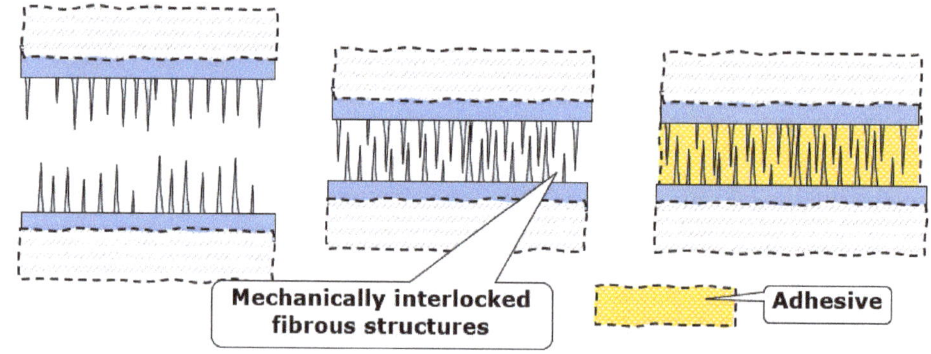

Figure 6. Schema of the combination of mechanical interlocking and adhesive attachment.

Assuming that the shear modulus of the adhesive is increased by α times, and using the formula by Rosen for shear buckling [23], one observes that such increase in the shear modulus of adhesive leads to the proportional increase in the buckling stress:

$$\sigma \sim \alpha/(1 - v_f) \quad (2)$$

Thus, the combination of fibrous and chemical interface joining increases the material properties of adhesives and reduces the likelihood of the blade buckling.

The engineered internal structures of the new adhesives will realize the local damage prevention mechanisms, local stress redistribution (more homogeneous stress distribution instead of high stress concentration, thanks to fibrous surfaces), and the damage tolerance mechanism (when the nanofibers or nanograss prevent the crack growth even when defects are formed, for instance, by so-called "fiber bridging" mechanism). Figure 7 shows the region of adhesive adjacent to the laminate, with a plane defect.

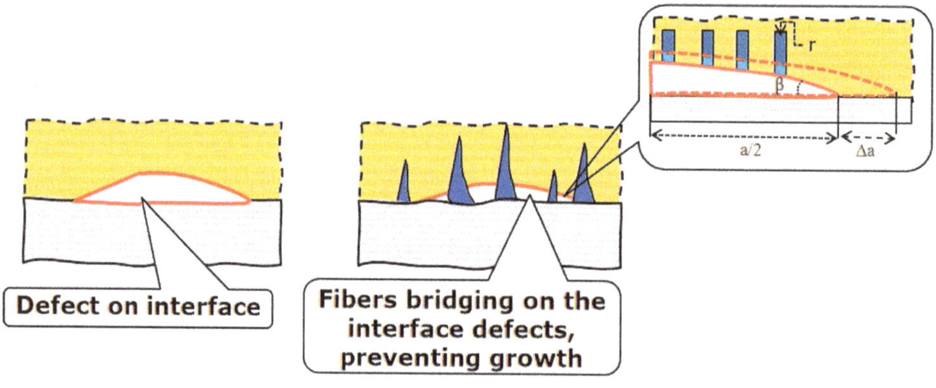

Figure 7. Schema of glued interface with defect (**left**), versus glued and fibrous interface with defect (**right**). Fiber bridging prevents defect growth.

In order to estimate the effect of the fibers available on interfaces on the energy balance of defect growth, we consider the fibers as vertical cylinders, with radius r and N fibers per unit area. The defect is considered a disc-shaped crack, with radius a and angle β at the tip. Assuming the defect extension by small value Δa, the surface of the unbridged crack increases from $\pi(2a)^2$ to $\pi(2a + \Delta a)^2$ and the new surface is:

$$\Delta S = \pi \Delta a(2a + \Delta a) \quad (3)$$

where ΔS—new formed surface; Δa—defect extension; and a—radius of disc-shaped crack.

However, in the case that the crack is bridged by fibers, which are attached chemically to the polymer, and the crack extension causes detachment of fibers, the additional surface is $2\pi rh$, where h–the length of fiber part which is detached from the polymer; $h = \Delta a \tan \beta$. Assuming that the extension of crack detaches only fibers located on one half of the disc-shaped crack, one can estimate the additional created surface as:

$$\Delta S = 2\pi r \Delta a \; N \tan \beta \frac{\pi a^2}{2} = 9.87 r N \Delta a \; a^2 \tan \beta \tag{4}$$

As mentioned above, the value N is given in unit per unit area. For the crack angle 5 to 30 degrees, the value $\tan\beta$ varies from 0.1 to 0.55. If one takes the new surface energy as constant value both for crack growth and fiber detachment, the fracture energy increases due to the fibrous surface, as compared to only polymer adhesive, using:

$$\eta = 1 + 9.87 \frac{r N \Delta a \; a^2 \tan \beta}{\pi \Delta a (2a + \Delta a)} = 1 + \frac{\pi r N \; a^2 \tan \beta}{2a + \Delta a} \tag{5}$$

For crack angle 10 degrees, $\Delta a = 0.1$ a, $a = 1$ mm, $N = 1000/\text{mm}^2$, $r = 0.001$ mm, and $\eta = 1.25$. Thus, by adding the fibrous mechanism to the adhesion of surfaces, the quite high increase in the adhesive toughness can be achieved.

In order to demonstrate the idea, test samples with probabilistic fastener structures were 3D printed and tested. Abaqus software v. 6.14 (SIMULIA, Johnston, RI, USA) was used to develop the 3D models of bonded cubes with different adhesive structures: (i) two cubes with long pins on the bonding surfaces (pin-pin), (ii) a cube with short pins and a cube with matching holes (pin-hole), and (iii) cubes with flat bonding surfaces (flat-flat). While the arrangement of pins in the pin–pin model results in the high friction between the pins and a tight fit, two adjoining cubes in the pin-hole model fit together with a loose fit. The dimensions of the cubes are 12 × 14 × 20 mm, the radius of pins is about 0.5 mm at the base, and the radius of holes defined on the cubes is around 0.6 mm. Figure 8a illustrates the two- and three-dimensional (2D and 3D) views of the developed models. A fused deposition modeling (FDM) 3D printer (Raise3D E2, Raise3D, Irvine, CA, USA) and a polylactic acid (PLA) filament (Raise3D, Irvine, CA, USA) were used to manufacture the developed models (Figure 8b). The printing settings are given in Table 2.

Table 2. Three-dimensional printing settings.

3D Printing Settings	
Filament type	Polylactic acid (PLA)
Produced by	Raise3D, Irvine, CA, US
Filament diameter (mm)	1.75
Nozzle diameter (mm)	0.4
Extrusion temperature (°C)	205
Bed temperature (°C)	55
Layer height (mm)	0.2
Fill pattern	Gyroid
Fill density (%)	30

Using the UHU polyvinyl acetate copolymer all-purpose adhesive (UHU GmbH & Co. KG, Bühl/Baden, Germany), the printed cubes were glued to each other. Models with pin–pin bonding shape were first interlocked into each other and then the interlocked pins were glued. After 48 h, the tensile behavior of the adhered cubes was quantified using a ZwickiLine uniaxial testing machine (Zwick Roell, Ulm, Germany) equipped with a 500 N load cell (Xforce P load cell, Zwick Roell). While one cube was fixed, the other one was pulled with the loading rate of 0.1 mm/s (Figure 9, Video S1). For each type of bonding shape (i.e., pin–pin, pin–hole, and flat–flat), 40 samples were prepared and tested in the same situation (n = 40). To study the influence of friction in pin–pin models, 20 additional pin–pin samples were tested with no adhesive (n = 20).

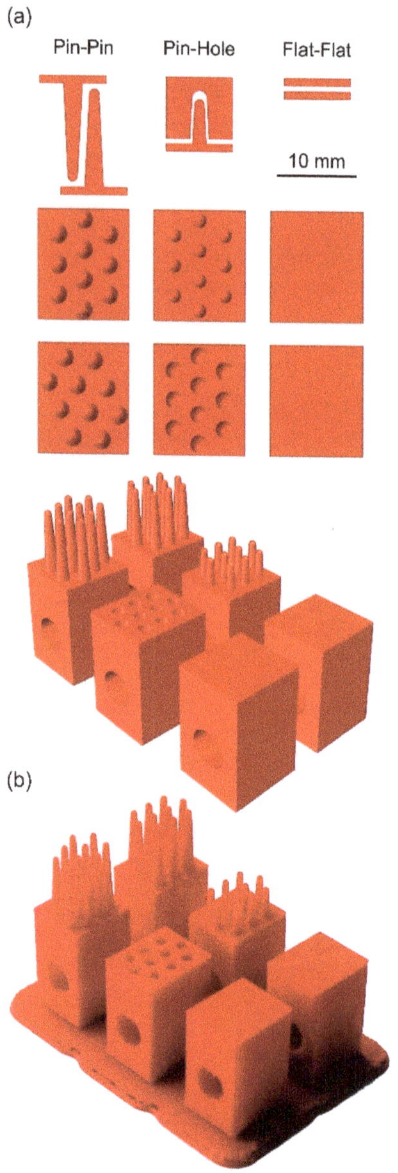

Figure 8. Modeling and manufacturing of the samples with various surface structures. (**a**) 2D schemas and 3D models of adjoining cubes with different bonding shapes, including long pins on the bonding surface (pin–pin), short pins and matching holes (pin–hole), and flat bonding surfaces (flat–flat). (**b**) 3D printed models.

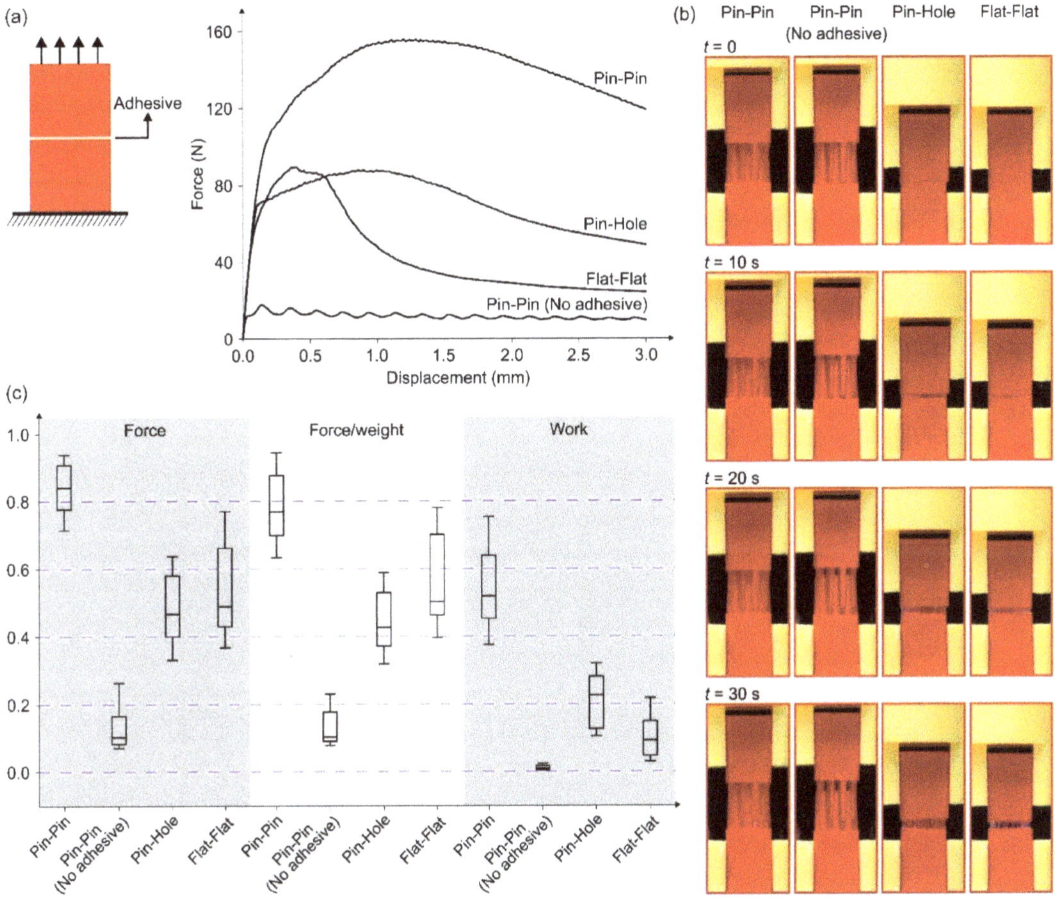

Figure 9. Testing results for the samples with four different bonding shapes. (**a**) Force-displacement curves, (**b**) stages of detachment, and (**c**) the boxplot comparing the maximum force, work, and force divided by the weights of the models.

The results of the experimental tests are available in Figure 9 and Table 3. Four random force-displacement curves related to four types of samples show the different performance of the cubes with different adhesive structures when subjected to tension (Figure 9a). The mechanical behavior of the models was investigated only up to their maximum load-bearing capacity. Besides the maximum force and work, the values of force divided by the weights of the samples were used to compare their behavior. The data were normalized by dividing the values of each comparing parameter (i.e., force, force/weight, and work) by the maximum value of that parameter and the boxplot was plotted (Figure 9c). The average values of the comparing parameters, besides their standard deviations, are presented in Table 3. In all tests conducted in this study, one cube was totally fixed in all directions, and the other cube was displaced vertically without any rotation.

It can be seen from Figure 9 that the pin–pin structure with additional adhesive ensures the largest fracture energy (work of detachment). This confirms the main assumption of this section: adhesive with both mechanical interlocking and chemical adhesion ensures much better attachment, than that with either only glue attachment (current common technology) or even with only interlocking. The pin–pin structure takes advantage of both friction and larger glued surface area to provide us with an efficient attachment mechanism.

Friction plays a crucial role by providing resistance to relative motion between the attached parts. The larger glued surface area further enhances the attachment by creating a greater adhesive bond, distributing the load across a larger surface, and improving the overall load-bearing capacity of the structure.

Table 3. Results of the experimental tests presented as mean ± standard deviation.

Model	Force (N)	Force/Weight	Work (N × mm)
Pin–Pin	155.7 ± 16.3	31,082 ± 4438	141.3 ± 36.5
Pin–Pin (No adhesive)	24.0 ± 11.6	5117 ± 2044	2.8 ± 1.9
Pin–Hole	89.1 ± 22.3	17,714 ± 3933	54.8 ± 23.3
Flat–Flat	100.0 ± 28.2	22,388 ± 5672	28.1 ± 19.45

Models with pin–pin bonding shapes were first interlocked with each other, and then the interlocked pins were glued together. A thin layer of glue covers the pins and fills the gaps. Considering the arrangement of the pins, there is some space in the middle of the two interconnected cubes that is not filled with glue. Both the thickness of the adhesive layer and the air trapped in the joints affect the performance of the structure. Future studies should focus on finding solutions that help control these parameters more precisely to increase the consistency and efficiency of the proposed strategy.

Furthermore, the type of glue used for adhesion is a factor that determines the functionality of the developed structures and their failure. In the pin–pin model, the pins act like cantilever beams in contact with each other. The maximum stress occurs at their base and is significantly influenced by the arrangement of the pins and the type of loading. While in this study, structures have been only loaded vertically for attachment and detachment, the risk of damage in pins was low. In pin–pin samples with adhesive as well, separation in the glued areas happened mostly before any damage to the pins. Future studies can investigate various failure modes in the pin–pin models by using different types of glues and adjusting structural parameters.

6. Detachable Interfaces of Composite Blades

As noted above, the problem of the end-of-life management of wind turbine blades is a critical problem of wind energy development [6,7]. The wind turbine blades are designed to sustain decades of service time and do not degrade. This means that they are very difficult to recycle (if recycling is considered as a material separation into constituents, or small parts). As discussed in Section 2, the main degradable parts of the blade are interfaces, adhesive layers, and other thin layers (e.g., coatings). The composite parts in the blades are in fact extremely durable and do not degrade sufficiently even after several decades of operation. This suggests that a blade can be disassembled and taken apart, the supposedly most durable parts (composite laminate parts) can be inspected, while the most damageable parts (adhesive layers, coatings) can be reapplied, thus allowing the further use of the blade materials.

The concept of structured, dual mechanism adhesive layer, presented in Section 4, opens the possibility of the disassemblable, separable blades. Figure 10 shows the schema of such separable adhesive, which includes the dissolution of matrix adhesive, and separation of mechanical interlocking surfaces of blade shall and spar. The matrix adhesive can be decomposed using reactive solvents, for instance, nitric acid, ammonia or glycol, below the critical temperature, or water or ethanol, near the critical temperature [7]. When the polymer matrix part of the adhesive is dissolved, only fibrous interlocking, plus friction remain as attachment mechanisms, and so the parts of a blade can be separated, inspected, and reused in future blades.

Comparing it with the results of previous section (Figure 9 and Table 3), one can see that the removal of adhesive from the pin–pin attached samples allows drastic reduction in the detachment work, almost to zero. Thus, the idea of strong and detachable adhesives is confirmed in the tests above.

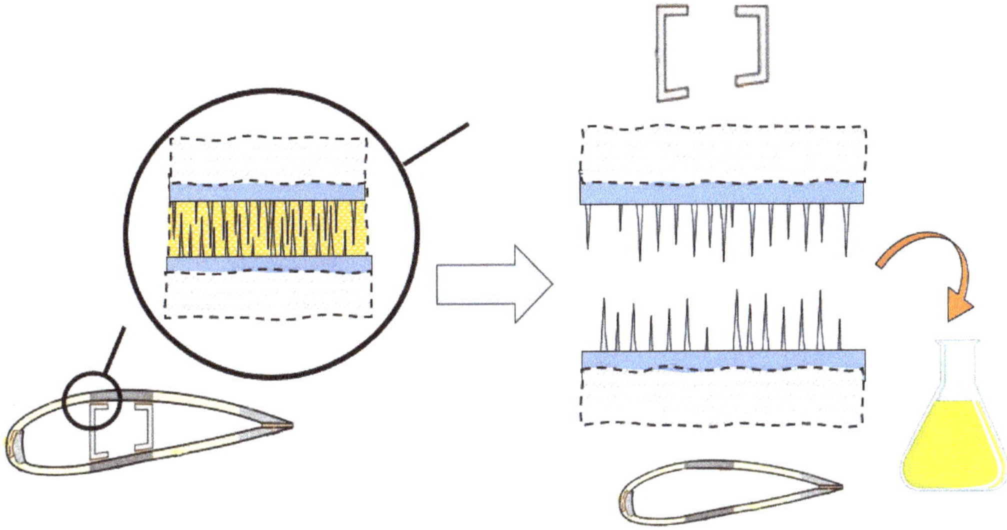

Figure 10. Schema of the separation of a wind turbine blade. After decommissioning with dissolution of adhesive, the fibrous surfaces become detachable allowing to separate the elements of blade structure. The blue arrow shows the transition from the "closed" interface to the debonded interface, with polymer matrix dissolved in yellow flask.

7. Conclusions

A new concept of extra-durable and sustainable wind turbine blades is presented based on a bio-inspired approach.

The analysis of most often observed damage mechanisms of wind turbine blades lead to the critical role of interfaces and adhesives for the durability and reliability of blades. Most damage mechanisms, e.g., buckling, trailing edge failure, coating detachment, are caused by weaker, damageable interfaces. Given the critical role of interfaces and adhesive layers for the durability of wind turbine blades, the development of extra-tough and separable adhesives should ensure the extended lifetime, reliability, and sustainability of wind turbine blades. The analysis of typical structural biological composites (e.g., nacre, shells) lead to the conclusion that their typical microscale structure includes layered, stiff platelets joined by tough, structured adhesive layers.

In this work, the idea is formulated to use a combination of mechanical and chemical attachment to develop new extra-strong adhesives. The high strength and toughness of new adhesives is achieved using mechanical interlocking of fibrous structures from each side, supported by chemical gluing. It is demonstrated that such adhesives show both higher shear modulus, lower likelihood of buckling, and higher damage tolerance. The potential of bio-inspired structured adhesives, which combine two adhesion mechanisms (i.e., the mechanical interlocking of fibrous surfaces and chemical adhesion) is discussed. The possibility of adhesive joint separation for re-use of blade parts, using two phase adhesives with solvable matrix component is presented.

Supplementary Materials: The following supporting information can be downloaded at: https://www.mdpi.com/article/10.3390/biomimetics8060448/s1, Video S1: Mechanical behavior of cubes with different bonding shapes under tension.

Author Contributions: Conceptualization, L.M.J. and S.N.G.; methodology, L.M.J., M.J. and S.N.G.; validation, M.J., J.K. and S.N.G.; investigation, L.M.J., M.J., J.K. and S.N.G.; writing—original draft preparation, L.M.J. and S.N.G.; writing—review and editing, L.M.J., M.J. and S.N.G.; visualization, L.M.J., M.J. and S.N.G.; supervision, L.M.J. and S.N.G.; project administration, L.M.J. and S.N.G. All authors have read and agreed to the published version of the manuscript.

Funding: L.M.J. acknowledges the financial support of the Innovation Foundation of Denmark in the framework of project "WiseWind: New generation of sustainable wind turbine blades", No. 2079-00004B, of the European Commission via Horizon project "Blades2Build: Recycle, repurpose and reuse end-of-life wind blades composites: A coupled pre- and co- processing demonstration plant", grant agreement101096437, and of the Ministry of Foreign Affairs of Denmark via Danida grant 19-M02-DTU "Maintenance and repair strategy for wind energy development". M.J. acknowledges the financial support provided by the "German Academic Exchange Service (DAAD)" (ID: 57684920).

Data Availability Statement: Data are available on request.

Conflicts of Interest: The authors declare no conflict of interest.

References

1. *Wind Energy in Europe: 2022 Statistics and the Outlook for 2023–2027*; WindEurope: Brussels, Belgium, 2022.
2. Frangoul, A. The Race to Roll out 'Super-Sized' Wind Turbines Is on. CNBC. 2022. Available online: https://www.cnbc.com/2022/04/13/green-energy-the-race-to-roll-out-super-sized-wind-turbines-is-on.html (accessed on 21 September 2023).
3. Williamson, R. Wind Turbine Failure Rates Are Rising—Has the Industry Gone Too Big, Too Fast? 9 February 2023, RenewEconomy. Available online: https://reneweconomy-com-au.cdn.ampproject.org/c/s/reneweconomy.com.au/wind-turbine-failure-rates-are-rising-has-the-industry-gone-too-big-too-fast/amp/ (accessed on 21 September 2023).
4. Mishnaevsky, L., Jr. Root causes and mechanisms of failure of wind turbine blades: Overview. *Materials* **2022**, *15*, 2959. [CrossRef] [PubMed]
5. Mishnaevsky, L., Jr.; Thomsen, K. Costs of repair of wind turbine blades: Influence of technology aspects. *Wind Energy* **2020**, *23*, 2247–2255. [CrossRef]
6. Mishnaevsky, L., Jr. Recycling of wind turbine blades: Recent developments. *Curr. Opin. Green Sustain. Chem.* **2023**, *39*, 100746.
7. Mishnaevsky, L., Jr. Sustainable end-of-life management of wind turbine blades: Overview of current and coming solutions. *Materials* **2021**, *14*, 1124. [CrossRef] [PubMed]
8. Liu, P.; Barlow, C.Y. Wind turbine blade waste in 2050. *Waste Manag.* **2017**, *62*, 229–240. [CrossRef] [PubMed]
9. Lichtenegger, G.; Rentizelas, A.A.; Trivyza, N.; Siegl, S. Offshore and onshore wind turbine blade waste material forecast at a regional level in Europe until 2050. *Waste Manag.* **2020**, *106*, 120–131. [CrossRef]
10. Mishnaevsky Jr., L.; Bendixen, B.; Mahajan, P.; Fæster, S.; Johansen, N.F.-J.; Paul, D.; Fraisse, A. Repair of wind turbine blades: Costs and quality. *J. Phys. Conf. Ser.* **2022**, *2265*, 032032. [CrossRef]
11. Mishnaevsky, L., Jr. Nanostructured interfaces for enhancing mechanical properties of materials: Computational micromechanical studies. *Compos. B* **2015**, *68*, 75–84. [CrossRef]
12. Song, F.; Soh, A.K.; Bai, Y.L. Structural and mechanical properties of the organic matrix layers of nacre. *Biomaterials* **2003**, *24*, 3623–3631. [CrossRef] [PubMed]
13. Heepe, L.; Xue, L.; Gorb, S.N. (Eds.) *Bio-Inspired Structured Adhesives: Biological Prototypes, Fabrication, Tribological Properties, and Novel Concepts*; Springer: Berlin/Heidelberg, Germany, 2017; p. 324.
14. Dwivedi, P.; Singh, K.; Chaudhary, K.; Mangal, R. Biomimetic polymer adhesives. *ACS Appl. Polym.* **2022**, *4*, 4588–4608. [CrossRef]
15. Rajabi, H.; Bazargan, P.; Pourbabaei, A.; Eshghi, S.; Darvizeh, A.; Gorb, S.N.; Taylor, D.; Dirks, J.H. Wing cross veins: An efficient biomechanical strategy to mitigate fatigue failure of insect cuticle. *Biomech. Model. Mechanobiol.* **2017**, *16*, 1947–1955. [CrossRef] [PubMed]
16. Aradhana, R.; Mohanty, S.; Nayak, S.K. High performance epoxy nanocomposite adhesive: Effect of nanofillers on adhesive strength and degradation kinetics. *Int. J. Adhes. Adhes.* **2018**, *84*, 238–249. [CrossRef]
17. Zhou, H.; Liu, H.Y.; Zhou, H.; Zhang, Y.; Gao, X.; Mai, Y.W. On adhesive properties of nano-silica/epoxy bonded single-lap joints. *Mater. Des.* **2016**, *95*, 212–218. [CrossRef]
18. Cho, H.; Wu, G.; Christopher Jolly, J.; Fortoul, N.; He, Z.; Gao, Y.; Jagota, A.; Yang, S. Intrinsically reversible superglues via shape adaptation inspired by snail epiphragm. *Proc. Natl. Acad. Sci. USA* **2019**, *116*, 13774–13779. [CrossRef] [PubMed]
19. Kaiser, I.; Tan, C.; Tan, K.T. Bio-inspired patterned adhesive single-lap joints for CFRP and titanium. *Composites B* **2021**, *224*, 109182. [CrossRef]
20. Gorb, S.N.; Popov, V.L. Probabilistic fasteners with parabolic elements:biological system, articial model and theoretical considerations. *Philos. Trans. R. Soc. Lond. A* **2002**, *360*, 211–225. [CrossRef] [PubMed]
21. Sánchez-Villagra, M.R. Suture closure as a paradigm to study late growth in recent and fossil mammals: A case study with giant deer and dwarf deer skulls. *J. Vertebr. Paleontol.* **2010**, *30*, 1895–1898. [CrossRef]

22. White, H.E.; Goswami, A.; Tucker, A.S. The Intertwined Evolution and Development of Sutures and Cranial Morphology. *Front. Cell Dev. Biol.* **2021**, *9*, 690. [CrossRef] [PubMed]
23. Rosen, B.W. *Mechanisms of Composite Strengthening, Fiber Composite Materials*; American Society of Metals: Cleveland, OH, USA, 1964; pp. 37–75.

biomimetics

MDPI

Article

Adhesive κ-Carrageenan Hydrogels by Polyphenol Intervention

Han-Yeol Yang [1,2], Jeongin Seo [1], Woongrak Choi [1], Eunu Kim [1], Sangho Yeo [1], Soeun Park [3] and Haeshin Lee [1,2,*]

1 Department of Chemistry, Korea Advanced Institute of Science and Technology (KAIST), Daejeon 34141, Republic of Korea
2 R&D Center, Polyphenol Factory Co., Ltd., Daejeon 34141, Republic of Korea
3 Department of Pharmacy, Kyung Hee University, Seoul 02447, Republic of Korea
* Correspondence: haeshin@kaist.ac.kr

Abstract

Kappa-carrageenan (κ-CRG) forms thermo-reversible physical hydrogels via a coil–helix transition and helix bundling, but its sulfate-driven electrostatic repulsion limits mechanical robustness and control over aqueous disintegration. Here, we show that plant-derived polyphenols reprogram κ-CRG gel through sulfate-directed binding in a structure-dependent manner. Tannic acid (TA) selectively engages κ-CRG sulfate groups, yielding transparent gels and a >5-fold increase in storage modulus, whereas the same TA triggers turbidity and precipitation in sulfate-free agarose, supporting sulfate-mediated specificity. Using monomeric pyrogallol as a galloyl analogue, we demonstrate that monovalent interactions partially reinforce κ-CRG but lack cooperative stabilization. Intervention timing further separates mechanism. Pyrogallol produces pathway-dependent mechanics and gelation temperature, while TA is stage-insensitive, consistent with multivalent network annealing. In simulated gastric/intestinal fluids, pyrogallol/κ-CRG gels retain morphology longer, whereas TA/κ-CRG ones disintegrate rapidly yet exhibit strong adhesion to rough substrates and human skin. These findings provide a fully food-grade route to tune κ-CRG mechanics, thermal behavior, adhesion and programmed disintegration.

Keywords: κ-carrageenan; tannic acid; pyrogallol; degradation; adhesive hydrogel

1. Introduction

Kappa-carrageenan (κ-CRG) is a sulfated galactan composed of repeating β-D-galactose-4-sulfate and 3,6-anhydro-D-galactose and extracted from red seaweeds that forms thermo-reversible physical hydrogels through a well-established mechanism [1]. At room temperature, κ-CRG exists as a self-supporting gel, whereas heating to around 90 °C disrupts the gel network and converts the polymer chains into a random-coil sol state. Upon cooling from this hot-sol state, individual chains undergo a random coil-to-helix transition, followed by aggregation of helices into junction zones that percolate into a three-dimensional network. This coil–helix transition and subsequently helix–helix association are highly sensitive to temperature, ionic environment, and polymer concentration, and they dictate the resulting gel stiffness, brittleness, and thermal hysteresis [2,3]. Owing to their biocompatibility and reversible gelation, κ-CRG hydrogels have been widely used as food thickeners and gelling agents [4], and have also been explored for drug delivery, wound dressings, and soft tissue engineering [5,6]. However, single-network κ-CRG gels are often mechanically fragile [7], show limited resistance to cyclic deformation, and offer only coarse control over degradation kinetics, which constrains their performance in demanding biomedical and cosmetic environments as well as in structured-food applications that require robust yet transient gels [8].

To overcome these limitations, considerable effort has been devoted to reinforcing κ-CRG-based hydrogels through chemical modification, double-network design [9], ion-mediated stabilization, and blending with synthetic or natural polymers such as poly(vinyl alcohol) (PVA) [10,11]. Recent studies and reviews continue to highlight the versatility of κ-carrageenan-based hydrogels in applications such as drug delivery, wound healing, and advanced biomaterial design, while also emphasizing the importance of controlling network stabilization and mechanical performance in these systems [12–14]. Although these approaches can enhance stiffness, toughness, and application-specific functionality, they may also require added salts, additional polymer networks, or complex formulations. This disconnect between mechanical reinforcement and formulation simplicity highlights the need for alternative strategies that preserve the naturally derived character of κ-CRG-based gels while enabling tunable mechanical and degradation properties. In this context, tannic acid (TA)—a plant-derived polyphenol composed of multiple galloyl units and widely present in foods and beverages—represents an attractive candidate crosslinker [15].

Beyond polysaccharide systems, TA has also been shown to act as a "molecular glue" for polyanionic biopolymers such as DNA. In a DNA hydrogel system termed "TNA gel" (TA + DNA), TA induces gelation by reversibly bridging the negatively charged phosphodiester backbone, without requiring complementary base-pairing or covalent modification of DNA [16]. In that design, the galloyl-rich architecture of TA mediates multivalent, non-covalent interactions with phosphate groups, yielding a hydrogel that is extensible upon pulling, adhesive to tissues, and intrinsically degradable because the ester linkages within TA gradually hydrolyze. This work demonstrated that a naturally derived polyphenol can directly engage an anionic backbone to construct mechanically robust yet degradable networks. Motivated by this precedent, we hypothesized that TA could similarly interact with the densely sulfated backbone of κ-CRG, forming sulfate-gallol physical crosslinks analogous to the phosphate–gallol interactions. On this basis, we set out to test whether such sulfate–gallol interactions could be harnessed to modulate both the mechanical reinforcement and degradation behavior of κ-CRG hydrogels.

Herein, we report a systematic investigation into how TA modulates the coil–helix transition, helix–helix association, and resulting network mechanics of κ-CRG hydrogels. We show that TA selectively interacts with the sulfate groups along the κ-CRG backbone, forming multivalent sulfate–galloyl physical crosslinks that tighten helix–helix junctions, increase junction-zone density, and substantially reinforce the three-dimensional gel network. To decouple the contribution of individual galloyl motifs from the multivalent architecture of TA, we further employed monofunctional pyrogallol as a comparison model. Unlike TA, pyrogallol alone is unable to generate extensive sulfate-bridging but instead partially inserts into nascent proto-helical regions, locally perturbing helix cohesion and altering thermal stability. This contrast highlights that the reinforcing capability of TA arises intrinsically from its multivalent galloyl organization rather than from galloyl chemistry alone. By integrating TA–κ-CRG interactions with pyrogallol-based comparisons, our study demonstrates that polyphenol–sulfate binding offers a previously underexplored and fully natural strategy to tune the microstructure, gelation kinetics, mechanical robustness, and thermal behavior of κ-CRG hydrogels without compromising their food-grade character.

https://doi.org/10.3390/biomimetics11040290

2. Materials and Methods

2.1. Materials

κ-CRG (Genugel κappa-carrageenan type AKW; reported Mw range, 200–800 kDa) was purchased from CP Kelco (Atlanta, GA, USA). Tannic acid (≥96%; JIURUI Biology & Chemistry Co., Ltd., Hunan, China) was supplied by COSFINE Co., Ltd. (Gunpo, Republic of Korea). Pyrogallol (≥96%) was obtained from Sigma-Aldrich (St. Louis, MO, USA), and agarose (certified molecular biology grade) was obtained from Bio-Rad (Hercules, CA, USA). All reagents were used without further purification and deionized water (18.2 MΩ·cm, Milli-Q system, Millipore, Burlington, MA, USA) was used in all experiments.

2.2. Preparation of κ-CRG Gels

κ-CRG gels were prepared by dissolving κ-CRG powder in deionized water to a final concentration of 1% (*w*/*v*). This concentration was selected because it reliably formed self-supporting hydrogels with a suitable mechanical property, clearly exhibiting the effects of polyphenol addition on gel mechanics, thermal behavior, degradation, and adhesion. The solution was heated in a water bath at 90 °C or 70 °C until complete dissolution, then poured into designated molds and allowed to cool at room temperature to form gels. For κ-CRG gels containing additives, 2% (*w*/*v*) κ-CRG solution and aqueous solutions of TA or pyrogallol at desired concentrations were prepared separately and heated to the same temperature. The two solutions were then mixed at a 1:1 (*v*/*v*) ratio to obtain a final κ-CRG concentration of 1% (*w*/*v*). The resulting mixtures were cast into molds and cooled at room temperature to form gels.

2.3. Optical Characterization of Gel Transparency

For optical characterization, native κ-CRG, agarose only, TA (15%)/κ-CRG, TA (0.2%)/agarose, pyrogallol (6%)/κ-CRG, and pyrogallol (6%)/agarose gels were prepared directly in disposable cuvettes (1 cm path length). After gelation, their optical spectra were recorded using a UV–Vis spectrophotometer (HP 8453, Agilent Technologies, Santa Clara, CA, USA). The 500–600 nm region was used to compare relative transparency among the gel samples. For baseline correction, κ-CRG only and agarose only gels were used as blanks for the corresponding polyphenol-containing gel systems.

2.4. Rheological Characterization of Gel Networks

Gels prepared under different conditions were loaded onto the plate of a rheometer (MCR 302, Anton Paar, Graz, Austria) using a syringe. A parallel upper plate (25 mm diameter) was mounted, the gap was set to 1 mm, and the excess solution was trimmed. Rheological measurements were performed in frequency sweep mode, and the elastic behavior of the gels was analyzed based on the storage modulus (G′) measured at an angular frequency of 10 rad s^{-1}. Prior to frequency sweep measurements, amplitude sweep tests were conducted to identify the linear viscoelastic region of each gel. A strain of 0.5% was then selected for all subsequent rheological measurements, as this value remained within the linear viscoelastic region for κ-CRG only, pyrogallol (6%)/κ-CRG, and TA (15%)/κ-CRG gels. To investigate synergy in mechanical properties when simultaneously adding TA (5%) + KCl (30 mM) compared with an individual addition either TA (5%) or KCl (30 mM), four κ-CRG (1%) gels (native κ-CRG, TA/κ-CRG, K^+/κ-CRG, and K^+/TA/κ-CRG) were prepared.

https://doi.org/10.3390/biomimetics11040290

2.5. Fourier-Transform Infrared (FTIR) Analysis

For FTIR analysis, κ-CRG and κ-CRG:TA mixtures were prepared for spectroscopic comparison. FTIR spectra were collected using a Bruker Equinox-55 spectrometer (Bruker Optics, Ettlingen, Germany). Spectral changes were analyzed with particular attention to the D-galactose-4-sulfate region (840–850 cm^{-1}).

2.6. Determination of Gel–Sol Transition Temperature

The gel–sol transition behavior was evaluated using temperature ramp measurements on the rheometer with a parallel-plate geometry. After loading the gel onto the lower plate, the gap was set and excess sample was trimmed. Measurements were conducted while heating from 25 °C to 90 °C at a constant rate (1 °C/5 s), with a fixed strain (γ) of 0.5% and angular frequency (ω) of 10 rad s^{-1}. The normal force was maintained at 0 N throughout the measurement. Changes in G' and G'' were recorded as a function of temperature, and the temperature at which G' and G'' intersected was defined as the gel–sol transition temperature.

2.7. Disintegration Behavior in Simulated Gastrointestinal Conditions

κ-CRG (1% *w*/*v*) solutions containing either pyrogallol (6%, *w*/*w* relative to the κ-CRG solution) or TA (15% *w*/*w*) were dispensed into a 24-well plate (1 g per well, well diameter 15.5 mm). The solutions were allowed to cool at room temperature to form gels, yielding gel discs approximately 15.5 mm in diameter and ~5 mm in height. For the disintegration test, the gel discs were immersed in 5 mL of simulated aqueous media and incubated at 37 °C under agitation (100 rpm). Two media were used: simulated gastric fluid (SGF, pH 1.2) and simulated intestinal fluid (SIF, pH 6.8). The SGF solution was prepared by dissolving NaCl (2 g L^{-1}) in deionized water followed by adjustment to pH 1.2 with 5 M HCl. The SIF solution consisted of 100 mM sodium phosphate buffer, with the pH adjusted to 6.8 using 5 M NaOH. The morphological changes and disintegration behavior of the gel discs were monitored over time. To quantitatively evaluate gel disintegration, the diameter of each gel disc was measured from top-view images taken at predetermined time points during incubation. The images were analyzed using ImageJ software (V1.54g, National Institutes of Health, Bethesda, MD, USA).

2.8. Adhesive Properties of Polyphenol/κ-CRG Gels

κ-CRG (1%), TA (15%)/κ-CRG (1%), and pyrogallol (6%)/κ-CRG (1%) gels were prepared using either cylindrical molds (35 mm diameter, 10 mm height) or rectangular molds (6545 mm^3: 35 mm × 55 mm × 7 mm). The gel specimens were placed on sandpaper (P100) or on the forearm skin, after which the substrates were inverted to assess adhesion retention. If the gel remained attached after inversion, additional mechanical stimuli were applied by tapping the substrate or by shaking the arm vertically to evaluate forced detachment. For quantitative evaluation of adhesion strength, lap shear tests were performed using sandpaper (P100) as the adherend substrate. Sandpaper strips were cut into 70 mm × 20 mm pieces. The gel sample was placed between two sandpaper strips, which were overlapped over an area of 20 mm × 20 mm so that the gel was positioned at the bonding interface. The adhered assemblies were then tested using a universal testing machine (EZ-SX, Shimadzu, Kyoto, Japan) at a crosshead speed of 10 mm min^{-1}. The maximum load at failure was recorded, and the lap shear adhesion strength was calculated by dividing the maximum load by the overlap area. The results were expressed in kPa.

All quantitative measurements were performed in triplicate, and the data are presented as mean ± standard deviation (SD).

https://doi.org/10.3390/biomimetics11040290

3. Results and Discussion

3.1. Sulfate-Mediated Interaction of TA Governs Divergent Gelation Behaviors of κ-CRG and Agarose

Figure 1 contrasts gelation behavior of κ-CRG and agarose because the two polysaccharides share a related galactan backbone structure yet differ critically in charge density. κ-CRG contains sulfate substituents that introduce strong anionic charges within networks, whereas agarose lacks sulfate groups. Thus, agarose gelation can rely effectively on helix–helix association supported by nonpolar (hydrophobic) interactions and dense hydrogen-bonding (Figure 1a). Although κ-CRG undergoes gelation through a similar helix-based association mechanism, electrostatic repulsion arising from its sulfate groups is expected to hinder close helix–helix packing and thereby reduce overall mechanical robustness relative to agarose. Consistent with the structural difference, the storage modulus (G') of a 1% agarose gel is substantially higher (4560 ± 223 Pa) than that of a 1% κ-CRG gel (294 ± 10 Pa) (Figure 1b), indicating that agarose forms a rigid percolated network under the same concentration. These baseline data establish κ-CRG as a mechanically weaker but highly tunable platform whose modulus is constrained by charge-driven repulsion.

To study how TA influences the gelation of the two polysaccharides, TA was introduced over a broad concentration range. For agarose, the incorporation of only trace amounts of TA (0.2–0.4 wt%) was sufficient to produce turbid gels, and a slightly higher TA content (0.6 wt%) completely inhibited gel formation, yielding precipitated aggregates instead (Figure 1c, lower row). This behavior indicates that TA interacts strongly with the agarose galactan backbone, and that even a modest increase in TA concentration can disrupt the intrinsic helix–helix domain formation required for agarose gelation, thereby driving macroscopic phase separation and aggregation. In contrast, κ-CRG retained its characteristic transparent gel state even in the presence of 5 wt% TA—approximately ninefold higher than the TA level that abrogated agarose gelation (Figure 1c, upper row). Further increasing TA up to 15 wt% did not measurably compromise gelation or optical clarity. The optical distinction between the two systems was further supported by UV–Vis measurements, which confirmed optical transparency in TA/κ-CRG ($A_{550} = -0.16$) and pronounced turbidity in TA/agarose ($A_{550} = 1.66$) (Figure 1d). The slight negative value in absorption might be due to light scattering of solid gels. Because the principal structural distinction between agarose and κ-CRG is the presence of sulfate groups tethered to the backbone, the large difference in gelation/aggregation outcomes upon TA addition is most consistently explained by preferential TA binding with the sulfate functionalities in κ-CRG (Figure 1e). Such sulfate-directed binding would effectively sequester TA away from extensive direct binding to the neutral galactan backbone, thereby suppressing the rapid turbidity increase and precipitation observed in the agarose/TA system. We further quantified the viscoelastic reinforcement of the TA/κ-CRG hydrogels by measuring the storage modulus (G') using an oscillatory rheometer (Figure 1f). All reported G' values were obtained at a strain of 0.5%, which belongs to the linear viscoelastic region for all tested gel formulations (Figure A1). Notably, G' increased in a TA concentration-dependent manner (5–15%), indicating progressive network strengthening with increasing TA content. In particular, at 15 wt% TA, the κ-CRG gel exhibited a G' value more than fivefold higher (1632 ± 31 Pa) than that of the negative control (κ-CRG only) (294 ± 12 Pa). This mechanical enhancement is consistent with the interpretation that TA preferentially associates with κ-CRG sulfate groups, thereby partially neutralizing the intrinsic electrostatic repulsion between anionic helices. Reduced inter-helix repulsion would facilitate closer helix–helix packing and/or enlargement of junction zones, effectively increasing the crosslinking density of the physically crosslinked network and manifesting macroscopically as a higher storage modulus.

https://doi.org/10.3390/biomimetics11040290

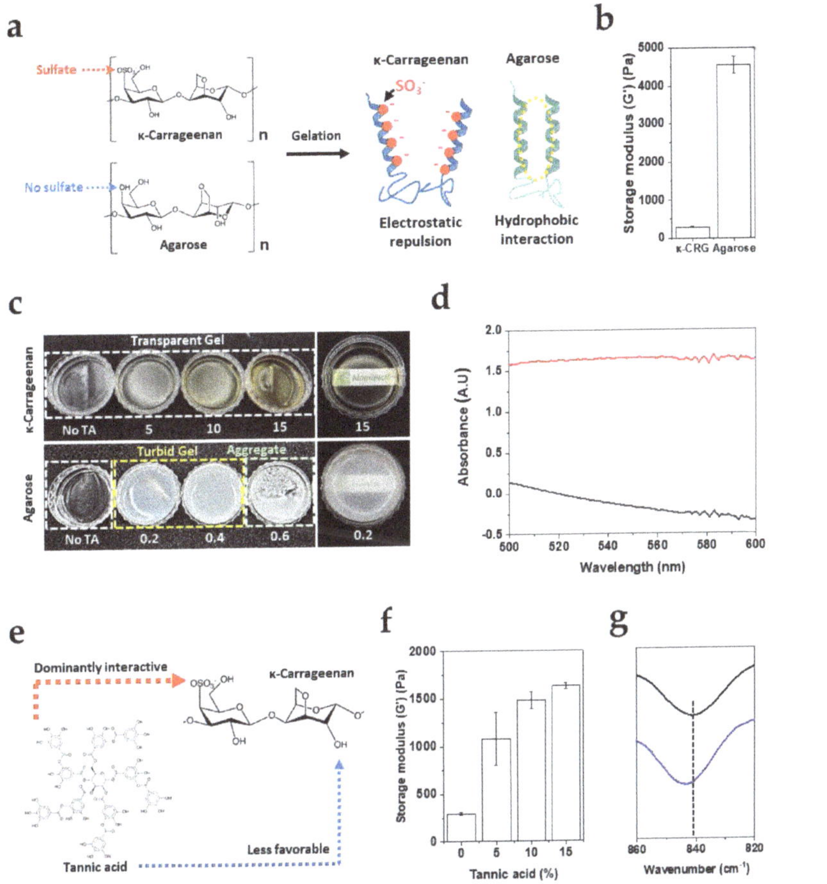

Figure 1. TA-mediated gelation behaviors of κ-CRG and agarose. (**a**) Schematic illustration comparing the molecular structures of κ-CRG and agarose. (**b**) Storage modulus (G′) of 1 wt% κ-CRG gels and 1 wt% agarose. (**c**) Optical appearance of κ-CRG and agarose gels prepared with increasing TA concentrations (wt%). (**d**) UV–Vis absorbance profiles of TA (15%)/κ-CRG (black) and TA (0.2%)/agarose (red) gels in the 500–600 nm region. (**e**) Proposed interaction scheme illustrating preferential binding of TA to sulfate groups in κ-CRG. (**f**) Storage modulus (G′) of κ-CRG gels as a function of TA concentration. (**g**) FTIR spectra of κ-CRG (black) and κ-CRG:TA ([κ-CRG]:[galloyl] = 1:14.7) (blue) in the D-galactose-4-sulfate region (840–850 cm^{-1}), with the dashed line indicating 841.3 cm^{-1}. Quantitative data in (**b**,**f**) are presented as mean ± SD (n = 3).

To provide evidence explaining the proposed sulfate-associated interaction between κ-CRG and the hydroxyl group of TA, FTIR spectroscopy was performed using κ-CRG and κ-CRG:TA mixture corresponding to [κ-CRG]:[galloyl] = 1:14.7. Under this condition, the peak position at 841.3 cm^{-1} was shifted to 843.2 cm^{-1} (Figure 1g). The range of wavenumber, 840–850 cm^{-1}, belongs to the well-known region assigned to the sulfate attached to D-galactose [17]. These results support sulfate-associated intermolecular interactions between κ-CRG and the hydroxyl groups of TA.

3.2. Monovalent Galloyl Interactions Partially Reinforce κ-CRG Gels but Lack Cooperative Network Stabilization

To distinguish whether the κ-CRG reinforcement observed with TA arises primarily from (i) additive, local charge screening at sulfate groups, or (ii) a multivalency-amplified,

https://doi.org/10.3390/biomimetics11040290

cooperative mechanism enabled by the clustered gallol units within TA, pyrogallol—a monomeric gallol lacking the multivalency of TA—was selected as a model small molecule and introduced at concentration of 2–6 wt%, chosen to provide gallol-equivalent inputs comparable to those used for TA. Qualitatively, both pyrogallol/κ-CRG and pyrogallol/agarose formed self-supporting, optically transparent gels across all tested pyrogallol concentrations (Figure 2a). Quantitatively, UV–Vis measurements showed $A_{550} = -0.13$ for pyrogallol/κ-CRG and $A_{550} = 0.19$ for pyrogallol/agarose (Figure 2b). Importantly, this result contrasts with the agarose/TA system shown in Figure 1c, where even trace amounts of TA induced turbidity and precipitation. The preservation of transparency and gel integrity in agarose upon pyrogallol addition suggests that monomeric gallol motifs, when presented without a multivalent scaffold, do not substantially disrupt helix formation or promote mesoscale complexation/phase separation in a neutral galactan matrix. In other words, the TA-driven turbidity and aggregation observed in agarose are consistent with multivalency-enabled bridging or condensation effects, rather than with a generic gallol–backbone interaction alone.

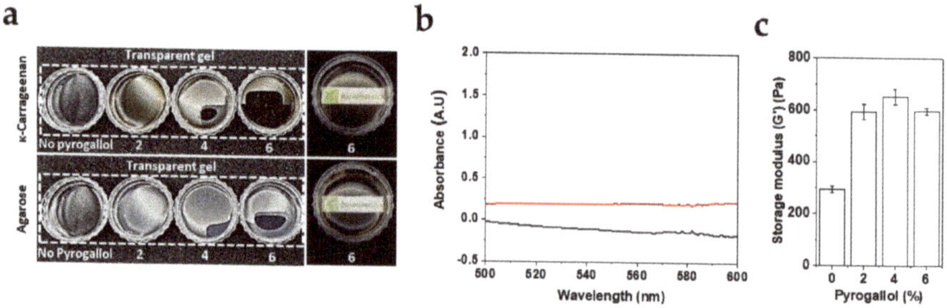

Figure 2. Effects of pyrogallol on the structure and mechanical properties of κ-CRG gels. (**a**) Optical appearance of κ-CRG and agarose gels prepared with increasing concentrations of pyrogallol. (**b**) UV–Vis absorbance profiles of pyrogallol (6%)/κ-CRG (black) and pyrogallol (6%)/agarose (red) gels in the 500–600 nm region. (**c**) Storage modulus (G′) of pyrogallol/κ-CRG gels as a function of pyrogallol concentration. Data are presented as mean ± SD ($n = 3$).

Consistent with selective strengthening of κ-CRG networks, oscillatory rheology revealed that pyrogallol increases the storage modulus (G′) of κ-CRG gels in a concentration-dependent manner, followed by an apparent saturation at higher pyrogallol loadings (Figure 2c). Relative to the pyrogallol-free control (~300 Pa), G′ increased markedly at 2–4 wt% pyrogallol to 600–700 Pa (593.15 ± 28.32 for 2 wt% and 653.21 ± 29.55 for 4 wt%) and then remained near this level at 6 wt%, indicating that a finite population of reinforcing interaction sites becomes progressively occupied and approaches a plateau (596.89 ± 12.72 Pa). These results support the conclusion that gallol motif alone can reinforce κ-CRG, plausibly associating with sulfate-rich domains and partially mitigating inter-helix electrostatic repulsion, thereby facilitating closer helix–helix association. However, when benchmarked against TA at comparable gallol equivalents (Figure 1f), pyrogallol yields substantially only a modest modulus enhancement, implying that charge screening by individual gallol units is not sufficient to recapitulate the full strengthening by TA. We attribute this gap to TA's multivalency [18].

3.3. Polyphenol Intervention Stage Controls Pathway-Dependent vs. Pathway-Independent Gelation

Gelation of κ-CRG proceeds through a stepwise process in which a high-temperature random-coil state undergoes a coil–helix transition upon cooling, followed by the development of inter-helical associations that form a three-dimensional network (Figure 3a, top).

https://doi.org/10.3390/biomimetics11040290

Because κ-CRG reinforcement observed with polyphenols in Figures 1 and 2 was attributed to sulfate-associated interactions, we next examined whether the intervention stage—i.e., whether the polyphenol encounters κ-CRG before or during helix development—alters the final network architecture and properties. Temperature-ramp rheology revealed a narrow plateau for 1 wt% κ-CRG between 66.6 and 71.7 °C, which we define as a proto-helical regime, whereas ≥90 °C corresponds to a fully random-coil regime (Figure 3a). Accordingly, polyphenols (TA or pyrogallol) were introduced either at 90 °C or 70 °C while maintaining identical composition and processing otherwise (Figure 3b).

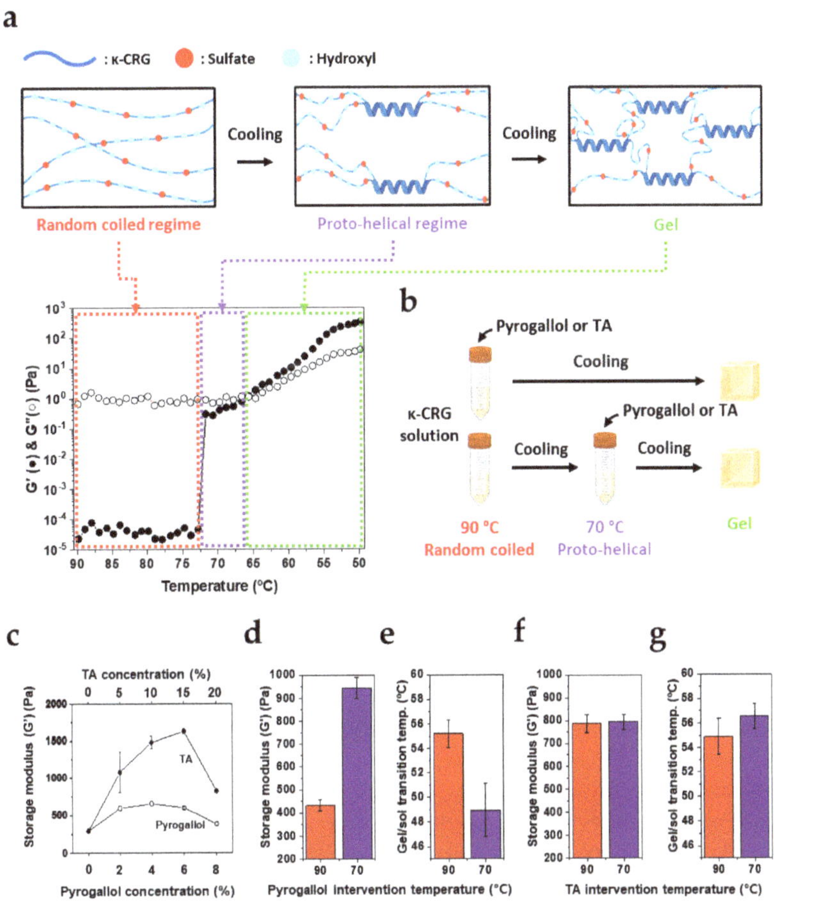

Figure 3. Stage-dependent effects of mono- and multivalent polyphenols on κ-CRG gelation. (**a**) Schematic illustration of κ-CRG gelation during cooling from a random-coil state to a proto-helical state and finally to a gel network, together with temperature-dependent rheological evolution defining the structural regimes. (**b**) Experimental design: polyphenols (Pyrogallol or TA) were introduced either in the random-coil regime (90 °C) or proto-helical regime (70 °C) prior to cooling and gel formation. (**c**) Storage modulus (G′) as a function of polyphenol concentration. (**d**) Storage modulus (G′) and (**e**) gel–sol transition temperature (°C) of pyrogallol. (**f**) Storage modulus (G′) and (**g**) gel–sol transition temperature (°C) of TA. Data in (**c**–**g**) are presented as mean ± SD (*n* = 3).

To select polyphenol concentrations for the temperature-dependent experiments (90 °C vs. 70 °C), we first monitored G′ as a function of polyphenol content. Across concentrations, both TA and pyrogallol exhibited a characteristic reinforcement-turnover profile. G′ increased at low polyphenol loadings, reached a maximum (~6 wt% for pyrogallol

https://doi.org/10.3390/biomimetics11040290

and ~15 wt% for TA), and then decreased at higher concentrations (Figure 3c). This trend is consistent with sulfate-mediated strengthening that eventually saturates, followed by structural perturbation of helix formation and/or junction-zone organization once excess polyphenol begins to frustrate helix–helix association. Based on these profiles, we selected 8 wt% for pyrogallol and 20 wt% for TA to clearly probe whether the final network depends on the stage of polyphenol intervention.

Strikingly, the intervention stage exerted a strong effect for pyrogallol but not for TA. When pyrogallol was introduced in the random-coil regime (90 °C), the resulting gel exhibited $G' = 432.5 \pm 23.7$ Pa after cooling to room temperature. In contrast, introduction of pyrogallol in the proto-helical regime (70 °C) produced substantially stiff gel $G' = 944.4 \pm 44.9$ Pa (Figure 3d). This result supports the view that κ-CRG is a weak gel, in which elasticity is governed by a subset of long-lived, load-bearing junction zones rather than network-average connectivity [19]. Importantly, the lower modulus upon pyrogallol addition at 90 °C (432.5 ± 23.7 Pa) can be rationalized by early-stage binding. Pyrogallol association with κ-CRG chains in the random-coil state likely perturbs subsequent coil → helix conversion and/or the organization of nascent helices into mechanically stable junction domains, thereby limiting the formation of load-bearing junction zones during cooling. By contrast, when pyrogallol is added after partial ordering has already emerged (70 °C), a fraction of proto-helical segments may be less accessible to disruptive binding [20], allowing junction-zone development to proceed more effectively and yielding higher G' (944.4 ± 44.9 Pa).

Thermal signature of gelation corroborates this stage-dependent effect. Neat κ-CRG gelled at ~67 °C (Figure 3a), whereas pyrogallol addition depressed the gelation temperature to ~55 °C when introduced at 90 °C and further to ~49 °C when introduced at 70 °C (Figure 3e). The larger decrease upon 70 °C intervention suggests that, once proto-helical domains have formed, pyrogallol preferentially associates with the remaining exposed/less ordered chain regions, effectively modifying backbone interactions that are critical for propagating helix formation; consequently, helix development becomes less favorable, and gelation shifts to lower temperature. In contrast, TA produced little change in storage modulus (G') and gelation temperature converged to a transition near 55 °C, comparable to the pyrogallol system mixed at 90 °C (Figure 3f,g). This stage-insensitive behavior is consistent with TA's multivalency, which can promote chain-to-chain association through multipoint sulfate-associated interactions and thereby facilitate network organization even when introduced at different structural states [21]. Considering all results, Figure 3 demonstrates that monomeric pyrogallol imprints pathway dependence—with both stiffness and gelation temperature sensitive to the intervention stage—whereas multivalent TA behaves as a pathway-independent organizer, yielding similar thermal behavior while preserving robust network formation in κ-CRG gels.

3.4. Valency-Dependent Disintegration of Polyphenol/κ-CRG Gels in Aqueous Conditions

Next, we assessed the macroscopic stability of the κ-CRG gel network in aqueous environments by monitoring the morphological evolution of gel discs over time. Gels were prepared by mixing pyrogallol (6 wt%) or TA (15 wt%) with hot 1 wt% κ-CRG followed by cooling, molding into discs (15.5 mm diameter, ~5 mm height), and incubating at 37 °C in simulated gastric fluid (SGF, pH 1.2) or simulated intestinal fluid (SIF, pH 6.8) with shaking (100 rpm) (Figure 4a). To quantitatively monitor the disintegration kinetics of the gel, the changes in gel diameter over time were additionally analyzed from top-view images using ImageJ (Figure 4b–d, right panels).

https://doi.org/10.3390/biomimetics11040290

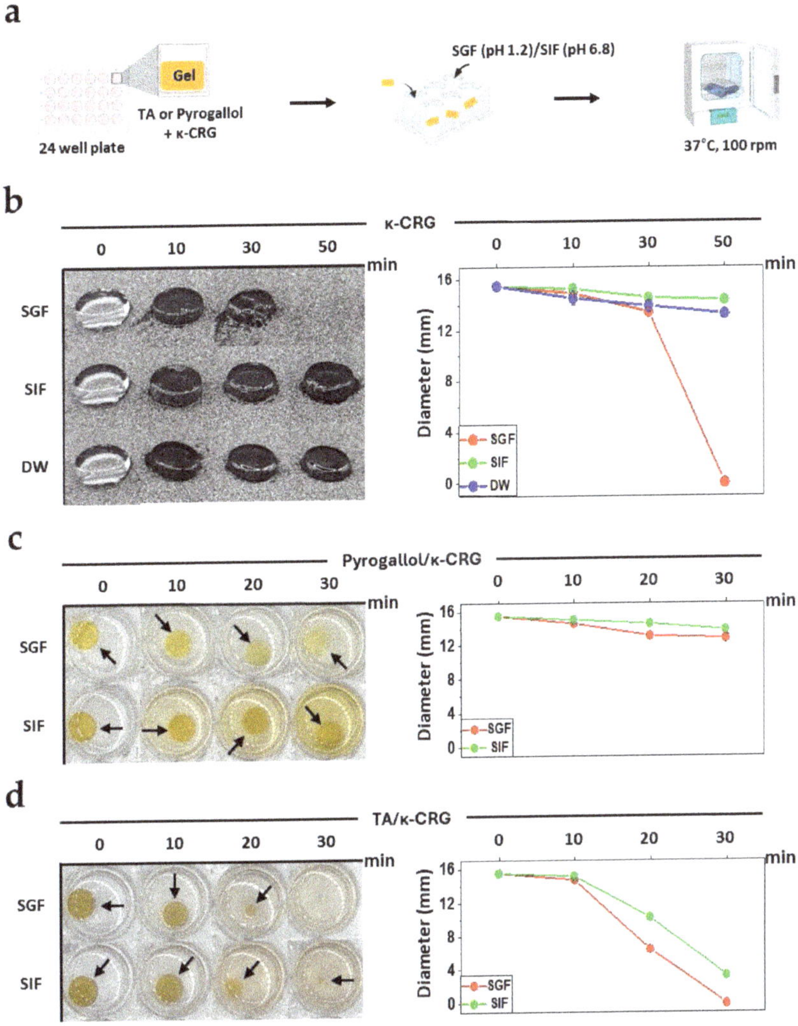

Figure 4. Disintegration behavior of κ-CRG gels in aqueous environments. (**a**) Schematic illustration of the experimental procedure. κ-CRG gels (15.5 mm diameter, 5 mm height) containing pyrogallol (6 wt%) or TA (15 wt%). Disintegration behaviors of κ-CRG only gels (**b**) in simulated gastric fluid (SGF), simulated intestinal fluid (SIF), and water (DW), pyrogallol/κ-CRG gels (**c**) in SGF and SIF, and TA/κ-CRG gels (**d**) in SGF and SIF. Arrows indicate the remaining gel discs at each time point. Quantitative results shown in (**b–d**) are presented as mean ± SD (n = 3).

We first established the baseline behavior of native κ-CRG gels without polyphenols. The rate of disintegration followed SGF > DW > SIF (Figure 4b). In SGF (≈35 mM sodium), gel discs rapidly disintegrated within ~30–50 min, whereas in SIF the gels retained their structure over the same time scale and were even more stable than in DW. This qualitative trend was also supported by the diameter analysis. In SGF, the gel diameter decreased to 13.4 ± 0.2 mm at 30 min, and no measurable gel disc remained at 50 min due to complete disintegration, whereas the gels in SIF and DW retained diameters of 14.4 ± 0.2 mm and 13.3 ± 0.1 mm, respectively, at 50 min (Figure 4b, right panel). This pH/ionic-strength dependence is consistent with reduced junction stability at low pH, where partial protonation

https://doi.org/10.3390/biomimetics11040290

of sulfate groups diminishes anionic coordination sites important for helix aggregation [22], while the sodium-rich environment in SIF (100 mM sodium phosphate buffer) can partially stabilize helix-bundle junctions.

Upon polyphenol incorporation, a clear valency-dependent divergence in stability emerged, with pyrogallol-containing gels being consistently more stable than TA-containing ones in both SGF and SIF (Figure 4c,d). In both media, pyrogallol/κ-CRG discs underwent only gradual reductions in diameter over the observation window (0–30 min), preserving macroscopic integrity throughout. The gel diameter decreased from 15.5 to 12.9 ± 0.3 mm in SGF (Figure 4c, right panel, red) and from 15.5 to 13.9 ± 0.1 mm in SIF over 30 min (Figure 4c right panel, green). In contrast, TA/κ-CRG discs exhibited rapid shrinkage immediately upon exposure to aqueous media, with pronounced size reduction. In SGF, the diameter of TA/κ-CRG gels decreased from 15.5 to 14.7 ± 0.3 mm at 10 min, to 6.4 ± 0.2 mm at 20 min, and no measurable gel disc remained at 30 min (Figure 4d, right panel, red). Similarly, the diameter of TA/κ-CRG gels was decreased from 15.5 to 10.3 ± 0.1 mm at 20 min and 3.3 ± 0.3 mm at 30 min (Figure 4d, right panel, green). The observed faster breakdown of TA/κ-CRG gels can be rationalized by the distinct network architectures inferred from Figures 1–3. Importantly, the increased initial G′ of TA/κ-CRG does not necessarily indicate greater resistance to aqueous disintegration. Pyrogallol primarily engages sulfate groups through localized, effectively monodentate interactions that partially reduce electrostatic repulsion yet preserve a substantial fraction of the native κ-CRG helix–helix junctions, allowing the network to retain global connectivity under hydrated conditions. In contrast, TA, owing to its multidentate architecture, promotes multipoint chain-to-chain bridging and dynamic bond exchange during gel formation [21], thereby markedly increasing gel stiffness while shifting network connectivity toward TA-mediated supramolecular junctions [23]. As a result, in aqueous environments, particularly in acidic SGF where native κ-CRG junctions are already destabilized, local dissociation or rearrangement of these TA-mediated junctions can directly disrupt percolation across the network and lead to rapid disintegration [24].

3.5. Multivalent TA-Mediated Surface Adhesion of κ-CRG Gels

The results presented in Figures 1–4 demonstrate that polyphenol/κ-CRG interactions can reorganize the gel network from local molecular binding to the supramolecular architecture of the entire system. In particular, TA forms multipoint interactions between κ-CRG chains through its multivalent galloyl groups, generating supramolecular junctions that simultaneously modulate the mechanical properties of the gel and its disintegration behavior in aqueous environments. Such interactions are not necessarily limited to the internal network structure but may also operate at the gel–surface interface. Indeed, polyphenols are known to form strong non-covalent interactions with various polymeric and biological surfaces, particularly through polyphenol–protein and polyphenol–polysaccharide interactions [25,26].

To test this hypothesis, the surface adhesion behavior of native κ-CRG gel, pyrogallol/κ-CRG gel, and TA/κ-CRG gel was compared. The gel discs were first placed on a rough sandpaper surface (P100), and their adhesion behavior was observed after the substrate was inverted so that the gels faced downward along the direction of gravity (Figure 5a). Both the native κ-CRG gel and the pyrogallol/κ-CRG gel detached immediately upon inversion. In contrast, the TA/κ-CRG gel remained firmly attached to the surface even when the substrate was completely inverted. The adhesion persisted even under repeated mechanical perturbations applied by tapping the substrate. These results indicate that the TA-containing gel forms stable interfacial interactions with rough solid surfaces. A similar trend was observed on human skin (Figure 5b). When gel discs of identical com-

https://doi.org/10.3390/biomimetics11040290

position were attached to the skin between the wrist and forearm, the native κ-CRG gel and pyrogallol/κ-CRG gel detached immediately. In contrast, the TA/κ-CRG gel remained adhered to the skin surface and maintained its attachment even after repeated arm movements. This observation demonstrates that the TA/κ-CRG gel exhibits strong adhesion not only on rigid substrates but also on compliant biological surfaces such as skin.

To quantitatively evaluate this adhesion behavior, lap shear tests were performed using sandpaper (P100) as the adherend substrate (Figure 5c). The shear strengths of native κ-CRG, pyrogallol/κ-CRG, and TA/κ-CRG gels were 3.4 ± 0.7, 2.2 ± 0.3, and 28.1 ± 5.0 kPa, respectively. Consistent with the qualitative adhesion results, TA/κ-CRG exhibited markedly higher adhesion strength than either native κ-CRG (~3.4 kPa) or pyrogallol/κ-CRG (~2.2 kPa). Inspection of the sandpaper surfaces after lap shear tests further revealed distinct failure modes (Figure A2). Native κ-CRG and pyrogallol/κ-CRG primarily exhibited adhesive failure at the gel–sandpaper interface, whereas TA/κ-CRG showed cohesive failure within the gel layer due to the aforementioned strong adhesion.

The origin of this adhesion can be attributed to the multivalent nature of TA. Even after forming the gel network with κ-CRG chains, a fraction of galloyl groups may remain exposed at the network surface. These exposed galloyl groups can interact with opposing surfaces through multipoint non-covalent interactions, thereby promoting stable interfacial adhesion. Such multivalent interfacial interactions allow the gel to maintain stable adhesion even under mechanical perturbations [27]. The large increase in lap shear adhesion strength observed for TA/κ-CRG is consistent with this interpretation and further supports that TA-mediated multivalent interactions operate not only within the bulk gel network but also at the gel–substrate interface. This behavior is consistent with the dynamic supramolecular network characteristics of TA-mediated junctions discussed in Figures 3 and 4.

The observed interfacial adhesion also suggests potential applications of TA/κ-CRG gels. When fabricated as thin sheets, the gels can form conformal coatings that adhere closely to skin surfaces. Through polyphenol-based interactions, the gels may effectively interact with biological macromolecules such as keratin in the stratum corneum. On this basis, the gels could potentially be utilized as exfoliating sheets that adhere to the skin surface (Figure 5d). In particular, the well-known protein affinity of TA [28] suggests that interactions with keratin may facilitate the removal of dead skin cells during sheet detachment. From a translational perspective, the TA content in skin-contacting κ-CRG gels should be selected by balancing interfacial adhesion and functional performance against possible irritation risk and cytocompatibility limitations, because tissue responses to TA can depend strongly on dose, formulation composition, and exposure conditions [29]. Although TA has been widely explored in skin-interfacing hydrogels owing to its polyphenol-mediated adhesion and additional antioxidant, antibacterial, and anti-inflammatory functions, these beneficial properties do not eliminate the need for careful dose optimization in topical formulations [15,28,30]. Prior dermatological literature has reported contact reactivity to topical tannic acid, with patch-test positivity observed at concentrations as low as 0.25% (aq.) in susceptible cases [30]. At the same time, the biological response to TA is likely formulation-dependent, and TA embedded within a polymeric network may behave differently from free TA applied directly to skin. Therefore, further studies will be required to define an appropriate TA dose window that preserves adhesion while minimizing barrier perturbation and irritation, ideally through standardized cytocompatibility assays, reconstructed epidermis or barrier-function testing, and human patch testing before biomedical or cosmetic translation.

https://doi.org/10.3390/biomimetics11040290

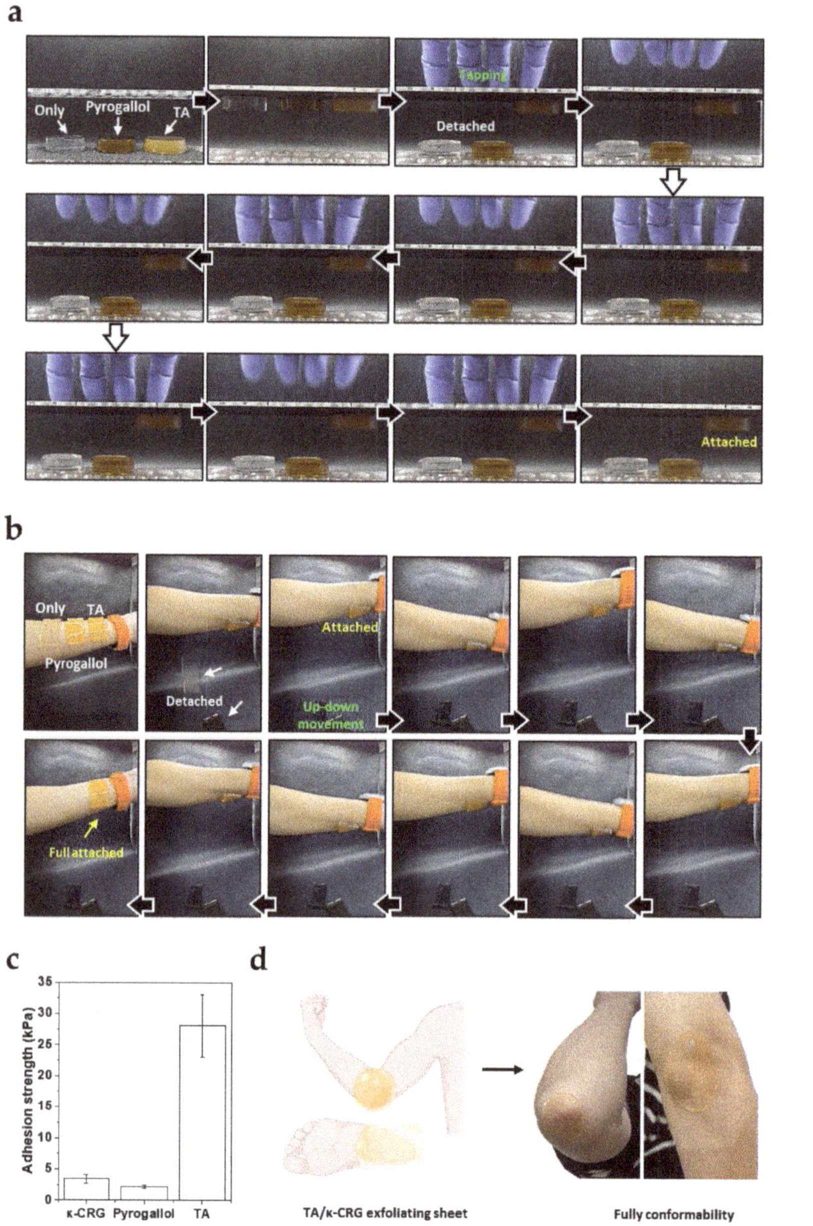

Figure 5. Surface adhesion and potential application of TA/κ-CRG gels. (**a**) Adhesion behavior of polyphenol/κ-CRG gels (35 mm diameter, 10 mm height) on a rough sandpaper (P100). (**b**) Adhesion behavior of gels (6545 mm^3: 55 × 17 × 7 mm) on human skin. (**c**) Lap shear adhesion strength of native κ-CRG (left bar) and polyphenol/κ-CRG gels measured on sandpaper (P100). Data are presented as mean ± SD (n = 3). (**d**) Schematic illustration of a potential application of TA/κ-CRG gel as a conformable exfoliating sheet.

In perspective of the context of existing κ-CRG reinforcement strategies, the present results show that polyphenol-mediated modulation of κ-CRG gels differs in character from conventional strategies used to control carrageenan gel networks. Existing approaches

https://doi.org/10.3390/biomimetics11040290

often rely on ion-mediated stabilization, such as K^+- or Ca^{2+}-assisted helix aggregation and junction-zone stabilization [31,32], or on polymer-based reinforcement through blending with additional polymers such as PVA, double-network design, or formation of an additional polymer-supported matrix [9–11]. By contrast, the present gelation strategies utilize small molecule polyphenols (pyrogallol = 126.1 Da and TA = 1701.2 Da) as new additives to form sulfate–polyphenol–sulfate bridges keeping the inherent helical domains intact [16]. Thus, the binding mode differences between K^+ and TA are allowed synergistic mechanical benefits in case of simultaneous addition of K^+ and TA. Upon the addition of potassium ion, the storage modulus of the κ-CRG gel was increased from 393.8 ± 81.6 to 1503.2 ± 156.1 Pa, which is similar to that of the TA/κ-CRG gel (1387.2 ± 12.7 Pa). Notably, additions of both TA (5%) + K^+ (30 mM) exhibited further increase (2366.1 ± 189.4 Pa) (Figure A3). This result indicates that TA-mediated modulation should be viewed not simply as another reinforcement additive, but as a distinct route for controlling the structure–property relationships of κ-CRG hydrogels.

4. Conclusions

In this study, we investigated how polyphenols with different valencies influence the structure, mechanics, and macroscopic behavior of κ-CRG hydrogels. By comparing TA and pyrogallol, we demonstrated that polyphenol valency plays a key role in regulating κ-CRG gel networks. Both polyphenols interact with κ-CRG through non-covalent interactions involving sulfate groups, which partially reduces electrostatic repulsion between helices and reinforces the gel structure. However, the extent and consequences of these interactions differ markedly depending on the polyphenol architecture. Pyrogallol produced moderate reinforcement and exhibited stage-dependent effects during gelation, indicating that its interaction with κ-CRG is sensitive to the conformational state of the polymer chains. In contrast, TA generated substantially stronger network reinforcement and showed stage-independent gelation behavior owing to its ability to form dynamic multivalent supramolecular junctions between polymer chains. These differences also resulted in distinct macroscopic degradation responses in aqueous environments. Pyrogallol/κ-CRG gels maintained structural integrity longer than TA/κ-CRG gels. Interestingly, TA/κ-CRG gels exhibited strong adhesion on rough, and biological surfaces, highlighting the interfacial functionality enabled by multivalent polyphenol chemistry. Overall, our findings reveal that polyphenol valency is a critical molecular parameter governing the mechanics, stability, and interfacial properties of κ-CRG hydrogels. This work provides new insights into polyphenol–polysaccharide interactions and offers a practical strategy for designing food-grade functional hydrogels with tunable mechanical and adhesive properties.

Author Contributions: Conceptualization and data analysis, H.L.; experiments and data analysis, H.-Y.Y., J.S., W.C., E.K., S.Y. and S.P.; writing—original draft preparation, H.-Y.Y. and H.L.; review and editing, H.L.; visualization, H.-Y.Y.; supervision, H.L.; project administration, H.L. All authors have read and agreed to the published version of the manuscript.

Funding: The manuscript was supported by Polyphenol Factory Co., Ltd. (Grant number G01250612 and G01250566).

Institutional Review Board Statement: The study was conducted in accordance with the Declaration of Helsinki, and approved by the Institutional Review Board of Korea Advanced Institute of Science and Technology (KAIST) (protocol code KAISTIRB-2025-182; date of approval: 25 July 2025).

Informed Consent Statement: Informed consent was obtained from all subjects in the study.

Data Availability Statement: The data presented in this study are available from the corresponding author upon reasonable request.

https://doi.org/10.3390/biomimetics11040290

Acknowledgments: This work was supported by Polyphenol Factory Co., Ltd.

Conflicts of Interest: Han-Yeol Yang and Haeshin Lee were employed by the company Polyphenol Factory Co., Ltd. The remaining authors declare that the research was conducted in the absence of any commercial or financial relationships that could be construed as a potential conflict of interest.

Appendix A

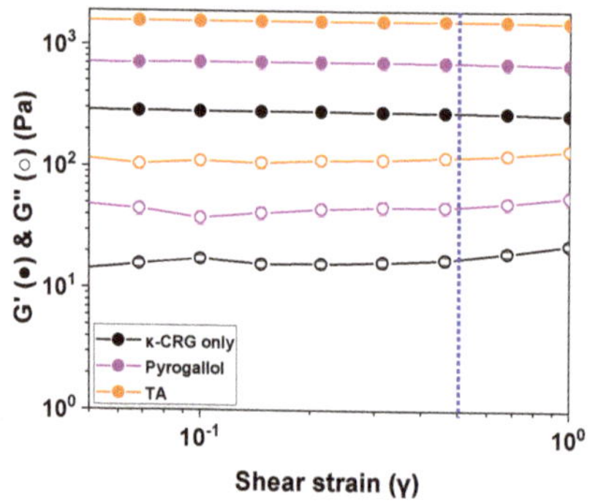

Figure A1. Amplitude sweep curves of κ-CRG only, pyrogallol (6%)/κ-CRG, and TA (15%)/κ-CRG gels to determine the linear viscoelastic region. Filled circles and open circles represent G′ and G″, respectively, for each formulation. The dashed blue line indicates the selected strain (0.5%) used for subsequent rheological measurements. Data are presented as mean ± SD (n = 3).

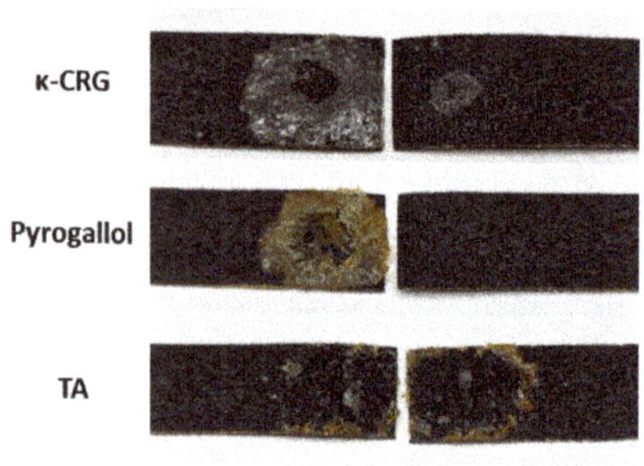

Figure A2. Failure mode after lap shear testing of κ-CRG, pyrogallol (6%)/κ-CRG, and TA (15%)/κ-CRG gels on sandpaper (P100).

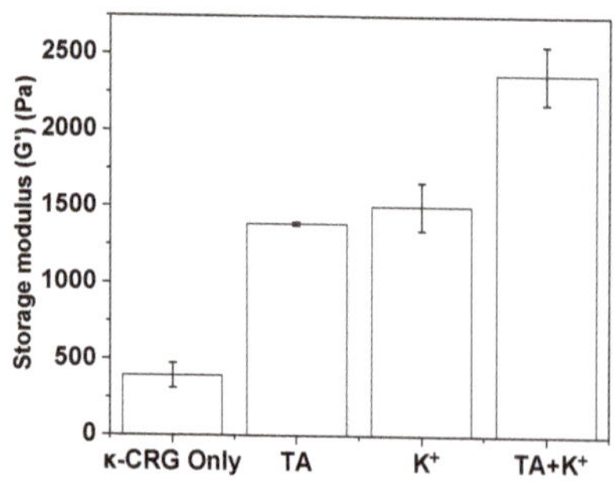

Figure A3. Storage modulus (G′) of κ-CRG only, TA (5%)/κ-CRG, K^+ (30 mM)/κ-CRG, and TA (5%)/K^+ (30 mM)/κ-CRG gels. Data are presented as mean ± SD (n = 3).

https://doi.org/10.3390/biomimetics11040290

References

1. Necas, J.; Bartosikova, L. Carrageenan: A Review. *Vet. Med.* **2013**, *58*, 187–205. [CrossRef]
2. Núñez-Santiago, M.C.; Tecante, A.; Garnier, C.; Doublier, J.L. Rheology and Microstructure of κ-Carrageenan under Different Conformations Induced by Several Concentrations of Potassium Ion. *Food Hydrocoll.* **2011**, *25*, 32–41. [CrossRef]
3. Clark, A.H.; Ross-Murphy, S.B. Structural and Mechanical Propcerties of Biopolymer Gels. In *Biopolymers*; Pearce, E.M., Ed.; Springer: Berlin/Heidelberg, Germany, 1987; Volume 83, pp. 57–192.
4. Saha, D.; Bhattacharya, S. Hydrocolloids as Thickening and Gelling Agents in Food: A Critical Review. *J. Food Sci. Technol.* **2010**, *47*, 587–597. [CrossRef] [PubMed]
5. Yegappan, R.; Selvaprithiviraj, V.; Amirthalingam, S.; Jayakumar, R. Carrageenan Based Hydrogels for Drug Delivery, Tissue Engineering and Wound Healing. *Carbohyd. Polym.* **2018**, *198*, 385–400. [CrossRef]
6. Pacheco-Quito, E.-M.; Ruiz-Caro, R.; Veiga, M.-D. Carrageenan: Drug Delivery Systems and Other Biomedical Applications. *Mar. Drugs* **2020**, *18*, 583. [CrossRef]
7. Dhand, A.P.; Galarraga, J.H.; Burdick, J.A. Enhancing Biopolymer Hydrogel Functionality through Interpenetrating Networks. *Trends Biotechnol.* **2021**, *39*, 519–538. [CrossRef]
8. Li, J.; Jia, X.; Yin, L. Hydrogel: Diversity of Structures and Applications in Food Science. *Food Rev. Int.* **2021**, *37*, 313–372. [CrossRef]
9. Yu, H.C.; Li, C.Y.; Du, M.; Song, Y.; Wu, Z.L.; Zheng, Q. Improved Toughness and Stability of κ-Carrageenan/Polyacrylamide Double-Network Hydrogels by Dual Cross-Linking of the First Network. *Macromolecules* **2019**, *52*, 629–638. [CrossRef]
10. Sabbagh, F.; Khatir, N.; Kiarostami, K. Synthesis and Characterization of K-Carrageenan/PVA Nanocomposite Hydrogels in Combination with MgZnO Nanoparticles to Evaluate the Catechin Release. *Polymers* **2023**, *15*, 272. [CrossRef]
11. Wang, P.; Liao, Q.; Zhang, H. Polysaccharide-Based Double-Network Hydrogels: Polysaccharide Effect, Strengthening Mechanisms, and Applications. *Biomacromolecules* **2023**, *24*, 5479–5510. [CrossRef]
12. Jabeen, F.; Ahmad, R.; Mir, S.; Awwad, N.S.; Ibrahium, H.A. Carrageenan: Structure, Properties and Applications with Special Emphasis on Food Science. *RSC Adv.* **2025**, *15*, 22035–22062. [CrossRef]
13. Ghasempour, A.; Naderi Allaf, M.R.; Charoghdoozi, K.; Dehghan, H.; Mahmoodabadi, S.; Bazrgaran, A.; Savoji, H.; Sedighi, M. Stimuli-Responsive Carrageenan-Based Biomaterials for Biomedical Applications. *Int. J. Biol. Macromol.* **2025**, *291*, 138920. [CrossRef]
14. Edo, G.I.; Ndudi, W.; Makia, R.S.; Ainyanbhor, I.E.; Yousif, E.; Gaaz, T.S.; Isoje, E.F.; Opiti, R.A.; Akpoghelie, P.O.; Igbuku, U.A.; et al. Carrageenan-Based Hydrogels for Advanced Wound Healing and Controlled Drug Delivery in Tissue Engineering. *J. Biomed. Mater. Res.* **2025**, *113*, e35594. [CrossRef]
15. Tang, Z.; Chowdhury, I.F.; Yang, J.; Li, S.; Mondal, A.K.; Wu, H. Recent Advances in Tannic Acid-Based Gels: Design, Properties, and Applications. *Adv. Colloid Interface Sci.* **2025**, *339*, 103425. [CrossRef] [PubMed]
16. Shin, M.; Ryu, J.H.; Park, J.P.; Kim, K.; Yang, J.W.; Lee, H. DNA/Tannic Acid Hybrid Gel Exhibiting Biodegradability, Extensibility, Tissue Adhesiveness, and Hemostatic Ability. *Adv. Funct. Mater.* **2015**, *25*, 1270–1278. [CrossRef]
17. Mendes, M.; Cotas, J.; Gutiérrez, I.B.; Gonçalves, A.M.M.; Critchley, A.T.; Hinaloc, L.A.R.; Roleda, M.Y.; Pereira, L. Advanced Extraction Techniques and Physicochemical Properties of Carrageenan from a Novel Kappaphycus Alvarezii Cultivar. *Mar. Drugs* **2024**, *22*, 491. [CrossRef] [PubMed]
18. Arunagiri, V.; Tsai, H.-C.; Darge, H.F.; Chiang, H.W.; Thankachan, D.; Mei, C.-J.; Lai, J.-Y. Preparation of Physically Crosslinked Polyelectrolyte Gelatin-Tannic Acid-κ-Carrageenan (GTC) Microparticles as Hemostatic Agents. *Int. J. Biol. Macromol.* **2021**, *191*, 324–334. [CrossRef]
19. Ikeda, S.; Nishinari, K. "Weak Gel"-Type Rheological Properties of Aqueous Dispersions of Nonaggregated κ -Carrageenan Helices. *J. Agric. Food Chem.* **2001**, *49*, 4436–4441. [CrossRef]
20. Bakó, I.; Wacha, A.; Pothoczki, S. Intramolecular H-Bond Properties in Solvated Cyclodextrins Elucidated by SAXS and Molecular Dynamics Simulation. *J. Mol. Liq.* **2025**, *421*, 126922. [CrossRef]
21. Wen, J.; Zhang, X.; Pan, M.; Yuan, J.; Jia, Z.; Zhu, L. A Robust, Tough and Multifunctional Polyurethane/Tannic Acid Hydrogel Fabricated by Physical-Chemical Dual Crosslinking. *Polymers* **2020**, *12*, 239. [CrossRef]
22. Madruga, L.Y.C.; Sabino, R.M.; Santos, E.C.G.; Popat, K.C.; Balaban, R.D.C. Carboxymethyl-Kappa-Carrageenan: A Study of Biocompatibility, Antioxidant and Antibacterial Activities. *Int. J. Biol. Macromol.* **2020**, *152*, 483–491. [CrossRef]
23. Hezaveh, H.; Muhamad, I.I. Modification and Swelling Kinetic Study of Kappa-Carrageenan-Based Hydrogel for Controlled Release Study. *J. Taiwan Inst. Chem. Eng.* **2013**, *44*, 182–191. [CrossRef]
24. Shikata, K.; Kikutsuji, T.; Yasoshima, N.; Kim, K.; Matubayasi, N. Revealing the Hidden Dynamics of Confined Water in Acrylate Polymers: Insights from Hydrogen-Bond Lifetime Analysis. *J. Chem. Phys.* **2023**, *158*, 174901. [CrossRef]
25. Sun, X.; Sarteshnizi, R.A.; Udenigwe, C.C. Recent Advances in Protein–Polyphenol Interactions Focusing on Structural Properties Related to Antioxidant Activities. *Curr. Opin. Food Sci.* **2022**, *45*, 100840. [CrossRef]
26. Xue, H.; Du, X.; Fang, S.; Gao, H.; Xie, K.; Wang, Y.; Tan, J. The Interaction of Polyphenols-Polysaccharides and Their Applications: A Review. *Int. J. Biol. Macromol.* **2024**, *278*, 134594. [CrossRef]

https://doi.org/10.3390/biomimetics11040290

27. Lee, D.; Hwang, H.; Kim, J.-S.; Park, J.; Youn, D.; Kim, D.; Hahn, J.; Seo, M.; Lee, H. VATA: A Poly(Vinyl Alcohol)- and Tannic Acid-Based Nontoxic Underwater Adhesive. *ACS Appl. Mater. Interfaces* **2020**, *12*, 20933–20941. [CrossRef] [PubMed]
28. Jafari, H.; Ghaffari-Bohlouli, P.; Niknezhad, S.V.; Abedi, A.; Izadifar, Z.; Mohammadinejad, R.; Varma, R.S. Tannic Acid: A Versatile Polyphenol for Design of Biomedical Hydrogels. *J. Mater. Chem. B* **2022**, *10*, 5873–5912. [CrossRef] [PubMed]
29. Ciriminna, R.; Li Petri, G.; Angellotti, G.; Fontananova, E.; Meneguzzo, F.; Luque, R.; Pagliaro, M. Tannin: An Insight into Its Cosmeceutical Properties and Uses. *Glob. Chall.* **2025**, *9*, 2500115. [CrossRef]
30. Calapai, G.; Minciullo, P.L.; Miroddi, M.; Chinou, I.; Gangemi, S.; Schmidt, R.J. Contact Dermatitis as an Adverse Reaction to Some Topically Used E Uropean Herbal Medicinal Products–P Art 3: M *Entha* × *Piperita*–S *Olanum Dulcamara*. *Contact Dermat.* **2016**, *74*, 131–144. [CrossRef] [PubMed]
31. Nickerson, M.T.; Darvesh, R.; Paulson, A.T. Formation of Calcium-Mediated Junction Zones at the Onset of the Sol-Gel Transition of Commercial κ-Carrageenan Solutions. *J. Food Sci.* **2010**, *75*, E153–E156. [CrossRef]
32. Russo Spena, S.; Grizzuti, N. κ-Carrageenan and Its Synergistic Blends: Next-Generation Food Gels. *Gels* **2025**, *11*, 976. [CrossRef] [PubMed]

https://doi.org/10.3390/biomimetics11040290

biomimetics

MDPI

Review

Bioinspired and Multifunctional Tribological Materials for Sliding, Erosive, Machining, and Energy-Absorbing Conditions: A Review

Rahul Kumar [1,*], Mansoureh Rezapourian [1], Ramin Rahmani [2,3], Himanshu S. Maurya [1,4], Nikhil Kamboj [1,5,6] and Irina Hussainova [1]

1 Department of Mechanical and Industrial Engineering, Tallinn University of Technology, Ehitajate Tee 5, 19086 Tallinn, Estonia; mansoureh.rezapourianghahfarokhi@taltech.ee (M.R.); himanshu.singh.maurya@associated.ltu.se (H.S.M.); nikhil.kamboj@utu.fi (N.K.); irina.hussainova@taltech.ee (I.H.)

2 CiTin–Centro de Interface Tecnológico Industrial, 4970-786 Arcos de Valdevez, Portugal; ramin.rahmaniahranjani@gmail.com

3 proMetheus–Instituto Politécnico de Viana do Castelo (IPVC), 4900-347 Viana do Castelo, Portugal

4 Department of Engineering Sciences and Mathematics, Luleå University of Technology, 97187 Luleå, Sweden

5 Department of Mechanical and Materials Engineering, University of Turku, 20500 Turku, Finland

6 TCBC–Turku Clinical Biomaterials Centre, Department of Biomaterials Science, Faculty of Medicine, Institute of Dentistry, University of Turku, 20014 Turku, Finland

* Correspondence: rahul.kumar@taltech.ee

Abstract: Friction, wear, and the consequent energy dissipation pose significant challenges in systems with moving components, spanning various domains, including nanoelectromechanical systems (NEMS/MEMS) and bio-MEMS (microrobots), hip prostheses (biomaterials), offshore wind and hydro turbines, space vehicles, solar mirrors for photovoltaics, triboelectric generators, etc. Nature-inspired bionic surfaces offer valuable examples of effective texturing strategies, encompassing various geometric and topological approaches tailored to mitigate frictional effects and related functionalities in various scenarios. By employing biomimetic surface modifications, for example, roughness tailoring, multifunctionality of the system can be generated to efficiently reduce friction and wear, enhance load-bearing capacity, improve self-adaptiveness in different environments, improve chemical interactions, facilitate biological interactions, etc. However, the full potential of bioinspired texturing remains untapped due to the limited mechanistic understanding of functional aspects in tribological/biotribological settings. The current review extends to surface engineering and provides a comprehensive and critical assessment of bioinspired texturing that exhibits sustainable synergy between tribology and biology. The successful evolving examples from nature for surface/tribological solutions that can efficiently solve complex tribological problems in both dry and lubricated contact situations are comprehensively discussed. The review encompasses four major wear conditions: sliding, solid-particle erosion, machining or cutting, and impact (energy absorbing). Furthermore, it explores how topographies and their design parameters can provide tailored responses (multifunctionality) under specified tribological conditions. Additionally, an interdisciplinary perspective on the future potential of bioinspired materials and structures with enhanced wear resistance is presented.

Keywords: bionic surface texturing; sliding wear; erosion; dynamic impact; cutting tools; friction; biomaterials; mechanical property; surface modification; biomimetics; nature-inspired tribology; sustainable development; green tribology; metamaterials

1. Introduction

Sustainable development is closely linked to improvement in tribology (low friction, low wear, and enhanced lubrication), which has a significant and beneficial impact on both the environment and society [1]. Global energy consumption is increasing year on year, putting pressure on both resources and the environment. Notably, the repercussions of friction and wear extend beyond substantial energy and economic loss (approximately 23% of global energy loss and 1–2% of a nation's GDP) and serve as a major source of CO_2 emissions [2,3]. Crucially, tribological losses (material degradation), comprising high friction and wear inefficiencies, expenses related to part replacement and remanufacturing, and maintenance costs, affect virtually all moving elements, ranging from industrial machinery to the intricate mechanics of natural bone joints and artificial implants integrated into the human body [4]. Any effort aimed at curbing these losses yields a direct and positive impact, not only conserving energy but also fostering economic stability, and enhancing the well-being of both individuals and the environment. Consequently, it is entirely valid to assert that tribology, with its inherent capability to diminish friction and wear, characteristically contributes to the noble pursuit of a sustainable and human-centric world, and directly or indirectly contributes to Sustainable Development Goals 3, 6, 7, 8, 9, 11, 13, 14, and 15 [5]. Figure 1 shows the triangle of sustainability, revealing nature, bioinspired materials, and bioinspired tribology as its three corners.

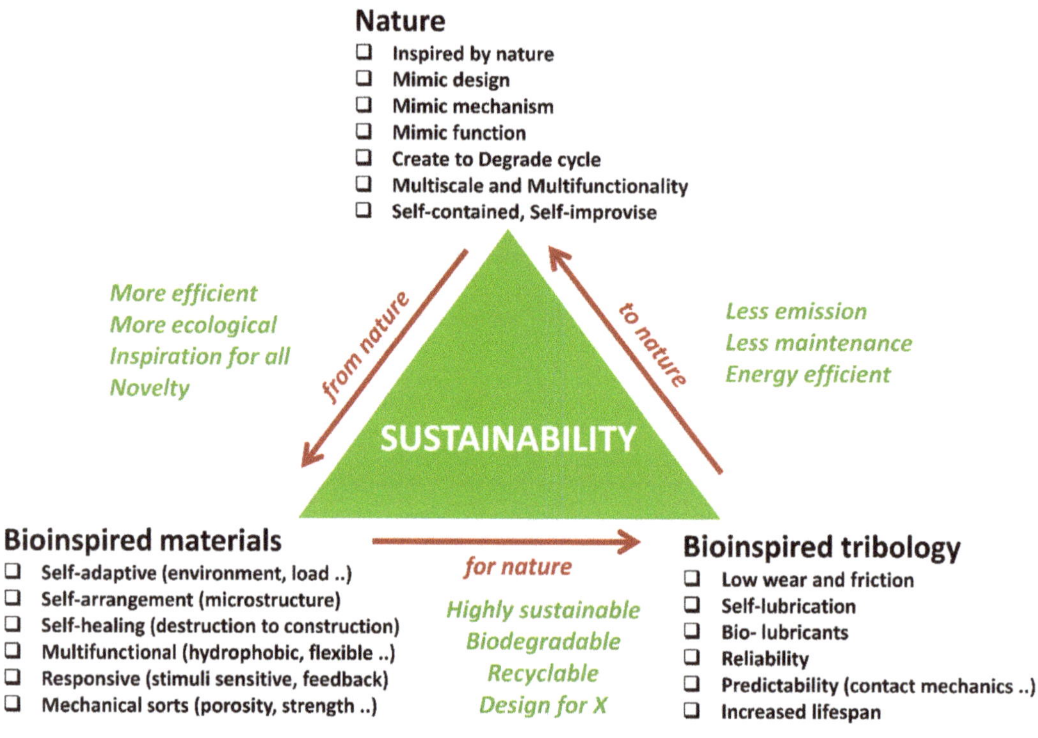

Figure 1. Sustainability triangle showing nature, materials, and tribology as its corners.

In this context, various methods can mitigate friction and its negative effects, such as wear. Those can be classified into (i) the mechanical approach (e.g., well-lubricated systems, novel composites, microstructural enhancements, and heat treatments, etc.) and (ii) the surface topography engineering approach (e.g., adjusting surface roughness or texture to reduce contact area). According to a model proposed by McFarlane and Tabor [6], friction force is the sum of two components: a mechanical contribution resulting from surface deformation and an adhesive contribution mainly influenced by surface energy. This model suggests that a minimum coefficient of friction (CoF) can be achieved at a critical value of real contact area [7]. Consequently, controlling surface topography can lead to CoF control. However, traditionally, the focus has been on mechanical approaches, overlooking the potential of engineered interfaces in friction control due to incomplete understanding of surface texturing (optimal choice and design of textures). Additionally, an innovative approach utilizes a combination of both approaches to enhance tribological properties and reduce friction and wear. The synergy may enable simultaneous tribological benefits, such as favorable effects of wear debris, continuous lubricant replenishment, formation of protective tribolayers or tribofilms, etc. [8,9].

Multifunctionality within tribological materials/surfaces represents a paradigm shift in engineering [10]. These materials not only mitigate wear and friction but also offer measurable benefits for efficient operation in specific applications. For instance, the wear rate of modern hip implants has been reduced by several orders of magnitude [11]. However, to enhance their longevity and minimize the necessity for revision surgeries, it is equally crucial for biomaterials to positively carry dynamic fluctuating high loads (during human movement), facilitate cell transfer and growth (biological interaction), and ensure seamless tribological integration within the human body (chemical interaction), even leveraging the positive effects of wear debris (e.g., hydroxyapatite-based biomaterials can generate wear debris that stimulates bone growth and integration around the implant, while optimized surface roughness/texture can lead to improved osteoconductivity or lubrication) [11–14]. Similarly, in the domain of photovoltaic (PV) solar cells, materials/surfaces are engineered to reduce wear (erosion) while simultaneously improving light absorption and self-cleaning properties [15]. This innovation can result in remarkable, up to 15%, enhancement in the overall efficiency and lifespan of solar panels [16]. In hydro-bearings and hydrofoils, meticulously designed surface textures that reduce drag lead to elevated energy conversion rates [17,18]. In certain instances, these materials have demonstrated substantial efficiency gains exceeding 20%, thereby translating into increased renewable energy production. In other cases, such as surgical instruments or soft robotics, it is advisable that the surface demonstrate higher friction or higher adhesion in order to improve the grip strength while accommodating high load transfer (pick up/drop) [19,20]. In space shuttles and satellites, the surfaces are designed to resist impact (wear) from space debris (meteors) while being lightweight, anti-weathering, etc. [21–23]. Space rover wheels are designed to navigate easily through harsh terrain, requiring enough flexibility while delivering tribological robustness [24]. Highly durable surfaces that demonstrate water-repellent properties, efficient water harvesting and spreading, or exceptional resistance to oils play a pivotal role in various applications [25–27]. These surfaces are essential for ensuring rapid lubricant dispersion, enabling rainwater harvesting systems, and powering nanogenerators, among other uses [28]. For instance, in the context of lubrication, surfaces with superoleophobic characteristics facilitate swift and uniform distribution of lubricants in machinery, reducing friction and wear. Moreover, surfaces engineered for efficient water spreading enable collection and storage of rainwater for various purposes while being resistant to enduring impact, erosion, etc. In summary, the creation of wear-resistant surfaces with specific functionalities, such as water harvesting and superoleophobic properties, has far-reaching implications across a range of applications, from industrial machinery to sustainable water management and energy harvesting technologies [29]. Frictional anisotropy is a critical tribological feature that spans from the molecular scale to the macroscale, characterized by directional asymmetry in friction response during sliding [30]. This feature is particularly valuable for microbots designed for friction-based locomotion, offering high maneuverability and precise targeting in confined three-dimensional (3D) spaces. Additionally, it ensures stress compatibility with soft living matter by limiting interface stresses [31]. As the demand grows for microbots in various applications like robotic assembly, drug delivery, lab-on-chip technology, sensing, microsurgery, and cancer treatment, there is a practical need to construct artificial prototypes [32]. These prototypes, exploiting geometric features to induce controllable anisotropic motion responses under external oscillating loads, serve as models for understanding and predicting the dynamics of future industrial microactuators [33]. For instance, in medical microbots navigating complex biological environments, anisotropic friction-based locomotion coupled with physical propulsion holds promise as an enabling technology for their development [34]. Table 1 shows the essential functional features required in key tribologically challenged fields.

Table 1. Multifunctional features required in different tribological materials/situations [13,24,35–61].

Applications	Key Functionality Required	Key Tribological Modes
Wind turbines and propellers	Anti-icing, Antifouling, Self-cleaning, Erosive resistance, Lightweight, Drag-resistance, Hydrophobicity	Sliding and Leading-edge erosion
Bearings, seals, actuators, gears, and shafts	Flexibility, Compactness, Self-lubrication, Adaptive response, Load carrying capability, Environmental adaptation, Vibration damping, Noise reduction, Friction reduction, Hydrophobicity	Sliding
Rail tracks, wheels, and braking systems	Self-healing smart materials, Self-adaptive wheels, and brakes (friction) to changing loads, speed, etc., Energy harvesting from vibration damping, Self-cleaning, Wear resistance	Sliding, Rolling
Piston rings, cylinder liners, camshafts	Self-lubrication, Adaptive response, Wear resistance, Heat dissipation, Friction reduction, Self-cleaning, Vibration damping, Noise reduction	Sliding
Cutting tools in machining, drill bits	Self-sharpening, Wear resistance, Heat dissipation, Anti-adhesion chip control, Lubrication enhancement, Adaptive control, Hydrophobicity	Sliding, Impact, Abrasion, Erosion
Prosthetic joints, dental implants, and surgical instruments	Biocompatibility, Antibacterial, Osseointegration, Flexibility and Range of Motion, Self-lubrication, Load carrying capability, Surgical precision, Responsive material to adapt tissue changes	Sliding, Impact, Tribocorrosion, Quasistatic compression
Ship hulls, hydrofoils, and swimsuits	Drag resistance, Antifouling, Hydrophobicity, Flexibility and maneuverability, Hydrophobicity, Air-filled cavities, or flexible fins for buoyancy	Hydro-drag, Erosion, Corrosion
Spatial satellites	Erosion and impact resistance from space debris, Adaptive or morphing structures, Energy harvesting solar panels, Radiation resistance, Bioinspired locomotion	Erosion, Impact
Material handling equipment (crushers, screens, conveyor systems, hoppers, chutes, etc.)	Erosion and impact resistance, Efficient material flow patterns, Abrasion resistance, Self-cleaning of debris or material build-up, Flexible components for easy throughput of feedstocks	Abrasion, Erosion, Impact
Pumps (impellers, casing, suction/discharge nozzles, etc.) and pipelines	Drag reduction, Self-cleaning, Improved fluid handling, Antifouling, Anti-corrosion	Abrasion, Erosion, Corrosion, Cavitation
Hydro turbines and boilers	Adaptive blade design, Efficient heat transfer, Drag reduction, Cavitation resistance	Slurry erosion
Plows, tillage tools	Self-sharpening, Self-cleaning, Hydrophobicity, Efficient soil penetration, and tillage	Erosion, Abrasion, Impact
Ski edges, snowboards, and bicycle chains	Hydrophobicity, Anti-icing, Reduced friction, Enhanced grip adhesion, Resilience, Shock absorption, Lightweight	Abrasion, Sliding
Solar panels	Self-cleaning, Erosion resistance, Flexibility, Enhanced light harvesting surface	Solid particle erosion
Airframe structures	Lightweight, High strength, Flexibility, Natural sound-damping (owl wings-inspired), Vibration stability and maneuverability, Drag reduction	Air drag/friction, Trailing edge erosion
Cricket bats, golf clubs, and hockey sticks	Lightweight, Shock absorption, High strength, Resilience, Drag reduction	Impact, Air drag

Table 1. *Cont.*

Applications	Key Functionality Required	Key Tribological Modes
Automotive body, crash barriers on highways, and military armor	Energy absorption, Impact resistance, Flexibility for movement, Adaptive camouflage	Impact
Bulldozers, hydraulic hammers, and excavator teeth	Impact resistance, Self-sharpening edges, Material anti-adhesion, Erosion resistance, Abrasion-resistant, Hydrophobicity	Impact, Abrasion, Erosion
Bone implants for hip, knee, skull, etc.	Energy absorption, Lightweight, Porous, High strength, Adaptive lubrication, Antibacterial, Cell attachment	Sliding, Impact, Quasistatic, and dynamic compression
Robotic grippers, Microbots, bio-MEMS/NEMS	Adaptive adhesion, Variable stiffness, Biocompatibility, Friction anisotropy, Antibacterial	Sliding, Frictional grip

Recently, significant attention has been paid to the exploration of materials, surfaces, and architectures inspired by nature [25,26,62]. The fact that natural objects have survived the harshest conditions through multifaceted evolution inspires engineers to create or imitate the construction of similar structures. Nature-inspired materials are man-made materials that mimic the structure, properties, or functions of natural materials or living organisms and offer the potential for sustainable synergy between tribology and multifunctionality [63]. The properties of these materials and surfaces result from complex interplay between the surface structure and the morphology and physical and chemical properties. Moreover, many materials, surfaces, and devices with such designs provide multifunctionality. Various terms such as bioinspiration, biomimicry, biomimetics, nature inspiration, and nature mimicry are commonly used interchangeably by researchers [63]. Biomimetic materials imitate the evolutionarily developed structural features, leading to adapted architecture, especially in living species [64–66]. Often, the architectures exhibit a graded structure spanning multiple scales, including macro-, micro-, and nanoscales. For example, bone's complex porous structure, characterized by complex ligament shapes and variations in density, allows it to achieve superior mechanical (stiffness, energy absorption, stress distribution, etc.) and biological characteristics (e.g., porosity leads to facilitating nutrient exchange and cell proliferation) compared to most man-made materials (biomaterials) [65].

Friction and adhesion are common in nature. Notable examples involve leveraging nature-inspired surface textures found in animal scales and skins, such as those of sharks (for drag reduction and hydrophobicity), snakes (for erosion resistance), pangolins and turtles (for flexibility combined with erosion/abrasion resistance), and gecko feet (for improved adhesion) [24,63,67]. Some species feature adaptive systems that enable changes in color, pattern, or texture for defense, signaling, temperature regulation, or reproduction (also called the 'chameleon mimetic system') [68,69]. Nacre, an organic material found in mollusk shells, exhibits remarkable strength and resilience [70]. Spider silk, renowned for its exceptional mechanical properties and supercontraction abilities, holds great potential for various structural applications [71]. Examining beetle adhesion systems at the nanoscale has uncovered a diverse array of intricate multiscale architectures serving crucial functions like wing fixation, crawling, mating, and external protection, mainly utilized regarding locomotion for microrobots (anisotropic friction and earthworm-inspired) [30,34,72]. These adhesion systems utilize different mechanisms; some rely on liquid secretion (capillary force and lubrication), while others operate through direct interlocking of high-density microfibers or contact of mushroom-shaped hairy structures (van der Waals force) [70]. A comprehensive analysis of materials (structures) inspired by nature reveals their unique functionality, encompassing various domains of sustainable science [63]. Bioinspired structures combined with surface topography (or texturing) of these materials demonstrate promising outcomes in customizing friction, wear, and other captivating properties such as antifouling [73–75], self-lubrication [8,61,76], self-adaptation [77,78], hydrophobicity [27], self-cleaning [16,25,73], cell growth [12,79], drug delivery [80–86], antibacterial [79,87,88], color manipulation [68], anti-reflection [89], anisotropic friction in MEMS/NEMS/microrobots/microactuators [30,70,72], and enhanced adhesion [24,25,70,90]. Figure 2 shows various biological organisms (animals/plants) for inspiration regarding creating an efficient multifunctional tribological material.

Figure 2. Bioinspired (nature) inspiration influencing various tribological conditions: (**a**) drag reduction in shark skin; (**b**) adhesion switch in gecko foot; (**c**) club of mantis shrimp; (**d**) flexibility of pangolin scales; (**e**) scorpion back; (**f**) drag reduction in penguin skin; (**g**) turtle shell; (**h**) fibered hierarchical layer of bamboo; (**i**) dung beetle back; (**j**) adhesion shift in sea horse tail; (**k**) friction anisotropy in snakeskin; (**l**) layered structure of Venus flower basket; (**m**) antibacterial cicada wings; (**n**) seashell; (**o**) hydrophobicity of lotus leaf; (**p**) trabecular bone network; (**q**) honeycomb cells; (**r**) adhesion manipulation in tree frog paw; (**s**) self-sharpening of maize leaf; and (**t**) hierarchical structure of softwood. Images reproduced with permission from The Biomimicry Institute, USA [91]. Reported areas of desired functionality concerning tribology are mentioned.

Achieving such complex nature-inspired geometries in materials or on their surface through conventional materials and technologies often proves challenging and expensive [92–95]. Historically, progress regarding materials (tribological) has primarily depended on modifying their chemical composition to alter their tribological and mechanical properties. Although this approach has yielded positive results, the journey from discovering new tribomaterial to its commercial availability has typically been time-consuming [92]. To this end, additive manufacturing (3D printing or laser surface texturing) offers promising prospects, especially in terms of generating complex bioinspired surfaces/materials and also allowing for faster fabrication and up-scaling production [94–99].

Any bottom-up approach to developing complex, multifunctional, and multiscale mimetics that can provide multiple robust functionalities requires a multidimensional strategy. The research in this multifaceted tribology domain is still at an early stage. The purpose of this review is to provide comprehensive information on the latest advances and future prospects in the field of "nature inspired" or "biologically inspired" tribological materials, with particular emphasis on their design, manufacturing processes, sources of inspiration, as well as friction and wear, performance, and other additional functions. The key focus is to establish a correlation between various tribological scenarios, bioinspired material characteristics, and their resilience to specific environmental conditions, aiming to formulate guidelines for the creation of bioinspired wear-resistant materials and systems. We have attempted to sum up relevant tribological information to a large extent. Nevertheless, due to some missing parameter specifications from the relevant reference publications that do not clearly and unambiguously characterize the physical conditions or tribological stress or the fact that tribological behavior is a system response and mere representation of some value such as force, rotational frequencies, etc., it is not in itself (alone) suitable for characterizing the stress level of the tribological system in question. Also, mixed or incorrect use of data/units for mass and forces should be avoided or carefully analyzed. For this, we recommend that readers defer to the relevant cited source(s). Figure 3 illustrates the research methodology framework for the current work.

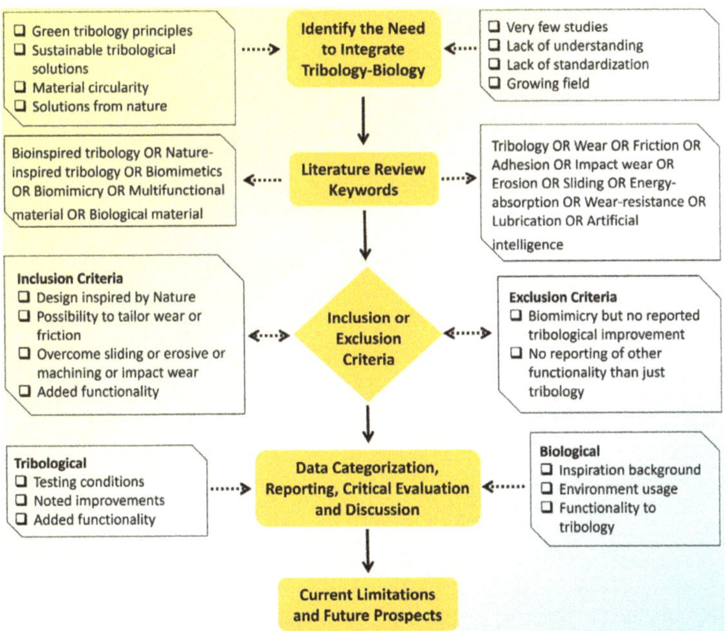

Figure 3. Research methodology framework. The arrow directions signify the flow direction of data.

2. Design and Fabrication Strategies for Bioinspired Tribological Surfaces

Bioinspired texturing techniques encompass a range of methods that mimic natural surface structures found in various biological systems. These techniques can be broadly classified into additive, subtractive, and deformation methods, each offering unique capabilities for creating surface textures with tailored properties and functionalities, benefiting a wide range of applications in tribology, biomedicine, materials science, and beyond. Figure 4 illustrates key texturing techniques as reported in the literature based on scales of textures and methods of fabrication.

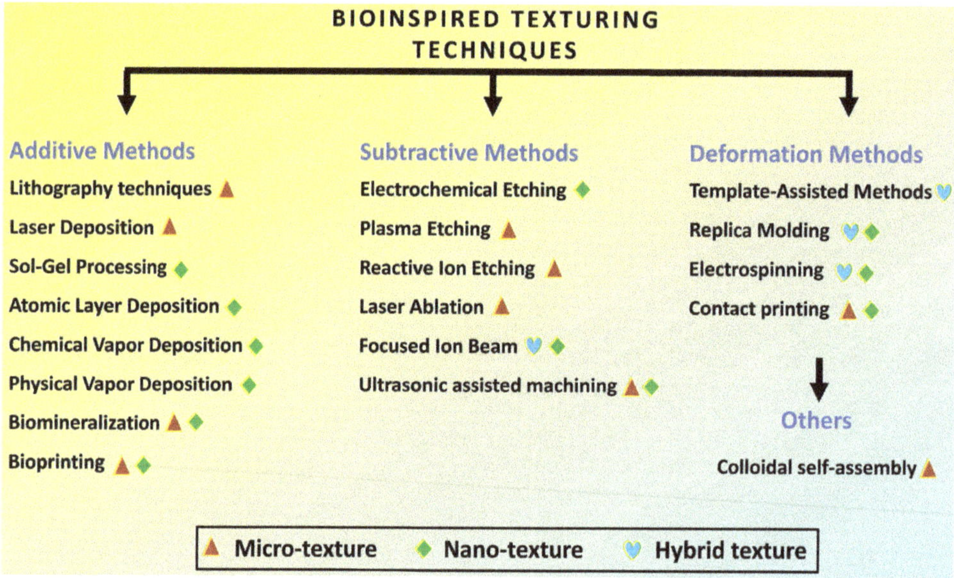

Figure 4. Bioinspired texturing techniques based on scales of textures and methods of fabrication.

Additive methods involve building up material to form desired textures. Lithography techniques, such as photolithography and electron-beam lithography, utilize masks or focused beams of light/electrons to selectively expose or remove material, creating precise patterns on surfaces [100–102]. Examples include microfluidic channels for lab-on-a-chip devices and photonic structures for optical applications.

Laser deposition involves using lasers to melt and deposit material onto a substrate, forming intricate structures layer by layer [97]. This technique is commonly used in additive manufacturing processes, including selective laser sintering and direct laser writing, to create complex 3D geometries with high resolution and accuracy [103].

Sol–gel processing involves the conversion of precursor solutions into solid materials through hydrolysis and condensation reactions [104,105]. By controlling the composition and processing conditions, sol–gel techniques can produce thin films, coatings, or bulk materials with tailored microstructures and properties. Applications include optical coatings, protective coatings for corrosion resistance, and biomedical implants.

Bioprinting utilizes CAD models to deposit biological materials, such as cells, growth factors, and biomaterials, layer by layer to create 3D structures. This technique is used in tissue engineering and regenerative medicine to fabricate customized scaffolds for tissue repair and organ transplantation [106,107].

Subtractive methods involve removing material to create textures on surfaces. Electrochemical etching utilizes electrochemical reactions to selectively dissolve material from a substrate, creating microscale or nanoscale features with high aspect ratios. This technique is widely used in semiconductor fabrication to pattern silicon wafers for microelectronics and MEMS devices [108].

Plasma etching involves using reactive plasma gases to remove material chemically or physically from a substrate, creating precise patterns through a process known as dry etching [108]. This technique is used in microfabrication processes to pattern semiconductor materials, polymers, and metals for electronic devices, sensors, and optical components.

Laser ablation utilizes high-power laser pulses to vaporize and remove material from a substrate, creating microscale or nanoscale features with high precision and resolution. This technique is used in materials processing, microfabrication, and surface modification applications, including micromachining, engraving, and thin-film patterning [84,109,110].

Ultrasonic-supported surface texturing is a specialized machining technique that utilizes ultrasonic vibrations to enhance the cutting action of the tool during surface texturing operations [111]. In this process, ultrasonic vibrations are applied to the cutting tool or workpiece, causing rapid small-scale movements that help to break up chips and improve material removal efficiency. This technique is particularly useful for creating microscale or nanoscale surface textures with precise control over texture depth, shape, and orientation.

Deformation methods alter surface textures through mechanical manipulation. Template-assisted methods involve using templates or molds with predefined patterns to transfer textures onto substrates through techniques such as nanoimprint lithography and soft lithography. Replica molding utilizes elastomeric molds to replicate surface textures from master molds onto different substrates, enabling large-scale production of biomimetic surfaces for various applications [34,100,112,113].

Electrospinning involves using an electric field to draw and elongate polymer solutions or melts into thin fibers, which are deposited onto a substrate to form textured surfaces with controlled fiber orientations and dimensions [114]. This technique is used in tissue engineering, filtration membranes, and functional textiles [115].

Contact printing utilizes patterned stamps or molds to transfer ink or functional materials onto substrates, creating textured surfaces with micrometer-scale or nanometer-scale features [116]. This technique is used in flexible electronics, sensors, and microfluidic devices for patterning conductive materials, biomolecules, and functional coatings.

Colloidal self-assembly involves the spontaneous organization of colloidal particles into ordered structures on a substrate through processes such as sedimentation, evaporation, or chemical interactions [117]. This technique is used in nanotechnology, photonic crystals, and surface engineering to create hierarchical structures with controlled properties for optical, electronic, and sensing applications.

3. Bioinspired Surfaces for Sliding Contacts

Sliding contacts bionic modifications and the bionic tribological surfaces thereof are designed to optimize lubrication (solid or liquid) and capture wear debris, such as in processes like honing cylinder liners. By creating bionic textures with controlled geometry, these surfaces retain lubricating oil and trap wear debris generated during rubbing. The material volume between the bionic textures acts as raised islands, cushioning the counterbody and reducing the actual contact area, thereby improving friction quality, and sometimes improving flexibility to accommodate sudden jerks. Properly structured cushioning islands can also reduce wear. The biotexturizing process often aims to adjust lubricating fluid film thickness, facilitating early separation between rubbing surfaces upon lubricant entry and reducing friction over a shorter distance [118]. It has been shown that biotextures can contribute to wear reduction by means of (i) enhancing hydrodynamic pressure to improve load carrying capacity and decrease shear–strain rate in the lubricant (hydrodynamic bearing effect), (ii) attracting more lubricant into the contact area (inlet suction effect), (iii) serving as a reservoir to store and supply lubricant to the interface (lubricant reservoir effect), (iv) capturing wear particles (debris-trapping effect), and (v) decreasing the actual contact area, thereby reducing friction [119,120]. It must be noted that

Ra values (roughness) are commonly reported for surface texture analysis in most of the published work. However, it is important to note that Ra, owing to its double averaging, does not inherently convey significant information.

Additionally, certain bionic textures such as snake or shark skin/scales, maize leaf, etc., are known to exhibit frictional anisotropy, adding to their unique features. For example, in Figure 5, a typical snake slithering mechanism reveals varying topography across different areas of the snake's skin. Pythons, known for constricting their prey before ingestion, must accommodate the prey's volume and shape by stretching their jaws upon inhalation. This process can result in multi-axial displacements of the skin's surface material [121]. To address this, the skin above line AA is composed of small patches of uniform octagonal shape connected by flexible strips. This design enables multi-axis stretching of the surface while potentially damping sudden jerks caused by prey resistance, as observed in the throat region. However, the frictional properties of the ventral surface (that touches the ground during slithering) of snakeskin vary widely, covering a broad range of values. Additionally, the degree of anisotropy, or directionality of friction, also varies within this range [122–126]. However, Filippov et al. [122] conducted a study comparing frictional experiments with anesthetized snakes on smooth and rough surfaces (with roughness values of Ra = 20 and 200 μm, respectively), revealing frictional anisotropy that diminished significantly on the smooth surface. They found that the macroscopic pattern of ventral scales contributes to frictional anisotropy at both macro- and microscales, with two hierarchical levels of structures, i.e., scales and denticles, playing a role. This explains why snakes experience decreased locomotory ability on smooth substrates and rely on a certain level of roughness for propulsion, in line with their numerical model's predictions.

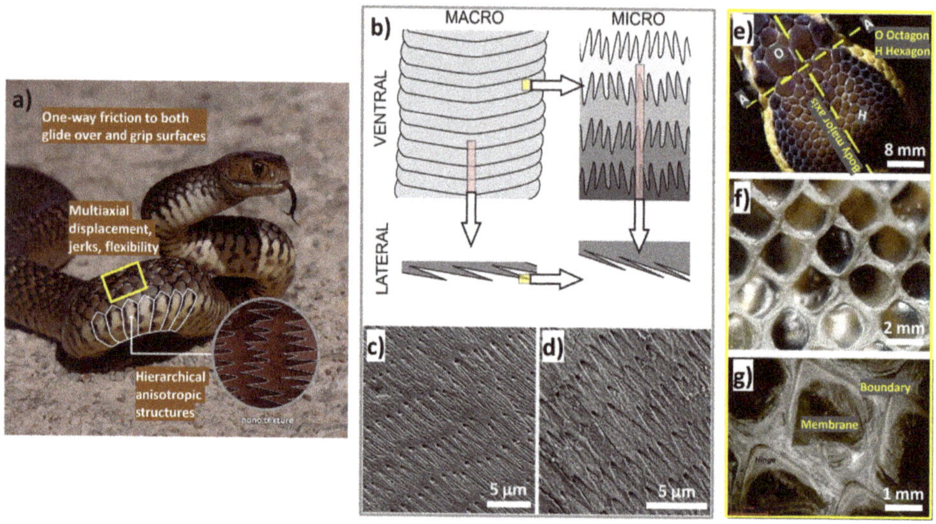

Figure 5. Images showing (**a**) snake slithering mechanism. Adopted from The Biomimicry Institute, USA [91]; (**b**) hierarchical anisotropic structures on the ventral snake surface [122]; (**c**) SEM of western diamondback rattlesnake [122]; (**d**) SEM of black-necked spitting python [122]; (**e**) details of dorsal (back) scale structure of a python [121]; and (**f**,**g**) magnified image of a python dorsal [121].

Mühlberger et al. [127] utilized a combination of powder injection molding and ultraviolet light-assisted nanoimprint lithography to replicate the intricate features of a Python/Cobra (*Naja nigrocollis*). This approach enabled the direct copying of the biological surface and imprinting of unit textural features three times the length of the actual features of the snake. Pin-on-disc experiments comparing the friction behavior of textured and polished ceramic samples showed favorable results for the textured surfaces. Snakeskin textured samples exhibited a 30% lower COF than polished non-textured samples, particularly when sliding in the parallel direction. However, the study emphasized the importance of employing an efficient replication method to achieve the required fine size texturization. Samples with features similar in length to the original skin did not perform well due to the inadequacy of the replication technique used rather than the tribological inefficiency of the geometric features themselves. In an another study by Greiner et al. [126], they used laser texturing to create scale-like morphologies on 100Cr6 bearing steel pins, inspired by snake and sandfish skin, for testing in both lubricated and unlubricated tribological contacts. The wide scale-like surface structure significantly reduced friction forces by over 40% in unlubricated conditions, while, in lubricated conditions, it increased friction by more than three times. In an another study, Kumar et al. [128] investigated the impact of lotus leaf-inspired dimple texturing (diameter 500 μm, depth 8 μm) on the dry sliding performance of bearing steel, yielding several notable findings. Two different types of dimple shapes were produced, namely circular- and bi-triangular-shaped dimples. Further, to study the effect of dimple density, two different types of arrangements in dimples were created, i.e., in radial and spiral arrays. It was observed that a spiral array has a higher and more stable number of microdimples than a radial array under the same contact area, so the density of dimples becomes important. Hence, a higher effective number of microcavities was observed in the contact zone for the spiral array compared to the radial pattern, leading to increased wear particle entrapment. Moreover, increasing dimple density from 7% to 20% enhanced wear debris entrapment, consequently reducing wear. However, once the cavity is filled, the texture effect diminishes. Thirdly, bi-triangular dimples with lower density were more effective than circular shapes in reducing both friction and wear, with 20% density exhibiting optimal performance. Finally, in dry conditions, the textured samples displayed lower coefficients of friction compared to the untextured ones. Several other studies conducted on lotus leaf-inspired dimpled surfaces point to its friction-diminishing character due to debris entrapment and lubricant storage effects [124,129–133].

Han et al. [105] developed a novel bionic anti-adhesive electrode inspired by maize leaves, known for their anti-adhesive traits. The study investigated tissue adhesion effects on electrosurgical electrodes under high temperatures. Using the bionic anti-adhesive electrode, liver tissue was electrically cut in animal experiments. The results showed that the single bionic electrode had slightly better anti-adhesive properties, while the coupled electrode exhibited lower adhesion. The coupled electrode, coated with TiO_2, gained superhydrophilic self-cleaning abilities. Later, Mzali et al. [134] studied the tribological behavior of five different surfaces' patterns bioinspired from maize leaf skin, shark skin, snakeskin, pitcher's structure, and lizard skin, with a focus on friction anisotropy and test conditions. Figure 6 depicts the apparent friction coefficient (CoF) evolution for various sliding cycles and contact pressures of 27 and 59 MPa [134]. Friction behavior significantly relies on surface texture and direction when interacting with anisotropic patterns. In the high-frictional direction (HFD), bioinspired patterns act as asperities, requiring substantial tangential load for deformation, correlating closely with plastic deformation and asperity sharpness. Asperities with sharper features induce higher plastic deformation, explaining the observed decrease in CoF (pitcher > lizard > snake). Conversely, in the low-frictional direction (LFD), the pattern effect is less pronounced, with an increase in pattern intensity resulting in higher absolute CoF values (0.8 to 0.4). When sliding in LFD, CoF decreases by 74%, 63%, and 61% for lizard, snake, and pitcher patterns, respectively. In contrast, isotropic patterns yield a large contact area (in relation to the apparent contact area), leading to relatively high CoF and increased adhesive wear. Increasing contact pressure reduces the friction coefficient, with a twofold pressure increase resulting in an 80% CoF drop. This phenomenon is attributed to frictional heat generation at the contact surface, affecting material strength and shear strength. In HFD, CoF ranges from 0.36 to 0.6, while, in LFD, CoF varies from 0.34 to 0.5. No significant difference in CoF between HFD and LFD is observed, indicating isotropic behavior among the tested specimens. However, at high contact pressures, flattening of surface asperities by the ball leads to early removal of high peaks, potentially reducing the microstructure's impact on the friction coefficient. The anisotropic nature of friction evolution concerning bioinspired surface designs was reported by a few other authors [27,30,34,125,130,135,136].

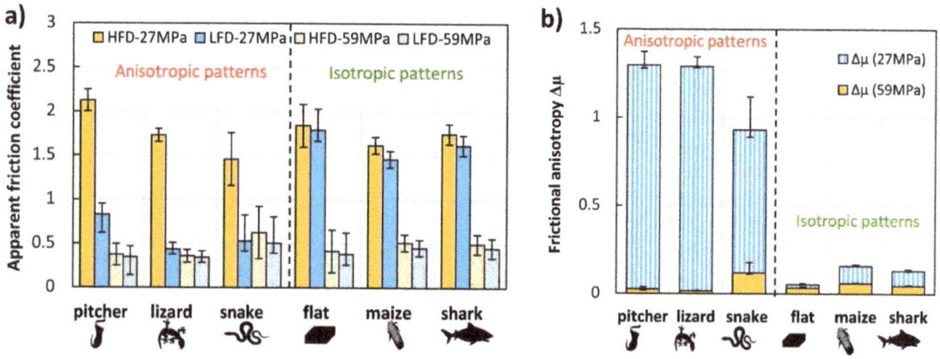

Figure 6. Images showing the progression of (**a**) CoF, i.e., μ, as function of the bioinspired surfaces for contact pressures and friction direction; and (**b**) direction-dependent CoF values of bioinspired surfaces for different contact pressure. The contact pressures were 27 MPa and 59 MPa. The friction directions are labeled as high friction direction (HFD) and low friction direction (LFD). Reproduced with permission from Ref. [134].

Furthermore, hexagonal textures exhibit noticeable anisotropy in various contact directions, influencing tribological properties. Nature's diverse organisms, serving as models for texture design, showcase the remarkable effectiveness of hexagonal patterns. Many creatures feature hexagonal textures for specific functions: stability (e.g., insect compound eyes and honeycombs), adhesion (e.g., frog toes and clingfish), wear resistance (e.g., snake scales), lightweight structures (e.g., dragonfly wings), impact resistance (e.g., insect tentacles), and superhydrophobicity (e.g., bubble rafts and cicada toes), among others [137–139], for instance, the toe pads of tree frogs, characterized by hexagonal columnar epithelial cells separated by mucous-filled channels as shown in Figure 7 [140,141]. Tree frogs' toe pads have garnered attention due to their ability to climb in humid environments without slipping. Similarly, hexagonal convex structures have been identified in the foot pads of katydids and bush crickets. Huang et al. and colleagues [132] engineered a surface structure inspired by tree frogs' toe paw for wet sliding contacts, featuring elastic microdimple and -pillar patterns on polydimethylsiloxane elastomers (PDMS), to investigate its tribological performance. The friction behaviors of these patterned surfaces, lubricated with deionized water, were assessed. The results showed that the PDMS surface with microdimples exhibited reduced friction with increasing pattern area density. The area density of pillar patterns had a negligible impact on friction at low sliding speeds but became significant as sliding speed increased. The relationship between friction force and area density was approximately linear for pillar-patterned surfaces, possibly explaining the evolution of polygonal columnar structures in newts' toe pads.

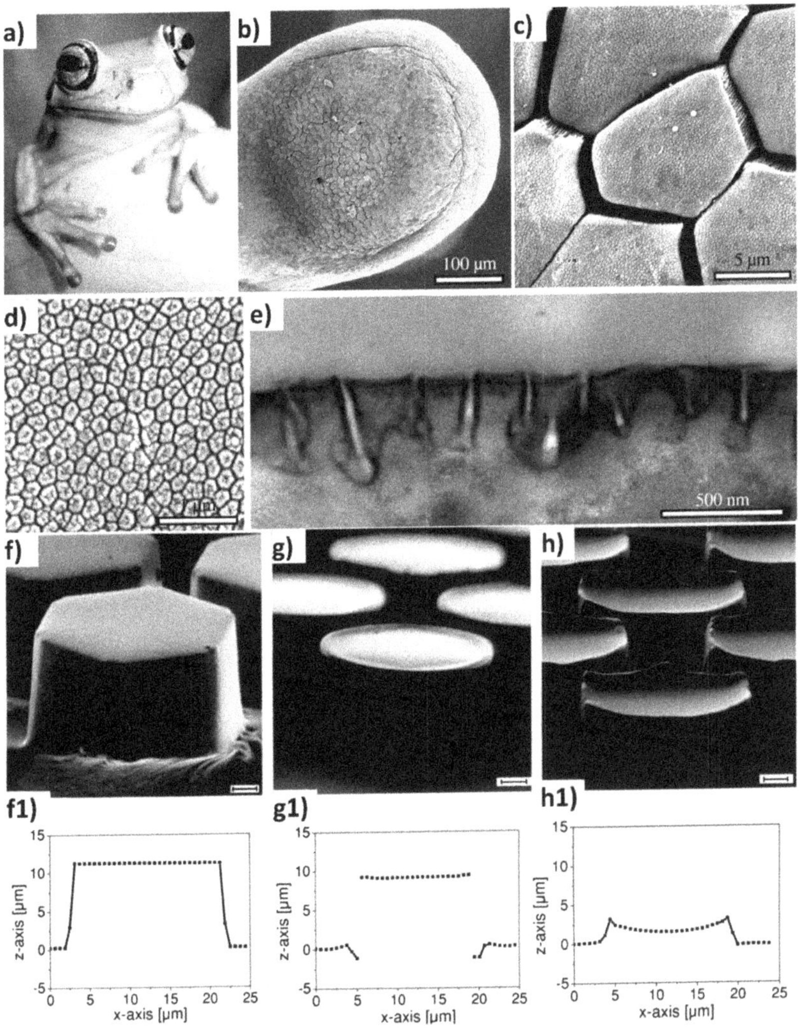

Figure 7. Morphology of tree frog toe pads. (**a**) White's tree frog (Litoria caerulea); SEMs of (**b**) toe pad, (**c**) epidermis with hexagonal epithelial cells, and (**d**) magnified view of the surface of a single hexagonal cell showing peg-like projections; (**e**) TEM of cross-section through cell surface. Reproduced with permission from Ref. [141]. SEM images depict arrays of hexagonal micropillars on PDMS surfaces, with diameters ranging from 15 to 19 μm. These pillars feature various tip shapes, (**f**,**f1**) flat tips, (**g**,**g1**) T-shaped tips with heights of 10 μm and channel widths of 3 and 5 μm, as well as (**h**,**h1**) concave tips with a height of 3 μm and a channel width of 5 μm. Scale bars in the images correspond to 2 μm; (**f1**,**g1**,**h1**) corresponding profiles of the pillars obtained with a confocal microscope are provided below. Reproduced with permission from Ref. [140].

Lu et al. [142] conducted a study inspired by Seashell to design Ni_3Al matrix bioinspired shell-like composite surface structures with excellent tribological properties. Wear tests were conducted to evaluate the effects of surface texturing and solid lubricants, revealing that the presence of solid lubricants significantly reduced friction coefficients and wear losses. The SAC305 alloy promoted self-compensating lubrication behaviors due to its ability to form a lubricating film under dry friction sliding. The combination of theoretical models and experimental measurements established a contact and collision model for microprotrusions between friction pairs, highlighting the roles of soft and hard materials in mutual sliding. The SAC305 alloy, with high surface energy and low hardness, facilitated material transfer and self-compensating lubrication compared to the Ni_3Al matrix.

In a separate investigation focusing on the combined effects of texturing and solid lubricants, researchers drew inspiration from the strong adsorption observed at the interface, known as the contact interface, which arises from the synergistic interaction between mucous (acting as a solid lubricant) and the surface microstructure of the frog paw, Huang et al. [137] investigated the tribological properties of AISI 4140 steel improved by biomimetic hexagonal textures (Figure 8) and incorporation of solid lubricants (Ag, Sn, and Cu) and TiC, and their optimization thereof [138]. Introducing TiC and SnAgCu showed significant antifriction and wear resistance. Compared to Ag and Sn, the modified material exhibited around a 65% decrease in average COF and around a 42% decrease in wear depth in addition to achieving a steady frictional condition (reduced fluctuations). The multi-solid lubricants formed a lubricating film and enabled self-repairing behavior for surface defects, contributing to excellent antifriction and wear resistance. Nano-TiC particles were gradually enveloped by ductile SnAgCu during wear, transitioning the contact mode from sliding to rolling for antifriction effects. Moreover, rolling lubrication particles were captured by wear defects, facilitating self-repair. In another study, Wang et al. [143] designed dung beetle-inspired multi-bioinspired hierarchical textures featuring superhydrophobicity to enhance AISI 440C steel wear resistance in water lubrication. It was shown that multi-bioinspired hierarchical textures significantly improved wear resistance under low loads and high frequencies (values are provided in Table 2). Superhydrophobic micro/nanostructures played a crucial role in enhancing wear resistance, resulting in wear reduction rates higher than single bioinspired textures. The geometrical parameters notably affected the wear properties, with the wear rates decreasing by 64% and 50% for textured configurations. The combined effects of multiple texturing and superhydrophobic micro/nanostructures boosted wear resistance by increasing surface hardness and improving hydrodynamic pressure. Several others have reported superior tribological performance (reduction in CoF/wear rate/wear mechanisms, etc.) upon synergistic application of biotexturing and solid lubrication [61,76,144,145]. These structures can trap lubricants (wear debris may lead to rolling friction), reduce contact area, improve hydrodynamic pressure, and alter the contact interface, leading to improved wear resistance, reduced friction, and contributing to increased efficiency, durability, and reliability of mechanical systems, making them valuable tools in various industrial applications.

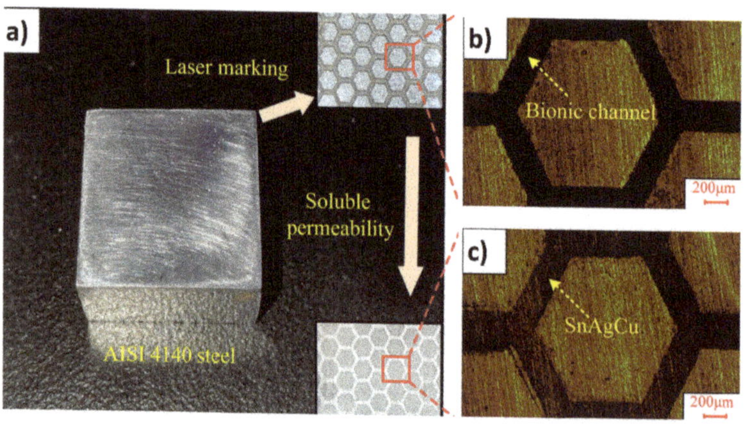

Figure 8. Showing the process of creating (**a**) AISI 4140 bionic self-lubricating material involves marking the specimen; (**b**) after laser processing and (**c**) then infiltrating it with solid lubricants. Reproduced with permission from Ref. [138].

Table 2. Results of different bioinspired structures under sliding wear conditions as reported in the literature. It is important to highlight that the absence of key tribotechnical parameters in many published findings prevents calculation of resulting tribological stress conditions. However, instances of tribocontact stress calculations have been observed assuming Hertzian contact and no wear to the counterbody in a few studies.

Bioinspired Texture and Material	Sliding Test Parameters	Friction Coefficient (CoF)	Conclusions/Texture Effect
Snake/Lizard scales-inspired (100Cr6) by laser texturing [124,126]	Velocity: 0.02–0.17 m/s Load: 2 N Pin-on-disc Counterbody: 100Cr6	~0.3 dry ~0.01 lubricated	50% low CoF in dry condition. 80% low CoF in lubricated condition for faster sliding speeds, i.e., >70 mm/s.
	Velocity: 0.02–0.17 m/s Load: 2 N Pin-on-disc Counterbody: PEEK	~0.12–0.18 dry ~0.1 lubricated	30–50% low CoF in dry condition. 40% low CoF in lubricated condition.
	Velocity: 0.02–0.17 m/s Load: 2 N Pin-on-disc Counterbody: Alumina	~0.6–0.9 dry ~0.05 lubricated	Reduction in friction strongly depends on the sliding speed. Low CoF (20%) at slow speed in dry condition. Low CoF (70%) at faster speed in lubricated condition.
Dung beetle-inspired multi-hierarchical micro/nanotextures (AISI 440C steel) by Laser ablation and coating technology [143]	Test: 15 min Load: 0.5, 1, 2, 4 N Hertzian contact pressure: 378–756 MPa Frequency: 3, 6, 9, 12 Hz Stroke: 0.004 m Ball-on-disc Counterbody: Steel ball Lubricant: Water	Decrease in CoF (~0.25) with increase in load or frequency for both single- and multi-bioinspired textures due to generation of high hydrodynamic pressure.	Possibility of developing superhydrophobic micro/nanostructures. Role of superhydrophobic in enhancing wear resistance. More than 50% wear rate reduction using multi-bioinspired textures than single-bioinspired.
Snake scales-inspired (AISI 52100) by Photochemical machining process [110,125]	Velocity: 1 m/s Load: 10 N Pin-on-disc Counterbody: AISI H13 tool steel pin	CoF Reduction of 37.5% was obtained for sliding parallel to lengthwise scales direction.	Anisotropic friction behavior observed as a function of the sliding direction. Friction depends on aspect ratio (scales) and percentage of texturized area on the surface.
Snake scales-inspired (Silicon/Epoxy) by Direct moulding [136]	Velocity: 0.002 m/s Load: 0.5–3.92 N Disc-on-disc Counterbody: Same as samples	~0.1–0.3 dry	Complex friction behavior may lead to the possibility of inversion of friction anisotropy depending on the materials' stiffnesses, the aspect ratios of the structural features, and substrate roughness.
Fish scale-inspired(MXene $Ti_3C_2T_x$ based fabric composites) by Impregnation and drying [146]	Velocity: 0.7–1 m/s Load: 25–40 MPa Pin-on-disc Counterbody: AISI-1045 steel pin	~0.04 dry	77% lower wear rate compared to pure fabric composite. 38% lower CoF compared to pure fabric composite. CoF and Wear rate increase with increasing sliding speed. Fish scale-like structure strengthens the mechanical interlocking and chemical bonding between matrix and MXenes.
Fish scale-inspired (Mold steel) by 3D Printing [147]	Velocity: 0.084–0.134 m/s Load: 5–25 N Pin-on-disc Counterbody: Bronze pin (HRC 45) Lubrication: Pharmaceutical white oil	~0.10–0.18 lubricated, tilt 45°	Tilt 45° resulted in the lowest CoF value. Squeezing out of lubricant observed. CoF decreases with the increase in sliding velocity. CoF decreases with the increase in textures number. CoF increases with the increase in sliding load.
Ocellated Skink Scale-inspired (polydimethylsiloxane) by 3D Printing [148]	Velocity: 0.002 m/s Load: 0.02, 0.05, 1 N Stroke: 0.004 m Ball-on-disc Counterbody: Chrome steel ball	~1.04 for 0.2 N, dry ~0.93 for 0.2 N, dry ~1.04 for 1 N, dry	For the sliding loads 2 and 0.5 N tests, the CoF reduced by 30% and 26%, respectively. Average wear tracks widths were reduced with 61%, 48%, and 44% reduction under 0.2 N, 0.5 N, and 1 N loading conditions, respectively.

Table 2. *Cont.*

Bioinspired Texture and Material	Sliding Test Parameters	Friction Coefficient (CoF)	Conclusions/Texture Effect
Pangolin scale-inspired (obtained from real animal) mounted on steel block [149]	Velocity: 0.42 and 0.84 m/s Load: 30–90 N Distance: 252 m Block-on-ring Counterbody: 0.04 m gray iron ring	~0.22–0.42 for 0.84 m/s dry ~0.28–0.49 for 0.42 m/s dry	The wear rate at higher velocity was larger than that at the lowest velocity. Transfer film at 0.84 $m{\cdot}s^{-1}$ velocity was enlarged a little as compared with that at 0.42 $m{\cdot}s^{-1}$ velocity. High loads led to lower CoF.
Lotus leaf-inspired nano pillars coated with PMMA (silicon wafers/PMMA) by Soft lithography [101]	Velocity: 2 μm/s, 0.001 m/s Load: 0–80 nN, Time: 15 min Ball-on-flat Counterbody: 1.25 μm Borosilicate ball	~0.025 untextured, dry ~0.01 textured, dry	Patterns show superior adhesion and friction behavior due to their hydrophobic nature and reduced contact area. Patterns show lower CoF than PMMA thin film owing to reduced contact area.
Lotus leaf-inspired micropillars coated with PMMA (silicon wafers/PMMA) by Soft lithography [101]	Velocity: 0.001 m/s Load: 0.003 N, Time: 15 min Ball-on-flat Counterbody: 1.25 μm Borosilicate ball	~0.65 untextured, dry ~0.2 textured, dry	Patterns show lower CoF than PMMA thin film owing to reduced contact area.
Lotus leaf-inspired circular and bi-triangular microdimples by 7 and 20% density (100Cr6) by Lasers [128]	Velocity: 0.2, 0.6, 1 m/s Load: 15 N Scar dia: 0.008 m Pin-on-disc Counterbody: 100Cr6	~0.56 for 7% and 20% dense dimples, 1 m/s, bi-triangular textured dimples ~0.55 for 20% dense dimples, 1 m/s, circular textured dimples	Dimples with higher density (20%) showed less wear and lower friction compared to untextured disks or disks with 7% dense dimples, especially at higher speeds. CoF in sliding pair with a textured sample is less than untextured one. Higher dimples increase the entrapment of a larger number of wear particles
Maize leaf-, Shark-, Snake-, Pitcher-, and Lizard skin-inspired surfaces (PLA) by FDM 3D printing [134]	Velocity: 0.01 m/s Distance: 200 m Load: 27, 59 MPa Time: 15 mins Stroke: 0.01 m Ball-on-plate Counterbody: 0.01 m AISI 52,100 balls Sliding directions: high friction direction (HFD) and the low friction direction (LFD)	At 27 MPa four times rise in the wear track width was reported when sliding against the anisotropic patterns. Lowest and highest values of wear track width (anisotropic) were found for pitcher and snake patterns, respectively. Lowest and highest values of wear track width (isotropic) were found for flat and shark patterns, respectively. Larger wear track width when increasing the contact pressure to 59 MPa. Both the bioinspired pattern and the sliding direction could be advised as the optimum test parameters for reaching the target frictional behavior.	The surface texture exhibited a considerable effect on friction anisotropy under a contact pressure of 27 MPa, specifically for the lizard, the snake, and the pitcher plant. Anisotropic pattern and sliding direction (LFD) could be advised as the optimum test parameters for reaching the best frictional behavior. The CoF decreases outstandingly when the contact pressure increases up to 59 MPa irrespective of the used patterns and particularly when sliding in the HFD due to the increased contact pressure; the morphology of the 3D-printed surface is damaged and compressed against the worn surface, causing the rise in surface temperature.
Tree frogs' toe pad-inspired elastic micropatterned dimple/pillar arrayed structures for wet sliding contacts (Polydimethylsiloxane elastomers) by FDM 3D printing [132]	Velocity: 0.003–0.2 m/s Load: 0.95 N Time: 15 mins Lubrication: Deionized water Pin-on-disc Counterbody: Spherical pins and disks made from Polydimethylsiloxane elastomers	High densities of microstructures are conducive to achieving stable friction from low to high sliding speed. Flat disk displayed higher friction than the microstructured disks at low speeds ($v < 0.04$ m/s). In the high-speed region ($v > 0.04$ m/s), all the friction forces were well above those of the flat disks.	Three types of patterns: round dimple, round pillar, and hexagonal pillar. Surface with microdimples displayed a reduction in friction with an increase in the pattern area density. The area density of pillar patterns had no significant effect on the friction property at low sliding speeds, whereas it became a dominating factor with the increase in sliding speed.

Table 2. *Cont.*

Bioinspired Texture and Material	Sliding Test Parameters	Friction Coefficient (CoF)	Conclusions/Texture Effect
Lotus leaf-inspired micropillars/channels coated with DLC or Z-DOL (silicon wafers/DLC/Z-DOL) by Photolithography and Deep Reactive Ion Etching [150]	Velocity: 0.001 m/s Load: 0.003 N Time: 15 mins Ball-on-flat Counterbody: 0.001 m soda lime balls	~0.5 for bare, untextured Si wafers ~0.15 for bare Si wafers, pillar-textured ~0.2 for DLC coated, untextured ~0.11 for DLC coated, pillar-textured ~0.15 for Z-DOL coated, untextured ~0.09 for Z-DOL coated, pillar-textured	Chemical modification by coating DLC/Z-DOL and topographical modification by micropatterns on Si surfaces lowers friction and wear. Topo-modification of Si surfaces renders a reduction in CoF due to the physical reduction in the contact area and debris removal. Micropillars show lower friction than microchannels due to lower contact area. A dual/combined surface modification significantly lowers friction and shows no observable wear of either material.
Lotus leaf-inspired circular microdimples (Ti-alloy coated with $CrN-MoS_2-Ag$) by Lasers and Sputtering [151]	Frequency: 20 Hz Load: 20, 30 N Stroke length: 0.001 m Ball-on-disc Counterbody: 10 mm SiC ball	~0.2, untextured and textured coated surface More than 50% reduced wear for textured coated surface in comparison to untextured coated surface.	Significant 74.4% and 60.4% improvement in wear resistance compared to substrate surface under sliding loads of 20 and 30 N. Dimples resulted in secondary lubrication and capture of debris.
Circular microdimples on bronze surface (CuSn6) by Macro indenter [133]	Velocity: 200 r/min Load: 5, 15, 25 N Scar dia: 0.008 m Ball-on-disc Counterbody: 0.004 m Graphite balls	~0.26, 0.22, 0.20 at 5, 15, 25 N for textured. ~0.35, 0.26, 0.24 at 5, 15, 25 N for untextured.	15%, 21%, 12% decrease in wear rate for textured samples after 5 N, 15 N, 25 N sliding.
Seashell-inspired composite surface pattern (Ni3Al matrix with Sn and Ag) by LMD and wire cutting [142]	Load: 25 N Amplitude: 5 mm Frequency: 1 Hz Sliding time: 1800 s Ball-on-disc Counterbody: 0.006 m high-carbon steel balls	Increasing texture density results in CoF and wear rate decrease first (until 20%) and later leads to increase in their values for textures without SL. For textures with SL, the decrease in their values was noted all throughout from 0–50% density of textures.	Textured surfaces with solid lubricants were always better than patterned without solid lubricant due to the wear debris entrapment of solid lubricants and could improve the anti-friction performance.
Mussel-inspired copolymer (MPC with surface induced Nanodiamonds by Copolymerization and stirring [152]	Load: 25 N Amplitude: 0.005 m Frequency: 1 Hz Sliding time: 1800 s Ball-on-disc Counterbody: 0.006 m high-carbon steel balls	~0.028 for non-lubricated textured surface. ~0.017 for ND-lubricated textured surface.	Significantly reduced the wear on the tribopairs but also further decreased the COF by approximately 40%. ND could be attributed to the rolling effect of the nanoparticles.
Honeycomb-inspired self/lubricating steel composite (AISI 4140 steel/SnAgCu–TiC) by Laser cladding [137]	Velocity: 200 r/min Load: 20 N Block-on-ring Counterbody: AISI 4140 steel	Compared with AISI 4140 steel, the average COF of self-lubricating textured composites was decreased by 67%, and the wear depth was decreased by 42%.	Self-repairing behavior of bionic textured AISI 4140 steel filled with multi-solid lubricants. Texturing enables wear debris trapping. Wear debris consists of solid lubricants.
Frog paw-inspired self/lubricating steel composite (AISI 4140 steel/SnAgCu) by Additive manufacturing [138]	Velocity: 0.016 m/s Load: 30 N Block-on-ring Counterbody: 0.063 m Si_3N_4 ceramic ball	Compared to untextured AISI 4140 Steel, AISI 4140-SnAgCu with optimized bionic texture reduced the average friction coefficient by 20%, fluctuation degree by 54%, and wear track depth by 65%.	Optimized parameters of bionic hexagonal type microtexture were reported. Optimized surface texturing resulted in uniformly dispersed lubricants on the worn surface. Superior friction and wear reduction after optimization.
Biomimetic coating (Calcium phosphate on Ti) by Biomimetic mineralization [153]	Velocity: 0.02–0.17 m/s Load: 2 N Pin-on-disc Counterbody: 100Cr6 pin Under dry and simulated body fluid.	~0.6 dry for pure Ti ~0.3 dry for coating ~0.01 lubricated for coating	50% lower CoF in dry condition in comparison to pure Ti. 80% lower CoF in lubricated condition for faster sliding speeds, i.e., >0.07 m/s.

Table 2. *Cont.*

Bioinspired Texture and Material	Sliding Test Parameters	Friction Coefficient (CoF)	Conclusions/Texture Effect
Cancellous bone-inspired ZrO_2 coating (ZrO_2/modified PTFE) by Thermal spraying/Laser texturing [77]	Velocity: 0.01–0.15 m/s Load: 1–30 N (654–2035 MPa) Ball-on-disc Counterbody: ZrO_2 ball, 0.005 m dia. Dry sliding.	Low CoF (<0.065) at ultrahigh contact pressure (10 N, 1411 MPa) with an extremely long lifetime (>1 × 10^6 cycles). CoF decreased from 0.070 to 0.048 as the load increased from 1 to 30 N. CoF decreased from 0.089 to 0.048 when the sliding velocity increased from 1 to 0.15 m/s.	Low CoF maintained a large range of sliding velocities and applied loads. Textured composites show zero-wear properties.
Biomimetic metal ceramic composite (Ti/Ti64 alloy and Calcium phosphate) by Laser deposition [154]	Stroke length: 0.01 m Stroke rate: 0.017 m/s Load: 5 N Ball-on-disc Counterbody: 100Cr6 ball Under simulated body fluid	~0.85–0.9 SBF lubricated	CaP addition to the Ti increased the strength, hardness, and wear resistance. Wear rate decreased by 92% when 10% wt CaP was added to CP-Ti. 5% wt CaP to Ti64 decreased the wear rate by 70%.

4. Bioinspired Surface Texturing of Biomaterials

Biomaterial surface modifications, regarding both chemical and topographic approaches inspired by nature, have been explored extensively [43]. Chemical modifications often involve the use of nature-derived compounds, such as biopolymers inspired by mussels, known for their exceptional biocompatibility and adhesive properties [155]. The current section addresses specific surface qualities targeted for reducing bacterial adhesion [88,156], improving cell adhesion, self-cleaning ability [25], including surface roughness, wettability [12], surface energy [147,157], biofouling resistance [113,158,159], and adhesion strength [100,160,161]. Utilizing natural surface topography for modifying these attributes offers advantages as it avoids the release of chemical compounds into the surrounding environment where such surfaces are employed [162]. This topographic alteration in biomaterial surfaces has the potential to reduce the reliance on antibiotics for treating implant-related infections, consequently mitigating the risk of antimicrobial resistance. Figure 9 shows the self-cleaning and hydrophobicity (water-repelling) properties of the lotus surface.

Another aspect is that bacteria can attach to the surface of a material through various mechanisms, resulting in the formation of biofilms when released into a biological environment [113]. To counteract this, various strategies have been devised to disrupt bacterial interactions with surfaces. Topographically textured surfaces offer a chemical-free approach to bacteria inactivation. The effectiveness of bacterial destruction depends on the interactions between bacterial cells and the surface's texture features (mechanical response), such as pillars, columns, or rods, as shown in Figure 10c. Chopra et al. [79] reported that, when bacteria encounter nanopillars, they undergo a brief downward movement that causes their cell wall to rupture, resulting in lethal damage, as shown in Figure 10d. Bandara et al. [163] suggested a mechanism that involves a combination of strong adhesion forces and shear forces in a mixed short–tall nanopillars topography presence. When bacteria move across taller nanopillars, it causes the nanopillars to bend, which in turn stretches the bacterial cell wall, leading to its rupture. Other attributes like height, width, diameter, and spacing of these features significantly impact the bacterial response upon contact [43,161]. Even more fascinating are the physical alterations that involve creating micropatterns on surfaces to manipulate the hydration layer, rendering them either superhydrophobic or superhydrophilic [102,157,164]. This has given rise to technologies that mimic the texture of shark skin (also observed on lotus leaves) and replicate the superhydrophobic properties, leading to water repulsion and preventing the adhesion of contaminants, contributing to self-cleaning and antifouling properties such as the textures developed on surgical instruments (knife, etc.) [57,113,165,166]. Moreover, superhydrophilic biomaterials, which attract water strongly and facilitate rapid wetting and spreading of liquids across their surfaces, are particularly valued for their ability to promote crucial processes, such as cell adhesion, tissue growth, and fluid transport. Examples of applications include biomedical implants, tissue engineering scaffolds, and microfluidic devices, where the enhancement of these processes is essential for optimal functionality and performance [113,158,167]. Studies have revealed that the interplay between surface topography and various cell types impacts cell morphology, behavior, alignment, migration, and proliferation, among other traits [168,169]. These interactions ultimately alter the dynamics between cells and surfaces. Table 3 lists several bioinspired surface textures fabricated on biomaterials for their added functionality.

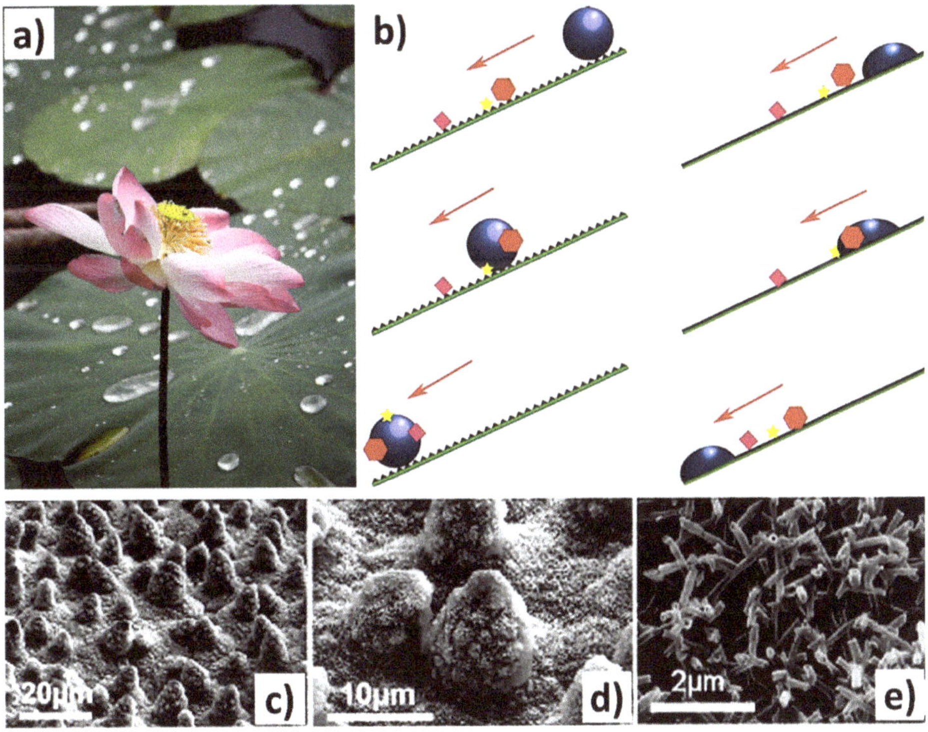

Figure 9. Images showing (**a**) hydrophobicity and self-cleaning characteristics of the lotus leaf surface; (**b**) a schematic representation of droplet movement (red arrow) on an inclined nanostructured superhydrophobic surface, known as the lotus effect. As the droplet rolls off the surface, it collects contaminating particles, effectively cleaning it (left). In contrast, a smooth surface merely redistributes the particles with the moving droplet (right); and (**c**–**e**) SEM micrographs of lotus leaf surface at various magnifications. Reproduced with permission from Ref. [25].

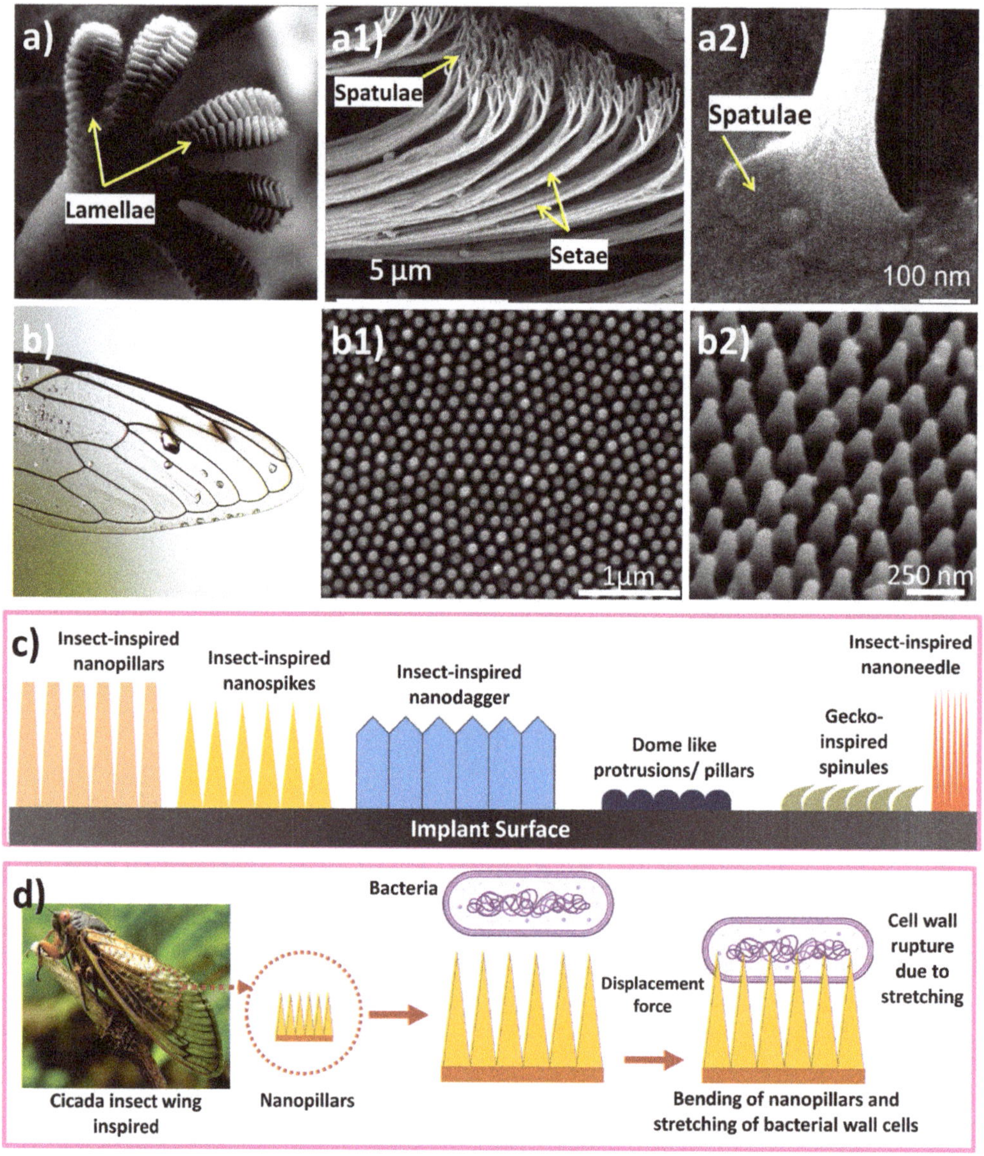

Figure 10. Illustration of (**a**) a gecko foot showing lamellae; (**a1**,**a2**) SEM of setae and hairlike spinules' structures on gecko toes [170]; (**b**) a cicada wing; (**b1**,**b2**) SEM showing nanopillar-like structure of cicada wing [171]; (**c**) schematic of different types of insect-inspired pillar-like nanotextures used towards fabricating antibacterial surfaces; and (**d**) schematic of bacterial killing/disrupting mechanism on a cicada wing-inspired nanopillar surface [79]. All images were reproduced with permission from respective publishers.

Table 3. Results of various bioinspired topographies on biomaterials for sliding wear conditions, as reported in the literature.

Bioinspired Texture	Biomaterial	Topography/Texture Feature	Functionalities
Lotus leaf-inspired [102,164,172]	Polymethyl methacrylate (PMMA) structures	Hierarchical surface featuring protrusions and valleys ranging from 3–10 μm, with nanometric particles (70–100 nm) of a hydrophobic wax-like material in the protrusions.	Antibacterial, self-cleaning
Sharkskin-inspired [165,166]	Polydimethylsiloxane embedded elastomer	Denticles refer to diamond-shaped scales with a raised ridge and concave groove, exhibiting some nanostructures. The Sharklet model comprises rectangular features.	Decrease in bacterial adhesion, whether independently or in combination with other chemical and photocatalytic substances
Cicada wings-inspired [109,173,174]	Poly (ethylene terephthalate), Titanium, Silicone-based elastomer nanopillars	Highly organized array of nanosized pillars or cones of varying dimensions.	Antibacterial
Rice leaves-inspired [156]	Polypropylene (PP)	Papillae in micron height on the surface.	Antibacterial, 53% reduction in bacterial adhesion.
Sea urchin-inspired [175]	Poly-lactic acid (PLLA)	Spiny finger-like structure.	Antibacterial, selfcleaning
Gecko skin-inspired [161]	Gecko skin	Dome-shaped scales are organized in a hexagonal pattern. These scales range from 100 to 190 μm in diameter and approximately 50 μm in height at the back. Spinules, resembling hairs reach up to 4 μm in length, with sub-micron spacing and a small radius of curvature usually ranging from 10 to 20 nm.	Antibacterial
Tree frog toe-inspired [160,176]	Polystyrene/Polydimethylsiloxane composites, Hydrogel	Hexagonal cells, separated by channels.	Improved adhesion
Butterfly wing-inspired [165]	Polydimethylsiloxane embedded elastomer	The wing surface features microscales, parallel ridges, and tile-like microstructures, along with nanoscale ribs and lamella-stacking nanostripe structures.	Self-cleaning coatings,
Nanoholes with atomic vacancies [177]	MoS_2	Nanohole-enabled nanomaterials	Anti-infection, enhanced biofilm formation.
Nanospikes [178]	Polypeptide-functionalized titania nanospikes	Titania (TiO_2) nanospike coating on the surface of a Ti substrate.	Antibacterial
Rose petals-inspired [100,179]	Polyethylene terephthalate glycol modified substrates	Hierarchical structures feature micropapillae.	Antibacterial and cell attachment.

Healthy articular cartilage in joints provides crucial lubrication for smooth movement of human hips or knees [180]. Inadequate lubrication of articular cartilage can lead to degenerative joint diseases, resulting in irreversible degradation of the cartilage. Hence, incorporating stimuli-responsive drug release into lubrication strategies can aid in repairing damaged articular cartilage. In terms of tribological studies, for instance, articular cartilage features a distinctive biphasic structure that incorporates both soft and hard materials to support heavy loads and minimize friction within joints [59,181–184]. To mimic this structure, Li et al. [59] developed a composite coating consisting of a dual-layer configuration comprising a thin PTFE layer atop a porous TiO_2 coating on Ti6Al4V alloy, as shown in Figure 11. The top self-lubricating layer reduces friction, while the underlying TiO_2 layer enhances wear resistance. This bioinspired bilayer coating demonstrates low friction (~0.1), exceptionally high load-bearing capacity (3.2 GPa under nanoindentation of 0.19 N, and a penetration depth of 600 nm), prolonged low-friction durability, and superior wear resistance during dry sliding. These remarkable properties are attributed to a self-repair mechanism, where PTFE is continuously replenished from the pores to the surface by frictional forces, effectively repairing scratches and combating abrasive wear. These findings offer insights into designing intelligent biomimetic coatings with enhanced mechanical properties, thus broadening the scope of applications for biomedical coatings in various fields.

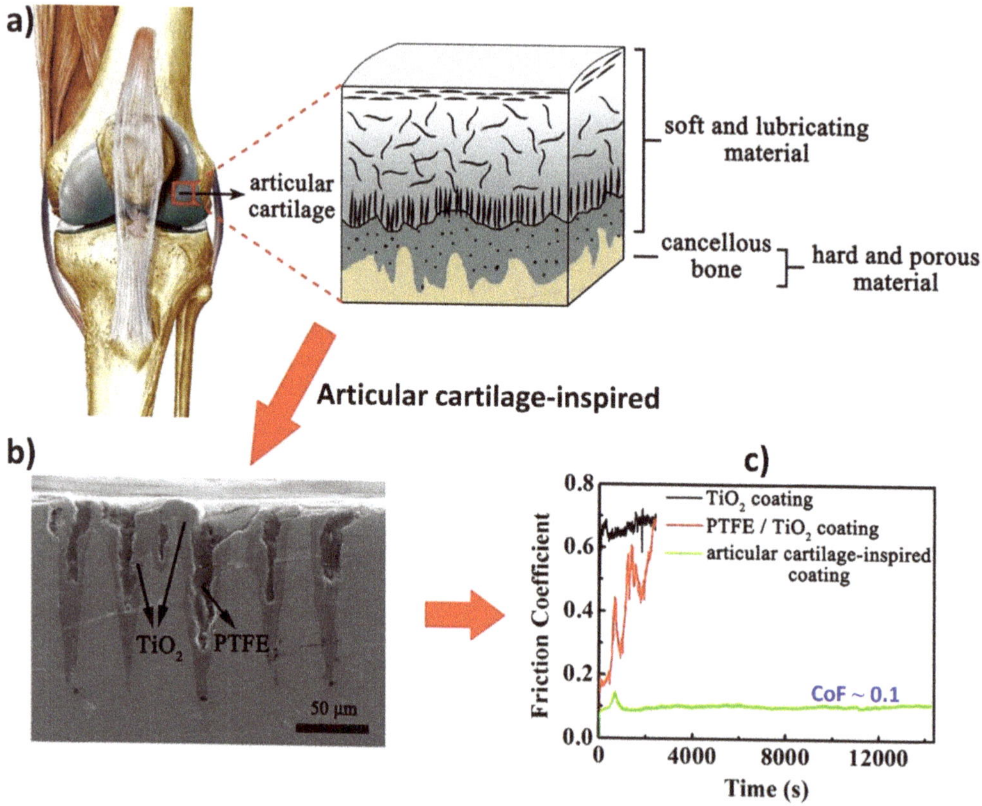

Figure 11. (**a**) Schematic illustration of cartilage organization in joints; (**b**) SEM image of the cross-section of the bioinspired bilayer coating; and (**c**) friction curves of the TiO_2, PTFE/TiO_2, and bioinspired bilayer coating under a ball-on-plate reciprocating configuration (counterbody 0.006 m dia. GCr15 steel ball, frequency of 1 Hz, stroke of 0.005 m, and a load of 20 N). Adopted with permission from Ref. [59].

5. Bioinspired Surfaces for Erosive Wear Resistance

Interest in anti-erosion characteristics and mimicked material design inspired by living creatures inhabiting harsh erosive environments, like desert lizards, sandfish, tamarisk, desert scorpions, and dung beetles, has grown gradually [24]. For instance, despite harsh desert conditions where sand particles travel at velocities over 100 km/h, *Androctonus australis*, a north African desert scorpion, remains resilient on the surface, enduring sand-laden harsh winds. Researchers have discovered that desert scorpions resist erosion through their body design. A team from Jilin University in China used 3D laser scanning to create a cloud point representation of the scorpion's back [185] as shown in Figure 12e. Their analysis revealed that the surface's convex and grooved design aids in erosion resistance. The grooves cause the air striking the scorpion's surface to rotate, creating a low-speed reverse flow zone. Essentially, this air flow zone acts like an air cushion, allowing some particles like sand to be harmlessly blown away rather than hitting the surface directly. The particles that do manage to pass through strike the surface with significantly lower velocity and impact angle, reducing the rate of erosion (Figure 12f).

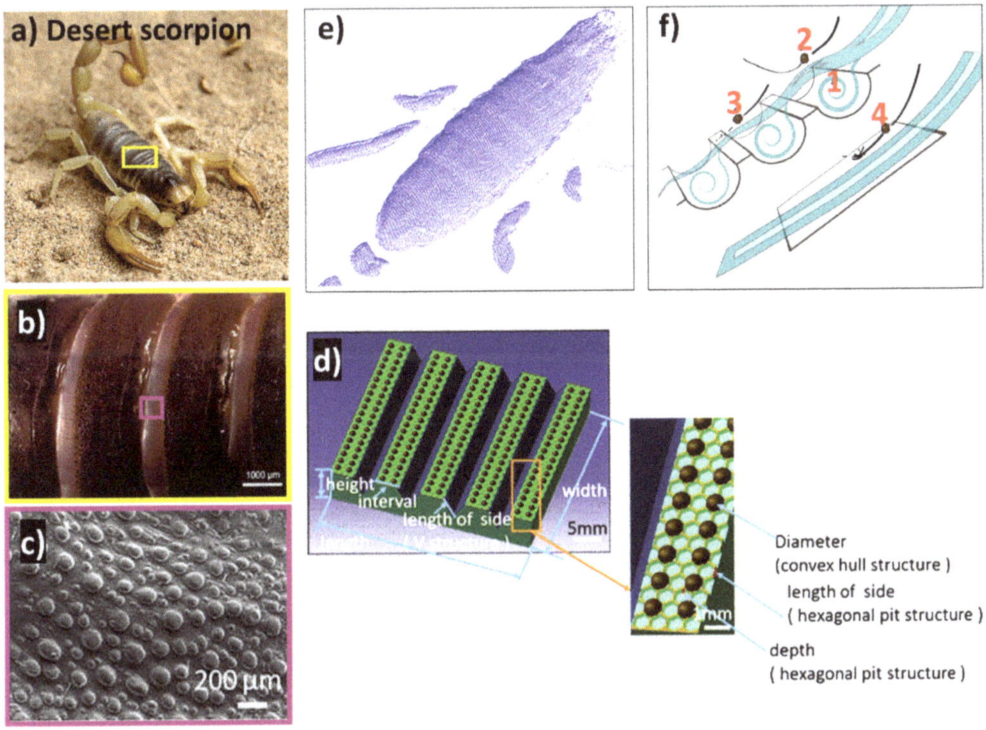

Figure 12. Images showing (**a**) back of an *Androctonus australis* (desert scorpion); (**b**) high-magnification image of desert scorpion back revealing V-groove-like pattern [186]; (**c**) scanning electron microscope image of scorpion back revealing outward bumps on the grooved surface [186]; (**d**) schematic of structure sizes of biomimetic samples; (**e**) point clouds of scorpion back showing bumps on the skin surface, later utilized to model an erosion study [185]; and (**f**) point 1—the swirling air caught in the grooves creates air cushions that slow down the velocity of the wind. This phenomenon reduces the speed of airborne particles, leading to different outcomes; point 2—some particles bounce away without impact, while point 3—others hit the surface at a much slower velocity and with less impact. Point 4—in contrast, a particle hitting a flat surface where the wind velocity remains unaffected will experience a much higher impact [187]. All images are adopted with permission from respective publishers.

Inspired by the back of the desert scorpion, a few researchers have reported a hexagonal pit structure with excellent anti-erosion properties [186,188–190]. The biomimetic samples with microstructures as shown in Figure 12d resulted in reduced erosion wear due to reduced solid particle velocity caused by rotational low velocity (explained in Figure 12f). Similarly, Han et al. [188] explored the erosion resistance of bump and groove shapes inspired by the desert scorpion surface, demonstrating increased erosion resistance compared to smooth structures by approximately 10% and 25%, respectively. Another study by Zhiwu et al. [185] reported that the erosion resistance of the desert scorpion involves a biological coupling approach, combining convex, grooved, and flexible elements. Numerical simulations showed that the grooved surface exhibited superior erosion resistance compared to smooth and convex surfaces, attributed to particle impact velocity, angle, and frequency. Erosion tests confirmed that grooved surfaces performed best at a 30° injection angle. Analysis of groove dimensions revealed that groove distance had the greatest influence on erosion resistance, followed by groove width and height.

Researchers have replicated snakeskin texture features, demonstrating that snake-like skin samples experience significantly less wear than smooth/untextured samples [24,110, 124–126]. Effectively mitigating erosion wear depends on optimizing the interaction and interference between contributing factors such as geometry, material, and mechanics. For instance, inspired by the head and pronotum surface of the dung beetle, Yang et al. and others [41,97,191] constructed and optimized a dung beetle-inspired textured drill bit surface, achieving better wear reduction and efficiency. However, the exact size and location of the dome and pit structures on the drill bit demand optimization based on soil environment and dung beetle species as the solid particles hitting these structures change trajectory, reducing wear on other structures. However, the relationship between the position and volume of the domes/outside dimples was overlooked and missing.

Tamarisk-inspired V-shaped grooved solid surfaces have been quantitatively studied for erosion rates under particle impact [55,192,193]. These grooves effectively reduce erosion wear within an impingement angle range of 20–60°. However, these impingement angles (global impingement angle) vary due to the ductility of the eroded substance. Jung et al. [194] successfully explained the dependence of the grooved surface's erosion rate on the impingement angle by considering the diversification of the local impact angle caused by the presence of the grooves. Other factors, such as multiple impacts of particles and air swirls within grooves, have negligible effects on erosion [194]. This suggests both the promise and limitations of this biologically inspired anti-erosive approach. Grooves aggravate erosion when the ambient flow direction is nearly parallel or perpendicular to the surface, which is eroded in a ductile manner. However, one can benefit from the grooves if the surface direction can be adjusted to align with effective anti-erosion impingement angles (i.e., 20–60°). Cylindrical surfaces with grooves exhibit greater erosion resistance compared to smooth surfaces, regardless of wind direction, as observed in tamarisks [30,195]. Yin et al. [193] tested different bioinspired grooves (square, U-shaped, and V-shaped) and found that the V-shaped groove morphology displayed the lowest erosion rate, as shown in Figure 13. However, over time, the bionic structure on the surface degrades, leading to decreased erosion resistance. Other approaches must be explored to limit erosive wear in bionic surfaces as eddy currents on non-smooth surfaces are reduced, eventually leading to erosion resistance loss.

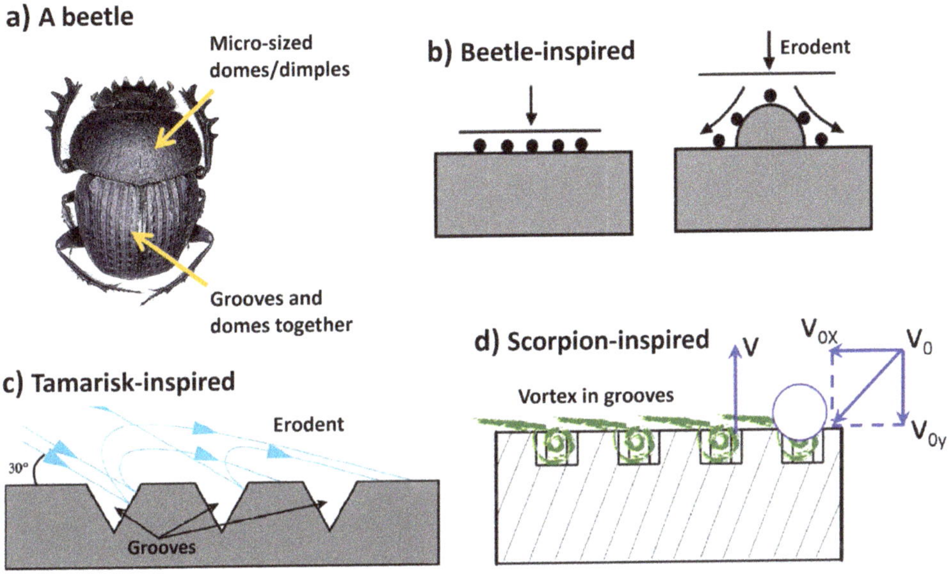

Figure 13. (**a**) Beetle skin surface revealing presence of domes/dimples and grooves [196]; (**b**) schematic diagram of smooth surface and beetle-inspired surface domes in erosion wear [41]; (**c**) schematic diagram of the interaction between tamarisk-inspired surface V-grooves and erodent particles [193]; and (**d**) schematic diagram of the interaction between desert scorpion-inspired grooved surface and erodent particles. Occurrence of rotational flow (vortex) for the impact velocity of particles (Vo), which forms the upward velocity vector perpendicular to the surface (V). The vertical upward velocity V is opposite to the velocity component Voy of the particle in the normal direction, which can slow the normal velocity of the solid particles. Studies show that, even if the rotational flow velocity is small, it still plays a very important role in reducing erosion wear [186]. All images were reproduced with permission.

Bionic tribological designs utilize the features of fast-swimming sharks that have skin with scaled structures aligned in the flow direction to minimize friction drag under turbulent conditions [24]. These structures, known as scales or denticles, resemble small dermal teeth and contribute to a 5 to 10% reduction in drag while swimming and are quite widely utilized to develop fluid flow erosive resistance [197]. Luo et al. [48] examined the morphology of shark skin scales and identified factors contributing to frictional drag reduction, including reduced wall friction from microgroove tips protruding into the viscous sublayer, decreased turbulence intensity near the wall due to backflowing microdroplets, a superhydrophobic effect from boundary slipping on the fluid–solid interface, and a nanochain of mucous increasing the thickness of the viscous sublayer to further reduce viscous resistance. Martin et al. [198] studied optimization of riblet geometries inspired by shark skin for minimal frictional drag. Their research emphasized that the reduction in drag was linked to vortices being lifted away from the surface and forming over the riblets, particularly under turbulent flow conditions. This phenomenon contributed to decreased overall shear stress. Fu et al. demonstrated that triangular-shaped riblets offered a more favorable balance between ease of manufacturing and effectiveness in drag reduction [46].

Many studies utilize various models such as V-shaped, U-shaped, I-shaped, scallop-shaped, and blade-shaped structures to replicate the characteristics of shark skin. Airbus applied grooved film to its A320 experimental aircraft, achieving an expected fuel-saving effect of 1% to 2% [199]. Similarly, China's Y7 aircraft experienced an 8% reduction in drag after being coated with a flow-grooved skin, while applying V-shaped trench film to the surface of the NACA 0012 aircraft resulted in a 6.6% drag reduction [199]. In contrast, the riblet surface exhibited a 9.9% decrease in adhesion resistance compared to the smooth surface [199]. The drag reduction effect was worse as the height of the riblet increased when h > 0.6 μm. Schulz et al. [200] investigated the application of shark skin riblet structures in high-temperature coatings for aerospace blade dynamics optimization. The experimental findings suggested that the zigzag groove spacing (s) should be approximately twice the height (h) of the zigzag groove, with h/s ratios ranging from 0.5 to 0.6 and s values between 15 and 17 μm. These findings offer valuable insights for establishing bionic riblet surface structures for erosive wear resistance. A study by Bechert et al. [201] reported the experimental results on frictional drag-reducing surfaces and their optimization with an adjustable geometry. The results of the study in terms of optimized structural dimensions are shown in Figure 14c.

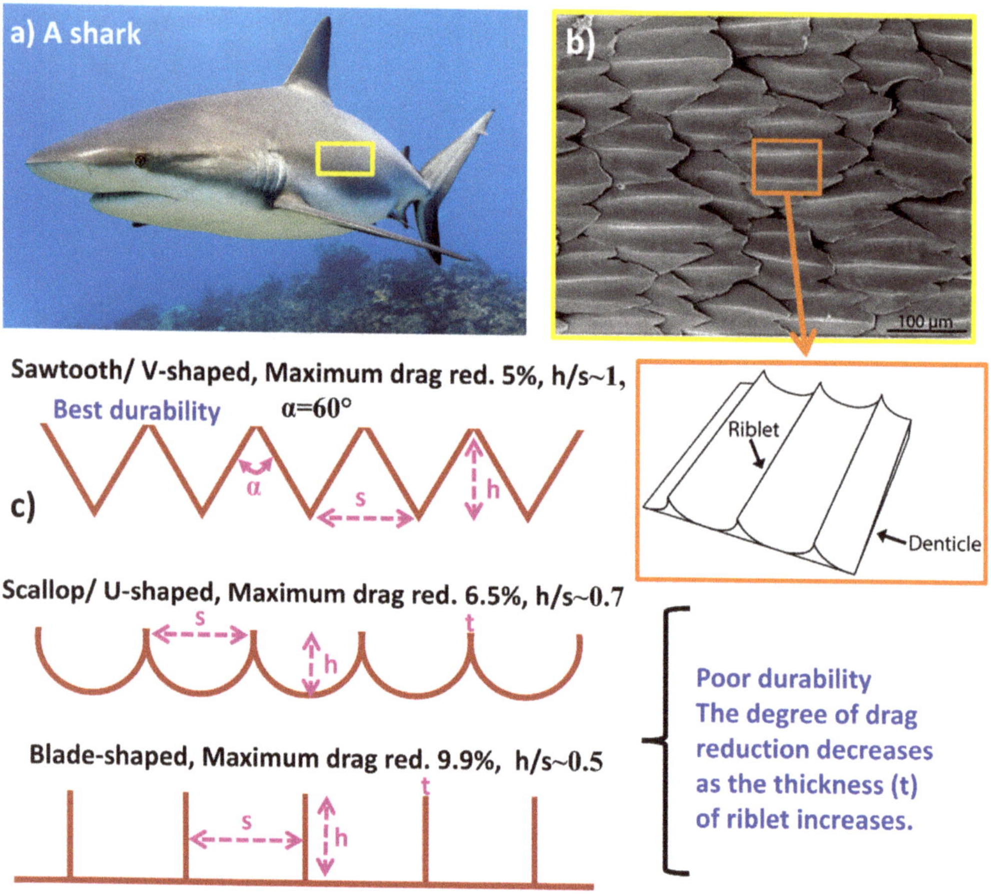

Figure 14. (**a**) Shark skin; (**b**) SEM image of shark skin showing microriblets structure [202,203]; and (**c**) proposed models of shark skin riblets' structures in the form of a sawtooth (V-shaped groove), scallop (U-shaped groove), and blade (I-shaped groove) [199]. Their optimal structural dimensions leading to improved performance are mentioned as per Ref. [201].

The surface effectiveness of the sharkskin or dorsal of the scorpion in resisting erosion is not solely due to its unique surface morphology but is also influenced by several factors related to coupling function. The protrusions and furrows on the surface serve as morphological coupling elements, while the adaptable linkage acts as a flexible coupling element.

The bionic model incorporates these coupling elements, as depicted in Figure 15. By eliminating the adaptable linkage from the model, it was divided into two layers. The upper layer comprises rigid material shaped into a bionic surface morphology, while the lower layer consists of pliable materials. This dual-layered arrangement creates an alternating composite structure of soft and hard materials, which is believed to offer superior erosion resistance, as shown in Figure 15d.

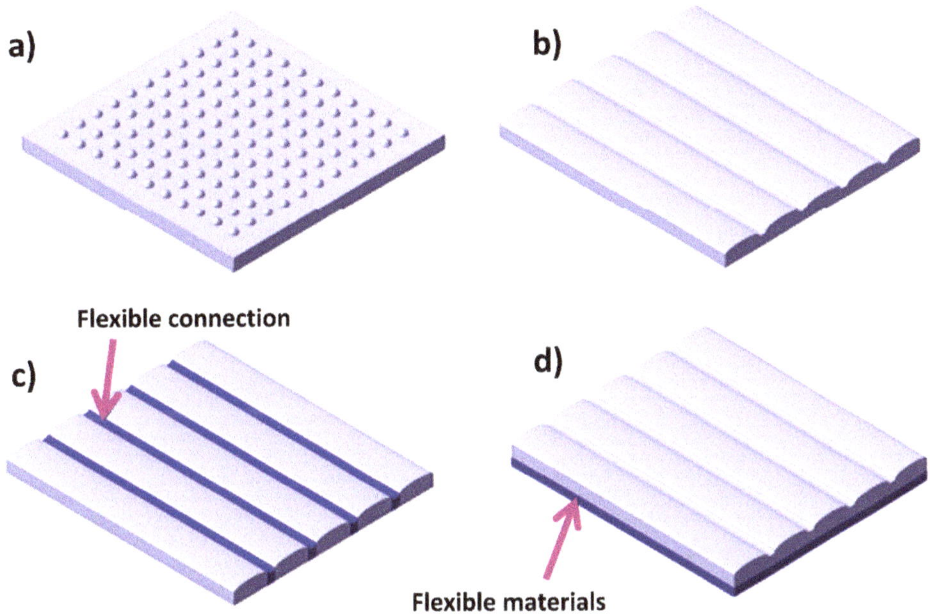

Figure 15. Bionic model of the desert scorpion's back with (**a**) dome pattern; (**b**) groove pattern; (**c**) coupling bionic models with flexible connection; (**d**) coupling bionic models with a composite structure alternating between soft and hard materials. Reproduced with permission from Ref. [204].

Erosive wear results from solid particles flowing along water or gas, impacting solid surfaces and causing repeated deformations [205]. The complexity of this phenomenon depends on various factors such as the size, shape, hardness, and concentration of erodent particles, as well as the elastic properties, surface hardness, and morphology of the eroded substance, and flow conditions like impacting velocity, angle, and location of impact [51,206–209]. Investigating erosion wear involves considering erosion rate, penetration depth, load spreading, and stress distribution [24,194]. The lack of a clear understanding of the erosion mechanism hinders the development of a simple, reliable, and universally quantitative erosion model. Despite the growing interest in anti-erosion characteristics and biomimicked material design, the current research in this area remains fragmented. Nevertheless, some significant results are documented in Table 4. Bioinspired surfaces offer erosion wear protection through various mechanisms: enhancing fluid turbulence along the wall, disrupting particle motion, reducing the number of particles impacting the eroded substance, and preventing particle sliding and rolling [194,210,211].

Table 4. Results of different bioinspired structures under erosive wear conditions, as reported in the literature.

Bioinspired Texture and Material	Erosion Parameters	Erosive Wear Rate	Conclusions/Texture Effect
Desert scorpion skin-inspired V-groove surface (AA6061 alloy) by Wire cut EDM [189]	Impact velocity: 30 m/s Erodent: SiC (irregular) Particle diameter: 100–150 μm Mass flow rate: 0.025 g/s Impact angle: 30° Nozzle diameter: 8 mm Impact distance: 200 mm Erosion time: 600 s	Stabilized erosion rates (200–600 s) for Textured: 4×10^{-8} g/mm^2·s Untextured: 9×10^{-8} g/mm^2·s	Anti-erosion property of the textured specimens increased by approximately 57.4%. The surface area due to V-shaped texture was increased by 67% compared to that of the smooth specimen. V-shaped textures changed the angle at which the erodent was incident to 90°, leading to reverse air flow and reduced erodent kinetic energy and erosion rate.
Desert scorpion skin-inspired V-groove surface (ABS polymer) by Simulation study [212]	Erodent: SiC (irregular) Solid particles density: 3.2 g/cm^3 Particle diameter: 105–830 μm Impact angle: 30°, 60°, 90° Erosion time: 10 s	Simulation by finite element software was applied to predict (regression analysis) the erosive wear behavior.	Bionic surface morphology can change the impact angle between particle and target and reduce the probability of impact angle of 20–0 to some extent and hence reduce erosion rate. Improved erosion resistance by 25–28% for textured samples for best conditions.
Desert scorpion skin-inspired Square, V, U grooved fan blades (Q235 steel) by Wire cut EDM [190]	Impact velocity: 30 m/s Erodent: SiO_2 (irregular) Particle diameter: 150 μm Mass flow rate: 0.025 kg/s Impact angle: 30° Nozzle diameter: 150 mm Nozzle length: 1000 mm Impact distance: 60 mm Erosion time: 450 s	Stabilized erosion rates (230–450 s) for Square-groove: 18×10^{-8} g/mm^2·s Square-groove: 15×10^{-8} g/mm^2·s U-groove: 12×10^{-8} g/mm^2·s V-groove: 12×10^{-8} g/mm^2·s	Blades with optimum parameters could effectively improve anti-erosion property by 29%. The formation of "air cushion" equips the V-type groove bionic surface with the best erosion resistance compared with other samples. V-groove surface, feature size of 4 mm, and distance of 2 mm
Androctonus australis skin-inspired V-groove, convex hull, and hexagonal pit grooved surface (Stainless steel) by 3D printing [186]	Impact velocity: 25 m/s Erodent: SiO_2 (irregular) Particle diameter: 260–941 μm (varied) Mass flow rate: 0.025 kg/s Impact angle: 30° Impact distance: 21 cm Erosion time: 350 s	Stabilized erosion rates (90–350 s) for Untextured: 10×10^{-7} g/mm^2·s V-groove: 9×10^{-7} g/mm^2·s V-groove + convex hull: 8×10^{-7} g/mm^2·s V-groove + hexagonal pit: 8×10^{-7} g/mm^2·s V-groove + convex hull + hexagonal pit: 7×10^{-7} g/mm^2·s	Hexagonal pit structure had best anti-erosion effect. The reason is that hexagonal pit slot can form the rotational flow, which can affect the motion direction of the particles and reduce impact velocity of the particles. A combination of groove + convex hull + large depth of hexagonal pit led to lowering the erosion rate.
Fish- and Snakeskin-inspired flexible structures (Ti6Al4V + rubber) by 3D printing [24]	Impact velocity: 30, 50, 80 m/s Erodent: SiO_2 Particle diameter: 0.1–0.6 mm Impact angle: 30°	10, 22, 96 mm^3/kg at 30 m/s, 50 m/s, 80 m/s.	Fish- and snake scale-inspired structures showed improved wear at 50 m/s and a slightly higher wear rate at 30 and 80 m/s compared to rubber.
Mole pelt-inspired flexible structures (AISI316L + Diamond + rubber) by 3D printing [24]	Impact velocity: 30, 50, 80 m/s Erodent: SiO_2 Particle diameter: 0.1–0.6 mm Impact angle: 30°	2, 16, 80 mm^3/kg at 30 m/s, 50 m/s, 80 m/s.	Mole pelt-inspired flexible structures exhibited around two times better erosion resistance during 30 m/s impact velocity compared with Hardox 400 and polymer while being compressed by 50% of its initial thickness without scarifying its wear resistance.

Table 4. *Cont.*

Bioinspired Texture and Material	Erosion Parameters	Erosive Wear Rate	Conclusions/Texture Effect
Dung beetle-inspired micro/macro surface (Z-ABS) by 3D printing [211]	Impact velocity: 56 m/s Erodent: Al_2O_2 Particle diameter: 212–300 μm Mass flow rate: 5.7 g/s Impact angle: 0–90° Impact distance: 20 mm Erosion time: 10 s	Erosion rate (g/g): 2.1×10^{-4} at the impact angle of 15°, 3.6×10^{-4} at impact angle of 30°, and 1.2×10^{-4} at impact angle of 90°. The erosion rate is 3.5×10^{-4} at impact angle of 45°.	Erosion rate of macrotextured surface in parallel (0°) direction was approximately 28% higher than microtextured surface. Erosion rate of macrotextured surface in perpendicular (90°) direction was approximately 71% lower than microtextured. Air cushion effect could be another mechanism of anti-erosive activity of the textured surface. Rebounded particles in perpendicular (90°) led to decreased incoming impact and final wear rate.
Tamarisk-inspired micro-V-groove surface (Original tamarisk trunk) [55]	Impact velocity: 30 m/s Erodent: SiO_2 Particle diameter: 40–70 mesh Mass flow rate: 25 g/s Impact angle: 90° Impact distance: 20 mm Erosion time: 10 s	Erosion weight loss (mg): Square groove: 93 mg U-type groove: 91 mg V-type groove: 75 mg	V-groove surface on centrifugal fan blades with optimum parameters can effectively improve anti-erosion property by 28.97% in comparison to other, i.e., with square groove surface, U-type groove surface, and convex dome surface, and the sample with smooth surface.
Tamarisk-inspired micro-V-groove surface (acrylic plate) [194]	Impact velocity: 30 m/s Erodent: SiO_2 Particle diameter: 230 μm Mass flow rate: 2.83 g/s Impact angle: 10°, 90° Impact distance: 20 mm Erosion time: 1 h	Erosion weight loss depends on global and local impingement angle.	Cylindrical surfaces with grooves are shown to be more resistant to erosion when compared to smooth surfaces, regardless of the wind direction. Erosive wear rate range of global impingement angles of a grooved surface changes according to the ductility of the target surface. Factors such as multiple impacts of particles and the air swirls within grooves were shown to exert negligible effects on erosion.
Scorpion skin-inspired bumped, grooved, and curvature structures (Stainless steel) by 3D printing [188]	Impact velocity: 30 m/s Erodent: SiO_2 Particle diameter: 230 μm Mass flow rate: 2.83 g/s Impact angle: 30° Erosion time: 180 s	Curved surfaces with bumps and grooves show a higher erosion rate than just curved surface. Smooth surfaces with bumps and grooves show lower erosion rate than just smooth surface.	Anti-erosion property due to bumps can be enhanced by 25%. Bumps have the better anti-erosion performance due to reduced area of impact and enlarge relative impact angle in a certain area, and, the smaller the relative impact angle, the lower the erosion rate.

6. Bioinspired Structures for Impact or Energy Absorption

Drawing inspiration from natural organisms such as beetles, woodpeckers, turtles, etc., researchers have explored novel structural designs and materials that exhibit remarkable impacts or energy absorption properties, as shown in Figure 16. By mimicking the unique features and mechanisms found in nature, such as hierarchical microstructures, multilayered compositions, and energy dissipation mechanisms, bioinspired structures offer promising solutions for mitigating impact forces, enhancing structural resilience, and sometimes improved flexibility.

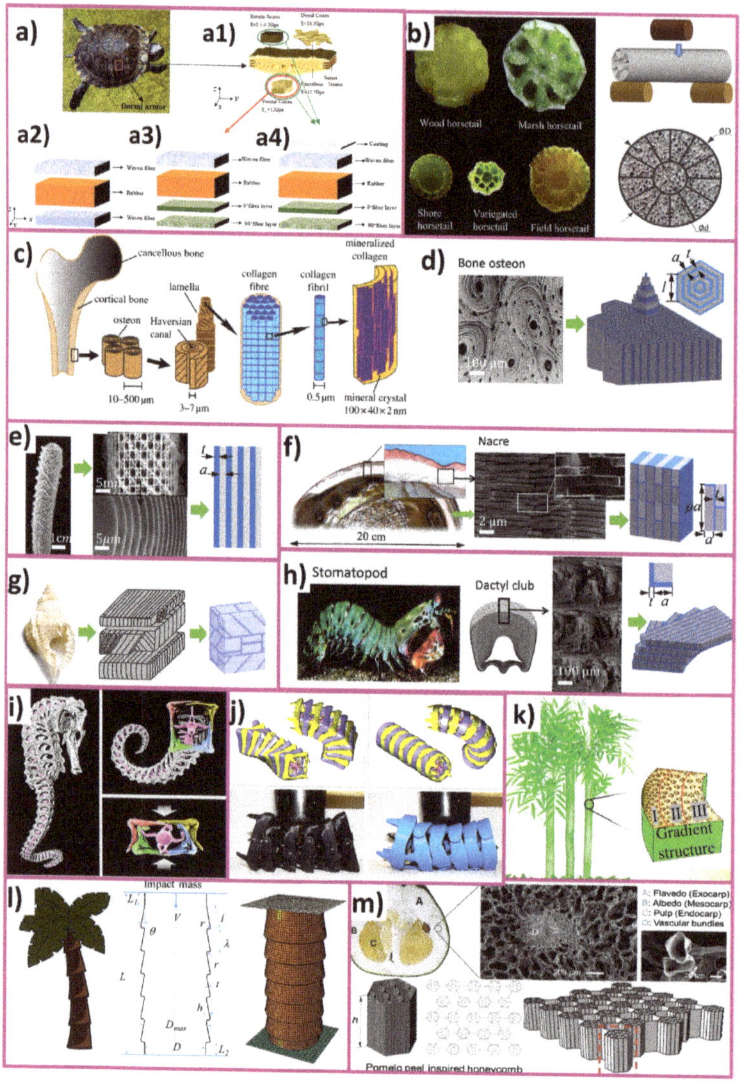

Figure 16. (**a**) The macroscopic morphology of the turtle shell is depicted [49]; (**a1**) illustrates a cross-sectional view of the turtle shell carapace, highlighting its composite layers; (**a2**) ordinary three-layer sandwich structure. Moving to the bionic sandwich structures, (**a3**) fine structure of the ventral cortical layer, resulting in a four-layer sandwich structure, and (**a4**) bionic sandwich structure that mimics the keratin scutes, leading to a five-layer sandwich structure. Reproduced with permission from Ref. [49]; (**b**) horsetail-inspired foamy structures under lateral loading. Reproduced with permission from Ref. [50]; (**c**) bone's architecture exhibits five levels of hierarchy; (**d**) SEM image illustrates the bone osteon along with a concentric hexagonal unit cell model, (**e–h**) optical image, SEM image, and microstructures of sea sponge, nacre, conch shell, and stomatopod dactyl club, respectively. Reproduced with permission from Ref. [213]; (**i**) seahorse skeletons are composed of highly articulated bony plates for flexibility. Reproduced with permission from Ref. [214]; (**j**) DNA-inspired helical structures and their deformations during compression tests. Reproduced with permission from Ref. [215]; (**k**) bamboo-inspired functional gradient structure. Reproduced with permission from Ref. [216]; (**l**) bioinspired conical corrugation tube. Reproduced with permission from Ref. [217]; (**m**) pomelo peel-inspired honeycomb. Reproduced with permission from Ref. [218].

The turtle shell abaxial armor, comprising lightweight and robust natural composite laminated material, typically consists of three to six layers. Its macro-mechanical properties are influenced not only by its multilayer structural characteristics but also by its intricate multiscale and multilayered microstructure and layering methods. The complex morphology of the turtle shell has been extensively studied by several researchers [49,219–225] to understand its superior protective capabilities against environmental penetration. Pangolin shells offer similar characteristics to turtle shells and have been studied by several researchers [24,135,149,226].

Beetles' elytra serve as lightweight yet rigid structures, effectively shielding beetles from external impact loads, capable of withstanding an impact velocity of up to 1 mm/min [227,228]. The superior mechanical properties of the elytra are attributed to the irregular cellular pattern observed in the laminated cross-section, featuring hollow pillars at the wall and intersections. This pattern includes circular cross-section fibers sparsely distributed around the outer edge, transitioning to rectangular structures with a denser distribution towards the middle lamination. Furthermore, the beetle's forewing is reinforced by trabecular structures between different laminas, significantly enhancing inter-laminar strength, approximately 30 times higher than that of pure chitin fiber laminas [228]. Some studies on beetle inspiration have shown its tremendous benefits regarding the impact-absorbing structure [41,70,97,227,229,230].

Woodpeckers demonstrate remarkable shock absorption capabilities when they drum rhythmically with their beaks. They can peck at a frequency of 18–22 times per second and around 12,000 times a day on average, with each peck lasting approximately 50 ms. Despite the high impact deceleration and a repeated impact velocity of 6–7 m/s during the pecking process, no head injuries are observed afterward [228,231]. Ha et al. [231] created a woodpeckers' beak-inspired novel wavy honeycomb sandwich panel and investigated its structural design approach by adjusting wave number and amplitude. This innovative design outperformed the traditional hexagonal honeycomb sandwich panel in energy absorption performance.

The mantis shrimp, known for its powerful raptorial projections, utilizes its sharp dactyl club as a biological hammer for smashing prey and defending against predators. With extreme accelerations facilitated by a unique power amplification mechanism, it can strike both hard-shelled and soft-bodied prey with remarkable force. Particularly, the peacock mantis shrimp can achieve accelerations of over 105 m/s^2 and impact velocity exceeding 20 m/s, making it one of nature's most formidable and fastest impact scenarios [228]. Most of the notable studies on shrimp-inspired energy-absorbing structures include [114,228,232–235].

Inspirations from various fruits and plants, such as nutshell, pomelo skin, bamboo structure, and horsetail, present promising opportunities for designing impact-absorbing structures. The pericarp, or fruit wall, of all angiosperms consists of three layers: the outermost exocarp, followed by the mesocarp, and the innermost endocarp. In nuts, the pericarp dehydrates entirely, resulting in a highly lignified, mechanically robust compact shell. On the other hand, in drupes, only the endocarp undergoes lignification, forming a resilient protective layer around the seed.

Yang et al. [236] studied pomelo peel and its bionic PEEK structures, finding that different varieties follow similar porosity trends: lowest near the endocarp and exocarp, highest in the mesocarp. The location of maximum porosity varies by pomelo variety. During quasistatic compression, deformation initiates in the highest porosity region, with a near-zero Poisson's ratio, while the region near the exocarp remains less compressed with a positive Poisson's ratio. Additively manufactured porous PEEK cubes and numerical simulations confirm the benefits of porous design from pomelo peels for efficient energy absorption, offering clear guidance for designing lightweight materials with high energy absorption.

Chen et al. [237] investigated three bioinspired hierarchical self-similar structures based on basic shapes (curve, circle, and hexagon). They assessed energy absorption rates of unit cells with varying structure ratios through experiments and simulations. Honeycomb and bamboo fractal structures outperformed snake-like fractals in energy absorption. The bamboo fractal structure displayed the highest energy absorption capacity and was used to construct carbon fiber-reinforced composite structures. The numerical findings showed that a seven-cell bamboo fractal structure effectively represents the energy absorption ability of a self-similar structure with an array of cells. Overall, the 3D-printed self-similar composite sandwich structure demonstrates significant potential as a lightweight energy absorption structure.

In terms of aircraft wing's navigation where energy absorption is as important as other mechanical parameters such as flexure strength, etc., Prakash et al. [238] developed a cashew nut-inspired bionic structure using 3D-printed high-stiffness lignin–ABS core and industrial hemp with aluminized glass fiber epoxy skin. The addition of 30 vol% Al-glass and hemp fiber with lignin-strengthened ABS core (60 vol%) significantly enhanced the mechanical properties of the composite. This composite demonstrated impressive tensile strength, flexural strength, Izod impact resistance, interlaminar shear strength, and compression strength. Table 5 records several bioinspired structures reported in recent literature for their superior impact or energy-absorbing capabilities.

Table 5. Results of different bioinspired structures under impact-absorbing (wear) conditions, as reported in the literature.

Bioinspired Structure	Impact/Mechanical Tests	Conclusions/Texture Effect
Pangolin scale (obtained from real animal) [135]	Microindentation: Load of 1 N holding for 15 s. Tensile testing: strain rates from 10^{-5}/s to 10^{-1}/s Compression testing: 10^{-3}/s	Microhardness around 220 MPa. Similar tensile behavior when stretched longitudinally and transversely, Young's modulus around 1.5 GPa, and tensile strength about 108 MPa. Able to absorb large energy when loaded at low strain rate (at 10^{-5}/s).
Pangolin scale (Al 6061-T6) by 3D Printing [226]	Ballistic experiment for armor. Projectile fired: Hard steel core Projectile mass: 0.785 kg Projectile velocity: 400 m/s Angle of attack: 0° Distance of attack: 10 m	Improved impact resistance through deflection of target due to the cantilever action (flexibility) offered by the single scale. Increased scales resulted in plugging of projectile due to absence of cantilever action.
Beetle wing-inspired honeycomb structure/column tubes (Aluminum) [239]	Impact velocity: 10 m/s Mass: 500 kg	More than 50% improved energy absorption under axial loading. Changing column thickness 0.5 mm–3.0 mm led to an increase in the total absorbed energy and energy absorptive effectiveness. Changing column height 50 mm–200 mm led to an increase in energy absorptive effectiveness first and then decrease due to buckling of the structure.
Oxhorn-inspired columnar tubes (Aluminum) [240]	Crashworthiness performance. Impact velocity: 10 m/s Mass: 500 kg	Specific energy absorption of bionic column was 46.2 kJ/kg and 1.3 times and 1.8 times higher than the four-cell conical tube. Energy absorption of bionic column increases with the increased thickness of the core.
Yakhorn-inspired columnar tubes (Aluminum foam) [241]	Loading rate: 5 mm/min Mass: 500 kg	Specific energy absorption of bionic tube was 51.23 kJ/kg and is 1.25, 0.89, and 1.02 times higher than that of circular tube, square tube, and tapered tube. Energy absorption capacity increases with the increase in taper, and then decreases.

Table 5. *Cont.*

Bioinspired Structure	Impact/Mechanical Tests	Conclusions/Texture Effect
Woodpecker's beak-inspired honeycomb sandwiched structure (AA6060 Al-alloy) [231]	Impact velocity: 6–7 m/s Mass: 1 kg	Specific energy absorption of the bionic structure increases by 125% and 63.7%, respectively, compared with that of the honeycomb sandwich panel with the same thickness core or the same volume core.
Nacre-inspired hierarchical structure (Photopolymer based soft matrix staggered composite) by 3D printing [242]	Strain rate: 0.156 and 0.178 min^{-1} Cyclical tensile strain loading at 1 Hz	Tests conducted on polymer composites revealed superior damping performance, with a loss modulus reaching approximately 0.5 GPa, significantly surpassing that of individual polymers. The damping enhancement is attributed to the large shear deformation of the viscous soft matrix and the large strengthening effect from the rigid inclusion phase.
Nacre-inspired composite (Stiff material—PLA; soft matrix—Nylon and Thermoplastic Polyurethane) by Fused deposition modeling [243]	Impact velocity: 3.13 m/s Mass: 20.41 kg Impact energy: 100 J	A comparative study of two different materials for soft matrix was carried out. Drop impact tests showed that there were increases of 25% and 120% in the energy absorption capabilities of the structure. Dimensions of the tablets and thickness of matrix play a major role in structural properties.
Conch Shell-inspired hierarchical structure (Stiff material—Veromagenta Soft material— TangoBlackPlus) by 3D printing [244]	Impact velocity: 2–3 m/s Load: 2500–7000 N Impact energy: 13–25 J	Impact tests demonstrate tortuous stress paths and even distribution within the bioinspired structures, leading to increased energy dissipation. Second-level hierarchy structures exhibit 70% and 85% increases in impact performance compared to single-level hierarchy structures and the stiff constituent, respectively.
Cybister Elytra-inspired sandwiched structure (Ti6Al4V alloy) by SLM [245]	Multi-objective optimization	Optimizing bioinspired sandwich structures involve exploring various core layers and arrangements. Two-layered core structures with vertically arranged configurations exhibit excellent mechanical properties, specific energy absorption of 9.16 × 10^3 J/kg, and energy absorption of 154.80 J.
Beetle Elytron-inspired sandwiched structure (Stainless Steel) by Direct metal laser sintering [246]	Tensile and compression experiment. Loading rate: 1 mm/min	The crash box test determined that the structure absorbed 375.5 J of energy, approximately 5 times more than the conventional structure.
Shrimp-inspired sandwiched structure (Photosensitive resin and nickel-coated carbon fibers) by Stereolithography 3D magnetic printer [112]	Compression and Charpy impact experiment. Loading rate: 1 mm/min	Magnetic 3D printing was utilized to fabricate Bouligand and herringbone architectures. The impact toughness and compression strength of the composites are contingent upon the orientation and distribution of the fibers. Bionic structures demonstrate enhanced fracture resistance and greater energy dissipation compared to traditional counterparts.

Table 5. *Cont.*

Bioinspired Structure	Impact/Mechanical Tests	Conclusions/Texture Effect
Turtle shell-inspired sandwiched structure (Photosensitive resin and nickel-coated carbon fibers) by Stereolithography 3D [247]	Tension and bending test. Loading rate: 2 mm/min and 0.15 mm/min. Impact wear test. Impact cycles: 10^3, 10^4, 10^5 Impact velocity: 184 mm/s, i.e., Kinetic energy: 7.6 mJ Force: 100 N	Studied hydration effects (12 h, 24 h, and 48 h) on mechanical and impacting properties of turtle shell. Under the same impact cycles, energy absorption and contact time increased with the extending of soaking time. The absorption rate is the lowest for shell without soaking.
Bamboo-inspired honeycomb structure (Nylon/Carbon fiber) by 3D printing [237]	Quasistatic compression test. Force: 10 kN Velocity: 2 mm/min until a displacement of 40 mm	Bionic structure absorbed 755.83 ± 39.6 J of energy. Bamboo-inspired structure shows better energy absorption than snake- or beetle-inspired structures.
Balanus-inspired thin-walled tube structure (AISI 304L) by Dep drawing [248]	Dynamic crushing test. Impact velocity: 4 m/s Impact velocity: 0.025 mm/s	Bionic shell absorbed more energy throughout the deformation history. High load carrying capacity of bionic structure due to hard outer core and soft hierarchical inner core. Crushing force of the balanus structure increased by 26%.
Horsetail-inspired thin-walled tube honeycombs structure (Aluminum alloy) by Simulation study [249]	Axial dynamic loading of different cross-sectional configurations (i.e., number of cells) in the bionic structure. Impact punch mass: 1000 kg Impact velocity: 15 m/s	Structures with 16 cells are recommended as energy absorbers. 16 cells were found to have the best overall crashworthiness for vehicle body applications.
Pomelo-inspired hierarchically structure (Actual pomelo peel) [250]	Free Fall. Impact height: 10, 13.5, 18 m	10 m height potential energy between 83.4 and 98.1 J. No sign of damage found. 13.5 m height potential energy between 100 and 117.9 J. No sign of damage found. 18 m height potential energy between 117 and 158.9 J. No sign of damage found.
Pomelo-inspired hierarchically structure (PEEK) by 3D printing [236]	Compression test. Loading rate: 0.02 mm/s Compression strain: up to 70%.	Guidance to design lightweight materials with high energy absorption. Crush force efficiency and dropping force efficiency are closer to 1 as compared with the non-pomelo peel-inspired structures. No abrupt changes in compression forces for bionic pomelo PEEK structures.
Cashew nutshell-inspired biocomposites (ABS/Hemp/Glass fiber) by 3D printing [238]	Tensile, Flexural, Compression, Izod impact, Low velocity drop load impact, Fatigue, Fracture toughness tests.	Best composition in vol.%: Resin: 60; Fiber: 30; ABS: 10; Lignin: 1 Maximum mechanical property achieved: tensile strength, flexural strength, Izod impact, and compression of 136 MPa, 168 MPa, 4.82 kJ/m^2, and 155 MPa, respectively. Maximum fatigue life counts noted for the composite about 33,709, 25,781, and 19,633 for 50%, 70%, and 90%. Maximum fracture toughness compared to pure epoxy resin, with value of 32.5 MPa·$m^{1/2}$. Highly toughened and marginally flexible composites could be potentially employed in the fabrication of high-endurance morphing wings for unmanned aerial vehicles and aircraft wings.

Table 5. *Cont.*

Bioinspired Structure	Impact/Mechanical Tests	Conclusions/Texture Effect
Turtle shell-inspired sandwiched structure (Rubber/Carbon fibers) by Vacuum assisted resin injection [49]	Drop hammer impact test. Impact mass: 7.29 kg Impact energy: 30 J, 90 J Impact height: 0.42 m, 1.26 m	Turtle-inspired structures show an improvement in impact resistance by 10–25%. Main impact failure modes include coating fracture, fiber fracture, compression damage deformation failure of the rubber core, fiber delamination, and interlayer degumming. Size of the hammer head radius has a significant effect on the energy absorption characteristics. Hammer head radius is 10 mm; the absorbed energy and specific energy absorption increase by 25% compared with hammer head radius of 8 mm and 83%.

A few promising examples of such inspiration applied to material structuring are described below and systematically compiled in Table 5.

Several features combine to influence the impact resistance of a material and its response to dynamic loading: hierarchical structure (discrete structural elements across multiple length scales); composition (distinct interfaces between a stiff phase for rigidity and strength and a softer phase for ductility); and porosity (gaps filled with air or fluid across all length scales).

Additionally, biological materials demonstrate viscoelastic and/or viscoplastic behaviors, characterized by time-dependent stress and strain responses and time-dependent permanent deformations, respectively [251]. These behaviors are attributed to the polymeric constituents present in biological materials, such as collagen, keratin, cellulose, hemi-cellulose, lignin, and chitin, which determine their ability to absorb and dissipate energy under dynamic conditions. These characteristics are ubiquitous across biological materials. Viscoelastic materials exhibit both elastic and viscous responses, where energy is stored during deformation and dissipated as heat upon loading, respectively. This time-dependent behavior allows viscoelastic materials to attenuate shock and dampen vibration, contributing to impact isolation. For instance, the viscoelastic response of muscle and tissues surrounding the woodpecker's hyoid apparatus reduces stress waves induced by pecking, while articular cartilage serves as a viscoelastic barrier against high-speed loading, limiting potential damage to surrounding tissues [252,253]. Despite the inherent viscoelasticity of all biological materials, alternative energy absorption mechanisms may prevail at high strain rates, influenced by structural factors such as layering and void distribution, as observed in wood's impact resistance mechanisms [231].

Furthermore, many biological systems exhibit specific impact-resistant structural elements predominantly on the micro- and mesoscale, including layered, gradient, tubular, sandwich, and sutured structures [251]. These elements, often found in combination, contribute to remarkable properties under dynamic loading conditions. While serving as foundational components for hierarchical impact-resistant structures, individual tests using computer simulations, 3D printing, and composite prepregs have shown that each of these arrangements can enhance a material's impact resistance independently [217,229,233,249,254].

A very good example is the woodpecker, as shown in Figure 17. When the woodpecker strikes an object with its beak, the immense force at the tip is cushioned by its beak structure and the resilient hyoid bone [255,256]. Consequently, the impact's stress is significantly diminished from the beak's tip to where it connects with the skull, reducing the force that reaches the bird's cranium. The cranial bone, consisting of dense compact bone enveloping layered plate-like structures, forms an efficient shock absorption system [257]. When encountering forces, this deeper bone disperses frequencies in various directions away from the central impact point. Although fragile on its own, the deeper bone, encased in compact bone, maintains flexibility within, absorbing shock during movement. Furthermore, the woodpecker's brain, proportionally small with a high surface area to weight ratio, further disperses impact force over a larger area, minimizing damage. Computer simulations suggest that only a minute fraction of the impact ultimately reaches the brain, bolstered by the bird's lengthy tongue, which envelops the brain for added protection [253].

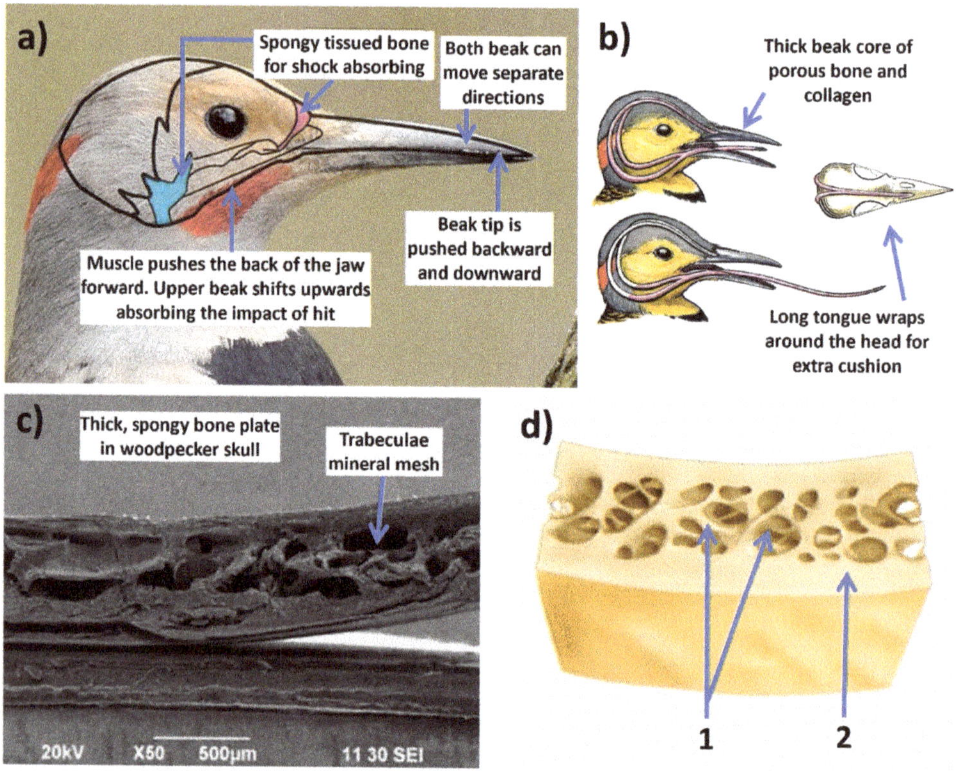

Figure 17. Images showing (**a**) a woodpecker's head and beak; (**b**) tongue wrapping around the brain to provide extra protection from impact; (**c**) SEM image of spongy bone situated in between brain and beak that absorbs impact (pink area in part (**a**)); and (**d**) schematic of bone showing 1—spongy bone; 2—compact bone. Images adopted CC BY-NC from Science China Press and asknature.org (Accessed on 12 February 2024) [255].

7. Bioinspired Design of Tools for Cutting or Machining

Bionic enhancements of cutting tools (soil, rock, metal, and biological) enhance tool durability, reduce friction, increase heat dissipation, and improve material/surgical removal rates. For example, soil adherence on soil excavation equipment poses a significant technical challenge due to soil characteristics. Metal cutting tools endure significant mechanical and thermal stresses due to the hardness and strength of metals, leading to considerable abrasion and chip adhesion [258]. This wear ultimately diminishes the tools' lifespan and compromises the quality of the workpiece during processing. This phenomenon elevates traction force by over 30% and concurrently diminishes productivity by the same margin [259]. In such cases, biomimetic designs inspired by the structure of animal teeth or plant leaves have been explored to increase wear resistance and reduce energy consumption during drilling operations. In rock-cutting tools or tunnel-boring machine applications, bioinspired surface modifications based on the properties of animal skins or plant surfaces have been investigated to reduce friction between the cutting tool and the tunnel surface, leading to greater mechanical traction, smoother excavation, and reduced tool wear. Moreover, features like anti-sticking properties for drilling the soft coal or earth [41,260], reducing sticking of drilling-breaks [41,261–265], reducing soil wear and adherence [41,259,266–268], self-cleaning [25,73,204,269,270], self-sharpening [271–273], self-lubricating [76,138,142,144], self-cooling [274], erosion protection (dry and slurry) [24,211,275,276], increased flexibility [24,277], reducing tool-chip friction [278], decreasing contact length between tool and chip [263,278], changing chip flow direction, improving adhesion resistance [88,266], stabilizing the built-up edge, lowering adhesive wear between tool and chip [105], etc., are shown to be attainable with specific biomimetic surface designs on machining/cutting tools. Additionally, some studies have suggested that a synergistic approach of solid lubricant and the bioinspired textures on tools (turning, etc.) can greatly enhance their performance due to considerable reduction in turning/milling/drilling force and temperature, resulting in major reduction in tool wear [138,270,279]. Some existing examples and nature inspirations reported are derived from python skin [123,274], sea urchin teeth [44,272,273], crab or dung beetle claw and legs [21,41,228,280], mole pelt claw [24,36,268,281,282], pangolin claw [24,52,226,277], wood wasp ovipositors [265,280], earthworm skin [262], seashell [283,284], human teeth [285,286], sunflower seeds [259], corn/maize leaf [287–289], badger teeth [36,44,271,273,285], etc. Figure 18 shows the self-sharpened serrated edges in different biological species.

Figure 18. Images showing self-sharpened serrated edges in (**a**) shark teeth [290]; (**b**) rock-eating sea urchin's teeth [291]; and (**c**,**d**) corn leaf [289]. Images were reproduced with permission.

In agricultural and other machining or cutting tools, a certain level of cutting resistance is expected and necessary for the tool to effectively perform its task. However, excessively high cutting resistance and soil adhesion can lead to increased working resistance, increased wear on the tool, and reduced cutting efficiency, significantly reducing service life [292]. Certain soil-burrowing animals, notably dung beetles, ants, and mole pelts, thrive in moist or sticky soil environments yet manage to keep their shells clean and free from soil adhesion. They achieve this anti-adhesive property through the evolution of unique non-smooth surfaces. Moreover, due to their claw's low soil-cutting resistance (a measure of the difficulty or effort required for the tool to move through the soil medium) and soil adhesion, they can easily dig long tunnels within a short period of time. Hence, the foreleg of the dung beetle and mole pelts serves as an effective bionic prototype for soil-cutting tools as well. For instance, Zhang et al. [35] fabricated a toothed gear inspired by the intricate outer contour curve of the foreleg end-tooth of a dung beetle (Figure 19a). The bionic toothed wheel significantly diminishes draft force by up to 16.5%. Finite element analysis demonstrated that the apex of the bionic peak tooth experiences the highest stress concentration at the contact point with the soil, leading to enhanced cutting tool performance and reduced adhesion and friction between the tool and soil material [35]. Ji et al. [282] created a bionic soil cutting model with ABS polymer, mimicking the foreclaw toes of a mole pelt through reverse engineering and 3D printing. The bionic model exhibited a 12.8% reduction in cutting resistance compared to a standard specimen.

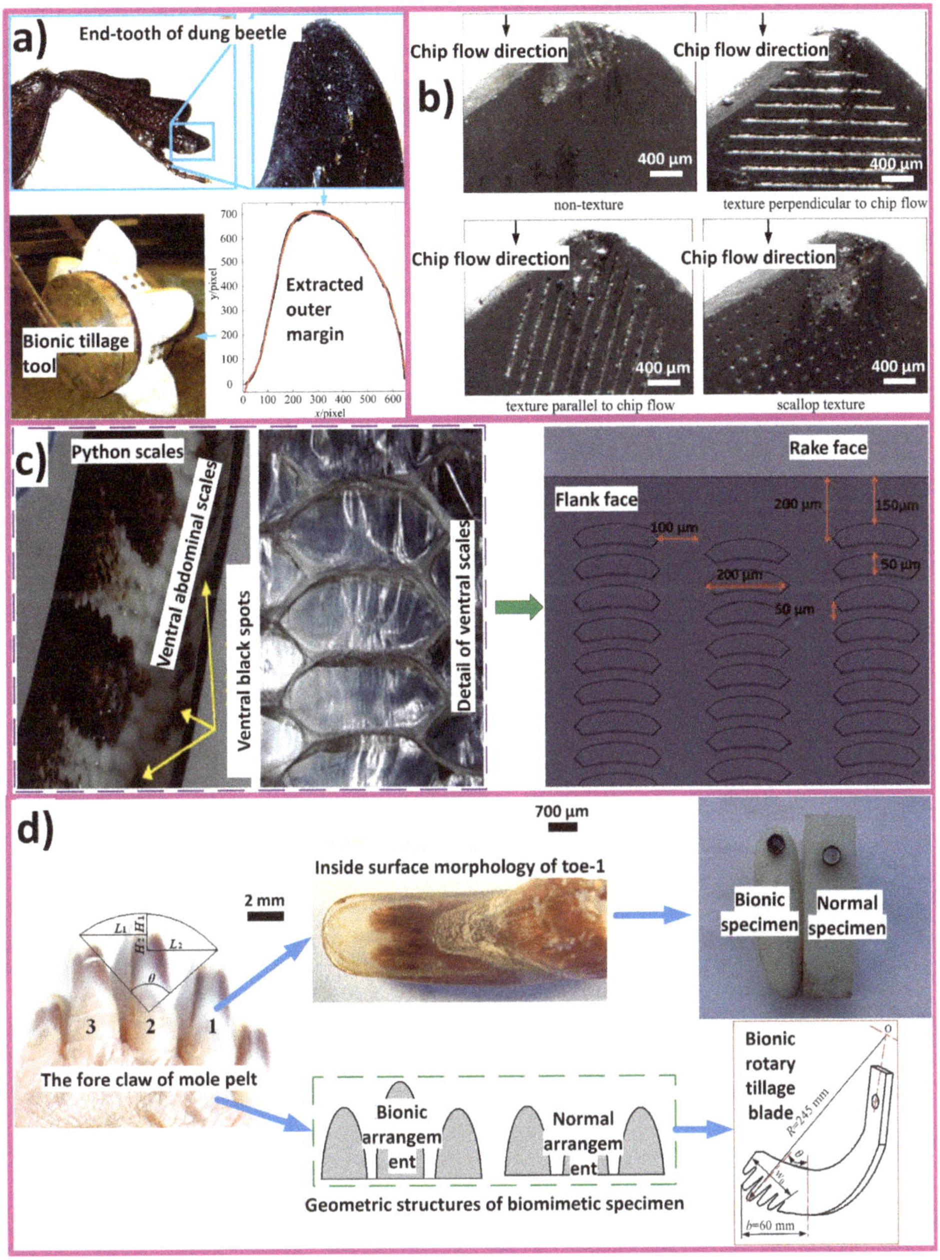

Figure 19. (**a**) Dung beetle-inspired design of tillage toothed wheel [35]; (**b**) various bionic texturing on rake face of a turning tool [259,293]; (**c**) snakeskin-inspired bionic turning tool [294]; and (**d**) bionic soil cutting tools inspired by claw toes of mole pelt [282]. All images were reproduced with permission.

In machine tools including turning, drilling, milling, etc., the bionic texture generation on the tool's surface enhances its tribological characteristics, including friction, lubrication, and wear, at the interfaces with the workpiece. This texture not only decreases friction by reducing the contact area between the chip and the tool interfaces but also entraps wear debris particles during cutting operations, reducing the plowing action of the tool, and further decreasing friction between the interfaces. Several other benefits, such as lubricant storage in the textures, etc., have already been discussed in the above sections. For instance, the rat, a prevalent rodent known for its strong incisive teeth, possesses remarkable metal cutting/machining abilities. Researchers discovered that the "rake face" of the rat's incisors featured a unique "crescent depression" morphology. This morphology, far from compromising structural integrity, enhanced the sharpness of the cutting edge, thereby bolstering cutting/machining efficiency. Drawing inspiration from this, it was shown through turning simulations and experiments that these bionic-gradient microtextured turning tools effectively reduce the primary cutting force while simultaneously enhancing structural resilience [295,296]. Another widely reported and effective inspiration comes from sea urchin teeth offering insights into self-sharpening mechanisms due to their ability to remain sharp when eating stone [272]. Three noteworthy aspects of these teeth are the following: (i) despite being composed of calcite, they can grind limestone, which shares a similar composition; (ii) the three structural elements of the teeth—plates, fibers, and the polycrystalline matrix between them—exhibit highly aligned orientation from the nanometer to the centimeter scale; and (iii) the sea urchin teeth exhibit self-sharpening properties with use, rather than dulling [273]. It was revealed that the tooth's structure determines the location of fracturing and renewal, resulting in self-sharpening. Fracturing is initiated by compressive or shear stresses, with cracks propagating through the plate, shedding off a portion along with its adjacent polycrystalline matrix [271]. In another inspiration taken from dung beetle skin, Siju et al. [297] employed laser biotexturing to create microgrooves, microdimples, square texture, and a combination of both on the carbide tool's rake face. These textured tools were then used for turning operations on Ti-6Al-4V alloy. The study found that the dual dimple and grooved textured tool exhibited superior performance in reducing friction, shear angle, and the chip reduction coefficient. Li et al. [298] replicated shark skin textures on the tool's rake face and nanotextures on the flank face of a carbide tool with WS_2/C coating. They observed that the shark-textured tool, parallel to the cutting edge, exhibited reduced flank wear, temperature, and surface roughness. Fatima et al. [294] imitated hexagonal textures from ball python skin on the tool's flank face during turning of AISI 4140 steel. They found that the bioinspired textured tool outperformed in reducing contact length, temperature, and cutting forces. Liu et al. [299] created microgrooves in parallel, perpendicular, and inclined orientations on the carbide tool's flank face, inspired from dung desert scorpion. They investigated the impact of these textured tools during the turning of green alumina ceramics and found that the parallel grooved tool displayed reduced flank wear compared to the conventional tool.

In earth drilling/tunneling works, a coal seam, being a soft rock layer, poses challenges in drilling due to hole collapses and the tendency of rock powder to stick to drill bits [300]. To address these issues, Yang et al. [41] developed a cemented carbide drill bit inspired by the dung beetle as shown in Figure 20. The domes on the bit minimize adhesion, while the pit accommodates detritus. Moreover, the curves facilitate detritus removal from the cutting area. With this bionic drill bit, the drilling speed rose by 45%, and the average abrasion rate dropped by 23%. Gao et al. [300] developed a bionic bit inspired by a pangolin claw to enhance cutting performance in drilling hot dry rocks with high hardness and poor drillability. Experimental drilling of hot dry rocks demonstrated around a 97% improvement in drilling rate and almost 1.2 times longer lifespan compared to the non-bionic bit (the statistical deviations in the reported results were missing). Similarly, Wang et al. [264] utilized the pattern from mole claws to design a polycrystalline diamond compact (PDC) rock cutter with a multistep microarc-structured diamond edge. This bionic PDC rock cutter exhibited remarkable anti-abrasion performance, increasing the service life by over three times and improving the drilling efficiency by approximately 200%. Going an extra mile, Wang et al. [191] introduced a novel bionic bit (Figure 21) inspired by the morphological traits of a dung beetle, featuring concave structures, and the abrasive wear (resistance) and other functional attributes of an earthworm, such as for a continuous lubricated surface texture. The bit was basically a composite of epoxy, graphite (70 wt.%), and diamond as reinforcement. The synergistic multitextured bionic bit was reported to offer high wear resistance and improved cutting capability (due to the continuous exposure of diamonds in the composite after epoxy worn out) in comparison to those with common surface. The graphite (solid lubricant and soft non-metal) composing the bionic unit possesses the ability to self-regenerate the concaves, providing a trapping effect for rock particles as shown in Figure 21. Additionally, the bionic surface enlarged the surface area, leading to enhanced heat dissipation and increasing the frictional contact interface.

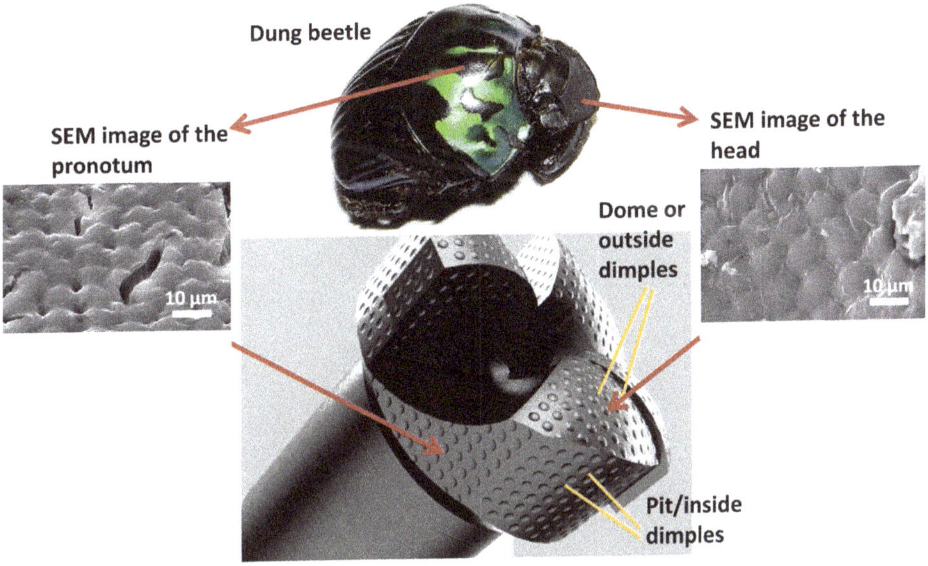

Figure 20. Bionic drill bit inspired by dung beetle. Reproduced with permission from Ref. [41].

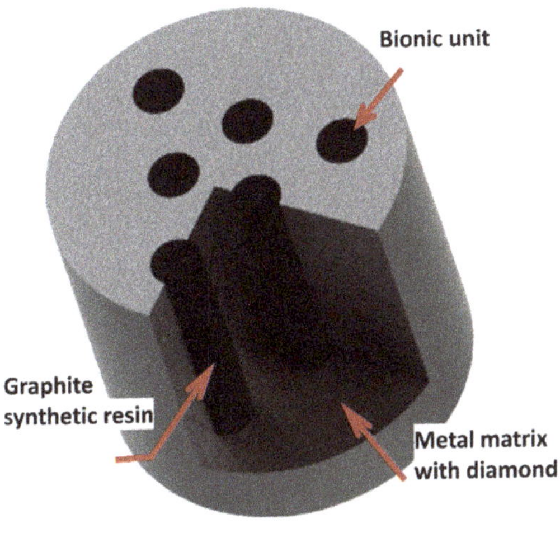

Figure 21. Bionic impregnated diamond bit inspired by dung beetle and earthworm to enhance lubrication mechanism. Reproduced with permission from Ref. [191].

In the biomedical field, surface modification of medical devices, particularly electrosurgical scalpels, has garnered considerable interest for its simplicity and minimal tissue trauma [106]. Introducing bionic texture to the tool's surface enhances hydrophobicity, increasing the contact angle and aiding faster fluid flow. This results in reduced temperature and adhesion during the surgical cutting process. Particularly, in orthopedic surgery, medical drill bits play a crucial role in pre-drilling. Incorporating bionic textures on these tools can effectively reduce drilling forces, optimize stress distribution, and mitigate stress concentration issues, ultimately enhancing the quality of pre-drilling procedures [263]. A promising method involves integrating harmless solid lubricants into the bionic texture of medical tools to improve cutting execution and ease patient pain [42]. For instance, Liu et al. [60] explored bionic designs inspired by the hydrophilic microstructure of Nepenthes openings to develop self-lubricating electrosurgical scalpels. By electrolytic etching, a wetting gradient was achieved on the scalpel's surface, featuring non-uniformly distributed cylindrical patterns/textures. This design facilitates rapid lubricant diffusion and coverage, effectively reducing adhesion. When used for biological surgeries, the surface texture of the bionic tool could retain a significant quantity of lubricant. This transformation changes the initial solid/solid contact to solid/liquid/solid contact, effectively reducing tissue adhesion by approximately 80%, tissue adhesion volume by about 88%, wound damage area by around 81%, and thermal damage area by about 70% [60,270,301]. Moreover, the slippery zone of Nepenthes has many crescentic arc structures, which have been shown to have drag reduction and hydrophobicity [269]. Figure 22 shows the Nepenthes-inspired design of an electrosurgical knife and its features.

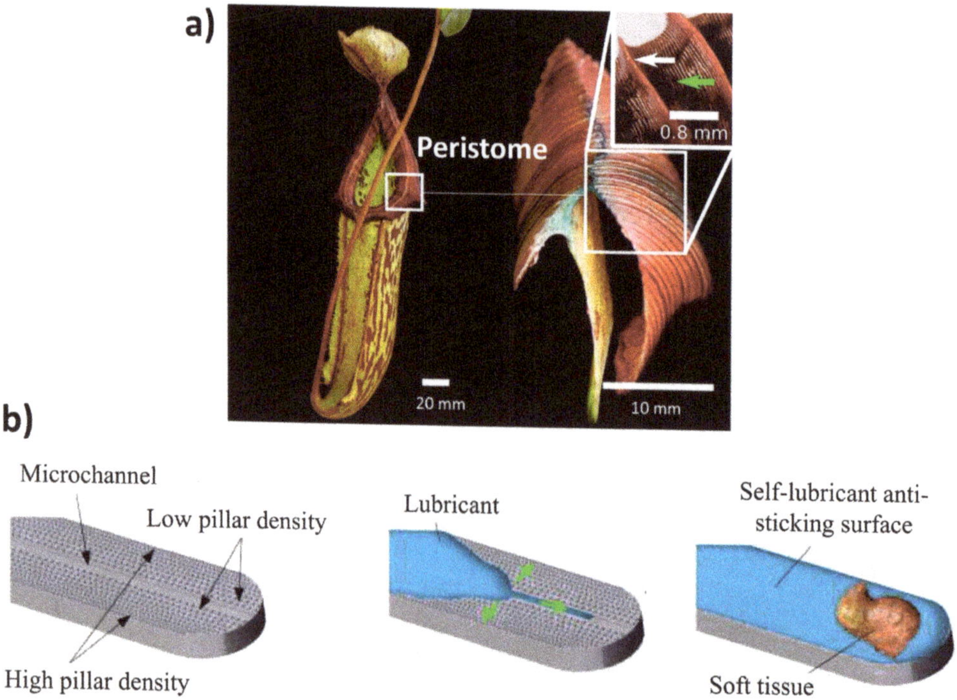

Figure 22. (**a**) A portion of the Nepenthes peristome showing both the inward (left) and outward (right) slanting regions and their microscopic grooves; and (**b**) self-lubricating mechanism of bionic design of electrosurgical knife inspired by Nepenthes [60,106,302].

Table 6 summarizes recent results from the literature concerning various bioinspired designs exploited for the enhancement of cutting or machining tools.

Table 6. Results of various bioinspired designs exploited for the enhancement of cutting or machining tools, as reported in the literature.

Bioinspired Structure	Conclusions/Texture Effect
Sea urchin teeth-inspired rock-cutting tool (Crystalline calcite teeth/epoxy composite) [271,272]	Self-breaking or fracturing leads to self-sharpening. Detailed description of self-sharpening phenomena.
Dung beetle-inspired Bulldozing plate (UHMWPE) [267]	Bionic UHMWPE could reduce soil adhesion and draught considerably. 30% reduction in cutting resistance.
Lotus leaf/Shark skin-inspired soil shovel/tillage tool (Steel-based) [266]	1.85–4% reduction in cutting resistance.
Dung beetle-inspired toothed wheel tool (Steel-based) [35]	16.5% reduction in cutting resistance. 11.8% reduction in stress concentration. 24% increased microbasin.
Shark skin-inspired subsoiler tillage tool (Steel-based) [303]	21.9% reduction in cutting resistance. 24.8% reduction in consumption energy. Improved crop stress resistance.
Mole pelt-inspired cutting blade for tillage tool (Steel-based) [36,282]	12.8% reduction in cutting resistance. 4% reduction in torque. 60% improved stubble breaking rate.
Pangolin-inspired drill bit for dry rock cutting (Impregnated diamond bit) [260]	97.5% increased drilling speed. 26.8% increased service life.
Mole claw-inspired drill bit (Impregnated diamond bit) [264]	230% increased drilling efficiency. 345% increased service life.
Beetle/Earthworm-inspired drill bit (Impregnated diamond bit) [191,261]	91% increased wear resistance. 27% increased grinding performance.
Bamboo/human tooth/annual ring/shell/mole pelt-inspired drill bit (Impregnated diamond bit) [285]	Improved cutting resistance. Improved wear resistance. 250% increased penetration rate.
Cat claw-inspired drill bit (Impregnated diamond bit) [304]	13% increased penetration rate.
Dung beetle-inspired drill bit (cemented carbide bit) [41]	45% increased drilling speed. 23% decreased wear rate.
Dung beetle-inspired microgrooves/pits turning tool for dry cutting (AISI 1045 steel) [305]	24% decreased cutting force. 29% decreased cutting temperature. Grooves and pits led to wear debris (graphite) storage and improved lubrication.
Dung beetle-inspired microgrooved tool for dry machining of Ti-alloy (PCD tools) byFiber lasers [278]	Reduced friction coefficient. Reduced tool-chip contact length. Reduced cutting force. Reduced adhesion.

as Mars or the Moon [334]. Furthermore, bioinspired surface coatings mimicking the self-cleaning attributes of lotus leaves could deter the accumulation of dust and debris on rover wheels or spacecraft bodies, ensuring optimal traction and minimizing frictional heating while reducing the likelihood of mechanical malfunctions. Notably, the combination of shape memory alloys and bioinspired surfaces presents synergistic prospects for developing advanced tribological solutions with adaptable microstructures and functionalities, allowing for customized friction and wear properties in response to the fluctuating environmental conditions encountered in space [335–337]. Figure 23 shows LEMUR with its gecko-imitating gripper technology climbing around (yellow dashed area).

Figure 23. Images showing (**a**) a gecko; (**b**) microscopic gecko's toepads with millions of microscopic hairs [170]; and (**c**) Limbed Excursion Mechanical Utility Robot (LEMUR, within a highlighted yellow border.) climbing around the outside of the International Space Station. JPL considered outfitting LEMUR with its gecko-imitating gripper technology. Images reproduced from NASA [338].

8.4. Renewable Energy

Bionic engineered materials/surfaces play a crucial role in advancing the transition to a net-zero economy by offering breakthroughs in various renewable energy sectors, including superior energy efficiency and sustainability through minimizing friction and wear while enhancing the energy produced. For instance, erosion wear of solar cells requires focus, especially in desert lands, and space where the phenomena can be catastrophic and lead to complete equipment failure [339]. In wind turbines, the wind blades often undergo erosion/impact through solid-particle or slurry processes or hurricanes (when installed in sea) [340]. In offshore wave energy converters, bearing failure by wear is a common issue resulting in downtime of the whole system. To this end, bionic enhancement of solar cell, wind turbine, or hydro bearing surface has been shown to facilitate more efficient capture of solar energy, enable the development of superior wind turbine blades, and enhance the energy efficiency of tribological generators, etc.

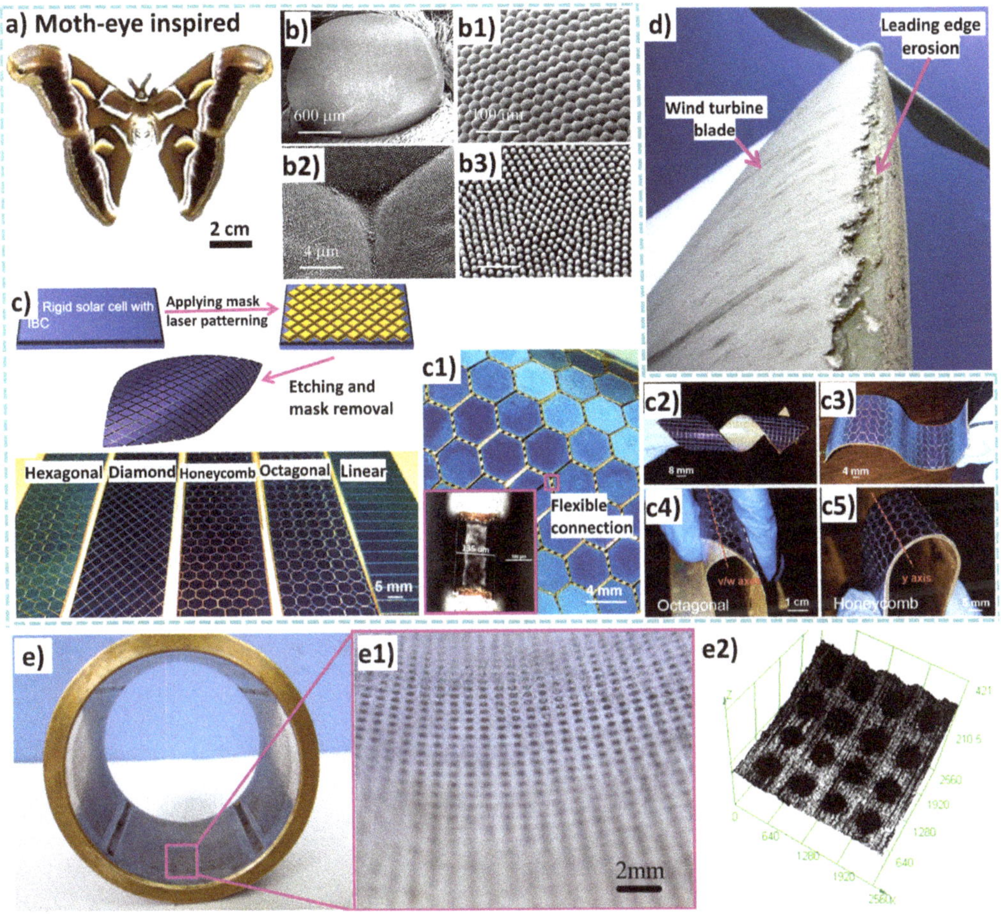

Figure 24. Images showing (**a**) a moth [341]; (**b**,**b1**,**b2**,**b3**) SEM images of moth eye at different magnifications [341]; (**c**) preparation step and optical image showing the five corrugation patterns that are studied, including hexagonal, diamond, honeycomb, octagonal, and linear structures [47]; (**c1**) optical and SEM images of the hexagonal corrugated solar cells for enhanced flexibility [47]; (**c2**,**c3**,**c4**,**c5**) optical images of the diamond, hexagonal, and honeycomb corrugated photovoltaic cells flexed in different axes [47]; (**d**) leading edge erosion in a wind turbine blade [342]; (**e**) a bearing [343]; (**e1**) enlarged photo of the texture on the inner surface of the bearing [343]; (**e2**) surface morphology of the texture as illustrated by laser scanning confocal microscopy [343]. All images were reproduced with permission from their respective publishers.

Especially in the area of solar cells, accumulation of dust (soiling) and repetitive impact of sand particles (erosion) affect the performance of solar power technologies. Replicating natural structures such as those found on lotus leaves or moth eyes can lead to enhanced light absorption, self-cleaning capabilities, hydrophobicity, and reduced reflection on solar panels while contributing to lower erosion rates [45,339,344–346], as shown in Figure 24a,b. By minimizing surface reflection, improving light trapping, and mitigating erosive wear, these textures ultimately enhance the efficiency and longevity of solar cells, resulting in higher energy yields. Furthermore, Atab et al. [47] showcased an innovative approach for creating ultraflexible, lightweight, and highly efficient (energy conversion) monocrystalline silicon solar cells. They achieved excellent reliability, mechanical resilience, and thermal performance by employing a corrugation method in conjunction with moth eye-inspired bionic laser patterning, as shown in Figure 24c.

Leading edge erosion (Figure 24d) is a significant issue for wind turbine blades, often caused by factors like insect impacts, dust accumulation, or ice formation [56,347,348]. These factors create surface roughness, affecting the aerodynamic performance and leading to energy production losses. Maintenance costs also increase as a result of this erosion [38]. To address this challenge, researchers have explored the potential of bionic textures on wind turbine surfaces to enhance longevity and energy efficiency. For example, Lv et al. [45] demonstrated that bionic-designed textures inspired by natural surfaces like lotus leaves, water strider legs, or owl feathers can trap air effectively [349], reducing the interaction with liquid water and preventing erosion from water droplets or ice adhesion (anti-icing functionality). These textures also offer additional benefits, such as self-cleaning capabilities, inspired by features like the lotus leaf, which help to keep dust off the blades and maintain aerodynamics, as well as noise reduction inspired by owl feathers [340]. Moreover, mimicking surface topographies from natural structures like shark skin or penguin skin can minimize drag and increase lift, resulting in significant improvements in energy output.

Bioinspired textures offer significant advantages in the bearing and sealing industry (Figure 24e), particularly for applications requiring long-term maintenance-free operation, such as wave energy converters or offshore triboelectric generators [350,351]. With the increasing emphasis on green tribology, bearings are designed to function in dry, water-based, or biolubricated conditions, often operating under boundary lubrication [352]. This requires surfaces to withstand heavy loads, vibrations, and start–stop cycles of hydrobearings while effectively entrapping debris or retaining lubricants through grooves or textures [353]. Bioinspired surfaces have demonstrated considerable benefits in this regard. Pattnayak et al. [354] developed fish skin-inspired microtextured bearings, showing improved friction, wear, and static and dynamic performance compared to conventional untextured ones. Ji et al. [355] investigated the effect of different surface textures on wettability and film formation during sliding wear of ceramic–steel pairs under fluid lubrication, proposing a model that suggests a bionic-feature tradeoff between the ball and race could customize the friction and wear properties of a bearing system. It can be understood that, under dry friction conditions, bionic texture pits play a crucial role in capturing abrasive particles and debris during friction, effectively reducing sample wear. In wet wear scenarios, these pits serve as reservoirs for abrasive particles and lubricants. The lubricant is dynamically released while capturing abrasives, aiding in wear reduction under boundary and mixed lubrication conditions. Computational fluid dynamics simulations by Xie et al. [356] revealed that surface textures retain lubricants within microgrooves during shaft rotation, enhancing local dynamic lubrication. Agglomeration of numerous micropits and microbearings improves bearing capacity and lubrication performance, as shown in Figure 25. Additionally, the wedge-shaped structures formed by textures create high-pressure zones during fluid flow, further enhancing bearing capacity and surface separation.

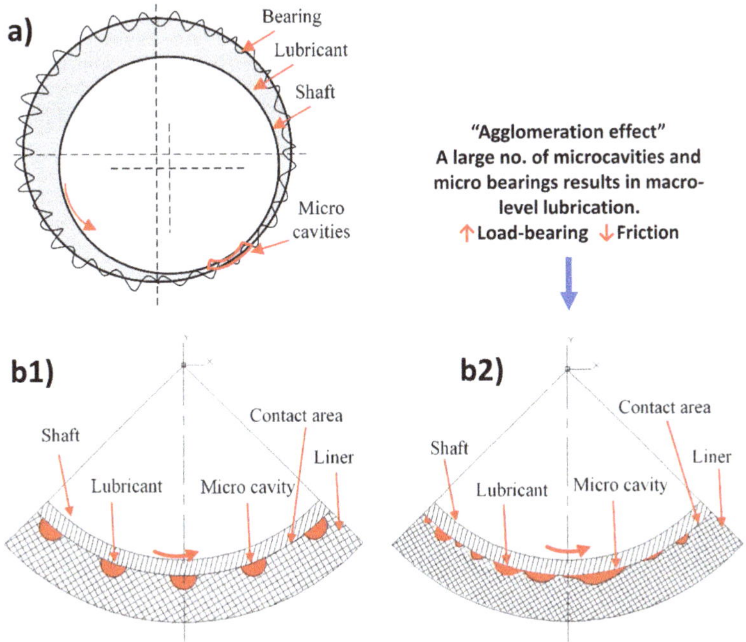

Figure 25. Schematic diagram of (**a**) bearing with rough surface; (**b1**) lubrication regime with fewer microcavities; and (**b2**) lubrication regime with a greater number of microcavities. This results in agglomeration effect and improved bearing life and tribology. Reproduced with permission from Ref. [356].

8.5. Bioinspired Metamaterials

Metamaterials, denoted by the prefix "meta", indicating "beyond", represent a novel class of synthetic materials characterized by intricate repetitive patterns at the micro- or nanoscale [357]. In contrast to conventional materials, their properties do not stem from inherent material characteristics but rather from precisely engineered structures or surfaces. These structures govern unique functionalities, such as the manipulation of mechanical stress or tribological stress waves, and are defined by their specific shape, geometry, size, orientation, and arrangement [358].

Metamaterials exhibit remarkable potential in reducing friction and wear, particularly under extreme tribological conditions involving high loads, temperatures, and stress. Reported tribological phenomena associated with metamaterials include adaptive interaction with non-equilibrium environments, surface reorganization, and efficient energy absorption or redistribution [359,360]. For example, Liefferink et al. [361] demonstrated the dynamic friction tuning capability of a kirigami metamaterial surface, enabling external friction adjustment without altering the sliding component. Similarly, Rafsanjani et al. [362] introduced the concept of "tune on-demand" texturing using a kirigami metamaterial, showcasing its application in bioinspired locomotion. They demonstrated the functionality of a bioinspired snakeskin, which, through the cyclic stretching and releasing of its kirigami skin, forms triangular scales when stretched, enabling forward crawling akin to a real snake [362]. The manipulation of scales through stretching and releasing allows for the external control of frictional energy detention or dissipation [362–364]. This "on-demand texturing" capability, based on kirigami scales, offers the ability to adjust friction by externally controlling surface roughness [362].

Moreover, friction serves as a crucial energy-absorbing mechanism in various engineering applications, such as automotive brakes, earthquake and bridge vibration absorbers, and prosthetics. Garland et al. [358] devised a honeycomb metamaterial unit cell with internal features capable of sliding past each other during loading and unloading cycles, leading to consistent Coulombic frictional energy dissipation during sliding contact. This technique of frictional energy dissipation serves as a key inspiration for the development of a novel metamaterial, aiming to achieve similar capabilities in dissipating repeated mechanical energy inputs through the movement of internal features.

In terms of energy absorption, lattice struts exhibit notable capabilities during impacts due to their ability to undergo plastic deformation or fracture, thus dissipating input energy [365]. Strategically placing features within a metamaterial design can enhance toughness by guiding cracks along predetermined paths. Additionally, grading the lattice structure can enhance energy absorption by ensuring sequential collapse of each layer rather than a single shear band failure [366]. The architecture and base material of unit cells play a crucial role in determining the macroscopic responses of the metamaterial, suggesting that modifying unit cell shape or topology could allow for tailored mechanical performance [367]. Another approach involves incorporating multiple morphologies within a metamaterial structure (Figure 26), with each morphology designed to fulfill specific functions, such as permeability, elongated stress range, elastic modulus, compressive strength, and more [368,369]. For instance, Rezapourian et al. [370] reported that triply periodic minimal surfaces (TPMSs) serve as bioinspired metamaterials due to their resemblance to biological structures like trabecular bone networks (Figure 26). It was shown that the highly interconnected porous structures of TPMS-based metamaterials can achieve improved strength-to-weight ratios, impact resistance, and structural integrity [371]. For example, the TPMS gyroid structure, resembling certain butterfly wing scales, contributes to their structural coloration and antibacterial functionality [367]. Similarly, the TPMS diamond structure mirrors the skeleton of the knobby starfish, featuring remarkable structural gradients and damage tolerance [372]. Meanwhile, TPMS primitive structures, found in sea urchin skeletal plates, are typically optimized (regarding design) and employed for their enhanced strength, flexibility, and delayed failure under impact [372].

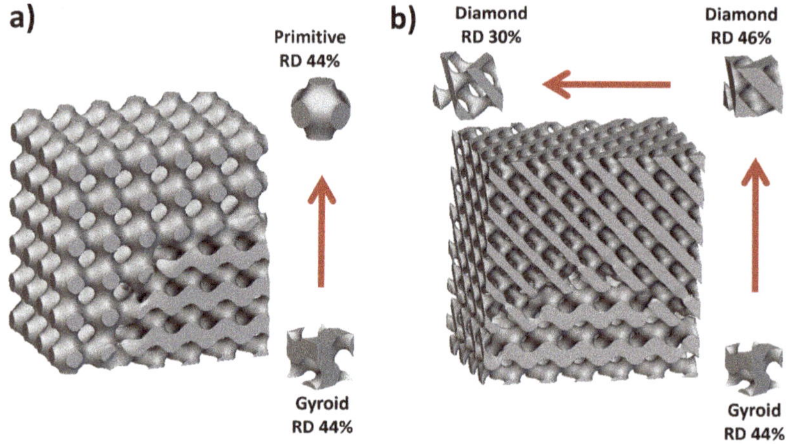

Figure 26. An illustration depicting a multi-morphology metamaterial architecture of triply periodic minimal surfaces (TPMSs), incorporating (**a**) gyroid and primitive unit cells with unit cell grading (printing direction); and (**b**) gyroid and diamond unit cells with unit cell (vertical direction and printing direction) and relative density (transverse direction) grading. The red arrow indicates the grading direction. The cell size for each model is $5 \times 5 \times 5$ mm^3. The relative density (RD) value is labeled. The model designs are inspired by trabecular bone networks. The 3D models were designed using nTopology software version 4.13.3.

9. Challenges and Future Prospects in Bioinspired Tribology

The concept of mimicking nature's design principles to create synthetic materials is gaining traction in tribological research. This approach is expected to remain pivotal in the quest for wear-resistant novel multifunctional materials in the years to come. To enhance comprehension and offer guidance for the initial selection and design of different bioinspired surface textures, Figure 27 was mapped. Figure 27 depicts a general schematic representation of various biologically inspired surfaces/textures along with their associated functionalities, potential tribological conditions of usage, and application areas, as discussed within this review. We think that graphical mapping could be a good tool for engineers and educationists. However, due to the length of information (acquired from published research) and statistical uncertainties regarding some topography, some data may have been mapped as per the authors' experiences. It is important to note that these biotextures may offer diverse functionalities beyond those mentioned in the schematic and can be customized to suit specific tribological conditions and applied across various domains. Moreover, in several applications, a combination of two or more textures is utilized to enable multifunctionality and superior performance. It is important to acknowledge that nature offers a wealth of features, and there may be additional characteristics that were not included in the graphical representation. The primary objective is to depict the functionalities and desirability of such materials and structures across different tribological scenarios and applications.

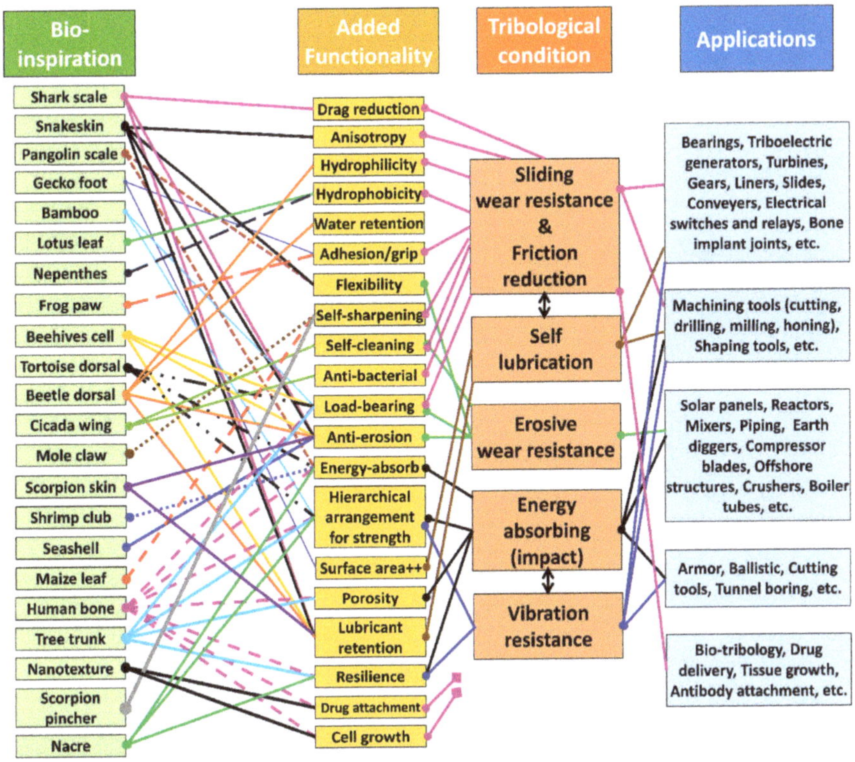

Figure 27. Schematic representation of various bioinspired textures linked to their associated key functionalities, further linked to their possible tribological use conditions and possible application areas. Different colors or styles of interconnected lines are employed to differentiate between bioinspiration types, which are linked to their specific attributes or functionalities, and the tribological conditions that may necessitate corresponding functionalities for specific applications.

Despite promising advancements, biomimicry in tribology remains in its early stages, presenting ample opportunities for further refinement and exploration. For instance, the study of superhydrophobic surfaces, inspired by the lotus leaf effect, has been extensive. Several studies have focused on replicating lotus leaf-like surfaces by combining surface roughness and low surface energy. However, the durability of superhydrophobicity remains a challenge. Recently, drawing inspiration from the Nepenthes pitcher plant, Aizenberg et al. [373] developed slippery liquid-infused porous surfaces (SLIPSs). Unlike conventional lotus leaf-like surfaces where air is trapped by surface roughness, SLIPSs feature pores infused with special liquids to create a solid–water–liquid interface, resulting in excellent repellence to various liquids, along with antifouling, anti-icing, and anti-frost properties, all with high durability, transparency, and even self-healing capabilities. The practical implementation of SLIPSs in tribology holds promise for enhancing the durability of lubricant or chemical-infused surfaces, along with providing additional benefits like antifouling properties. However, their widespread application in tribological systems is still pending realization.

Furthermore, bioinspired materials such as biomaterials must exhibit self-organization under mild conditions of relative motion in the form of debris usage, bone self-structuring, etc. While their sophisticated hierarchical structures (cellular) have never been perfectly replicated using modern technologies, precisely studying self-healing or self-organization behavior, especially at small scales, remains a significant challenge. In contrast, these cellular materials typically lose their function once their structure is destroyed. Therefore, understanding and replicating the dynamic and nanoscale molecular mechanisms of these systems, as well as emulating the adaptation and self-healing abilities of biomaterials under sliding tribology, are crucial.

Generally, biomimetics follows a top-down approach [374], where natural similarities are sought to address a challenge in certain tribological conditions, like the development of an anisotropic low-frictional surface inspired by snakeskin. Conversely, an advanced bottom-up process [375] entails analyzing natural systems to uncover potential tribological and mechanical applications, a field that is underexplored due to limitations in nanomechanical characterization, structural investigation, and manufacturing techniques, resulting in unelicited laboratory studies.

Another topic that requires focus is investigation regarding synergistic effects of multiple bioinspiration features [104]. By examining how various features within biological systems interact and complement each other to achieve specific tribological functions, researchers can gain deeper insights into designing more efficient and effective biomimetic materials and surfaces. For example, understanding how the combination of surface roughness, hierarchical structures, and material/lubricant composition in natural systems contributes to enhanced friction reduction or wear resistance can inspire the development of novel tribological solutions.

Large-scale production of bioinspired materials, especially in achieving simplicity and ambient processing conditions, has been a serious challenge [376]. Attempts to fabricate biomimetic materials often fall short of natural performance as biological materials exhibit multifunctionality. Replicating such multifunctionality remains a challenge. For instance, designing cellular materials inspired by nacre or bone or tree trunk, etc., involves incorporating microcapsules or vascular networks with transport/healing agents to enable spontaneous recovery from damage (self-healing).

Improvements in biomimetic designs through utilization of modeling hold imminent importance as they enable exploration of a wide range of design parameters, such as surface topography, material properties, and environmental factors, to optimize the performance of bioinspired tribological systems [198,377,378]. Furthermore, modeling enables us to gain insights into the underlying mechanisms governing tribological behavior in systems, leading to the development of more realistic and effective biomimetic designs. Although biological systems advocate the use of hierarchical multiscale surface textures, most of the published experimental and numerical works have mainly addressed effects induced by single-scale surface textures [379,380]. Multiscale surfaces offer advantages in high-load, low-speed conditions by reducing real contact area, supplying additional lubricant, and trapping wear particles. They also increase hydrodynamic pressure and provide a greater reservoir volume for lubricant and wear debris in mixed lubrication. Moreover, smaller densely distributed textures improve wetting behavior and lubricant distribution, reducing cavitation and flow circulation to enhance load carrying capacity. However, further fundamental studies are required to evaluate essential parameters in multiscale textures, including aspect ratio, area density, and their relationship to texture size, to optimize their tribological behavior. Additionally, machine learning algorithms have been increasingly utilized to analyze large datasets and identify optimal texture designs for improved tribological performance [381,382].

Another essential step in futuristic growth in bioinspired tribology is the utilization of materials-property databases in terms of comprehensive repositories of material properties, including tribological features [383,384]. Through data-driven approaches and computational modeling, materials-property databases enable the prediction and exploration of material behaviors under different conditions, facilitating the design of bioinspired surfaces with tailored tribological characteristics. Moreover, these databases facilitate knowledge sharing and collaboration within the scientific community, accelerating the pace of innovation in bioinspired tribology. Figure 28 shows a general circle of a bioinspired tribological material.

In general, the advancement of bioinspired wear-resistant materials depends on comprehending structure–composition–property relationships, building a comprehensive database for customizable design and optimization, as well as employing advanced manufacturing techniques to meet complex design requirements. Advancements in additive manufacturing enable the printing of a wider range of materials and the creation of hierarchical and composite structures [385]. Additionally, 4D printing can lead to integration of biological materials into synthetic components towards the development of novel bioinspired materials with superior properties [386,387]. Furthermore, establishing relationships between quasistatic and dynamic testing, as well as advancements in testing and modeling of bioinspired materials under dynamic loads, are crucial for understanding material responses and optimizing wear resistance [13,251]. Investigating the effects of multiple structural elements (multi-bioinspired) acting in synergy and understanding the role of self-healing and structural hierarchy on wear resistance are also essential for future progress. Finally, optimal utilization of materials, porosity, and fluids for superior lubrication, heat transfer, etc., is of interest in numerous applications [53,388,389]. Apart from this, the creation of efficient databases and computational models supported by a machine learning approach for ease in technology transfer will aid in broadening the range of knowledge and collaboration [382,384]. By addressing these key areas, bioinspired tribology can advance significantly, leading to the development of durable, efficient, and sustainable materials and structures with enhanced tribological properties. Figure 29 shows the aspects to be integrated for advancements in bioinspired tribology.

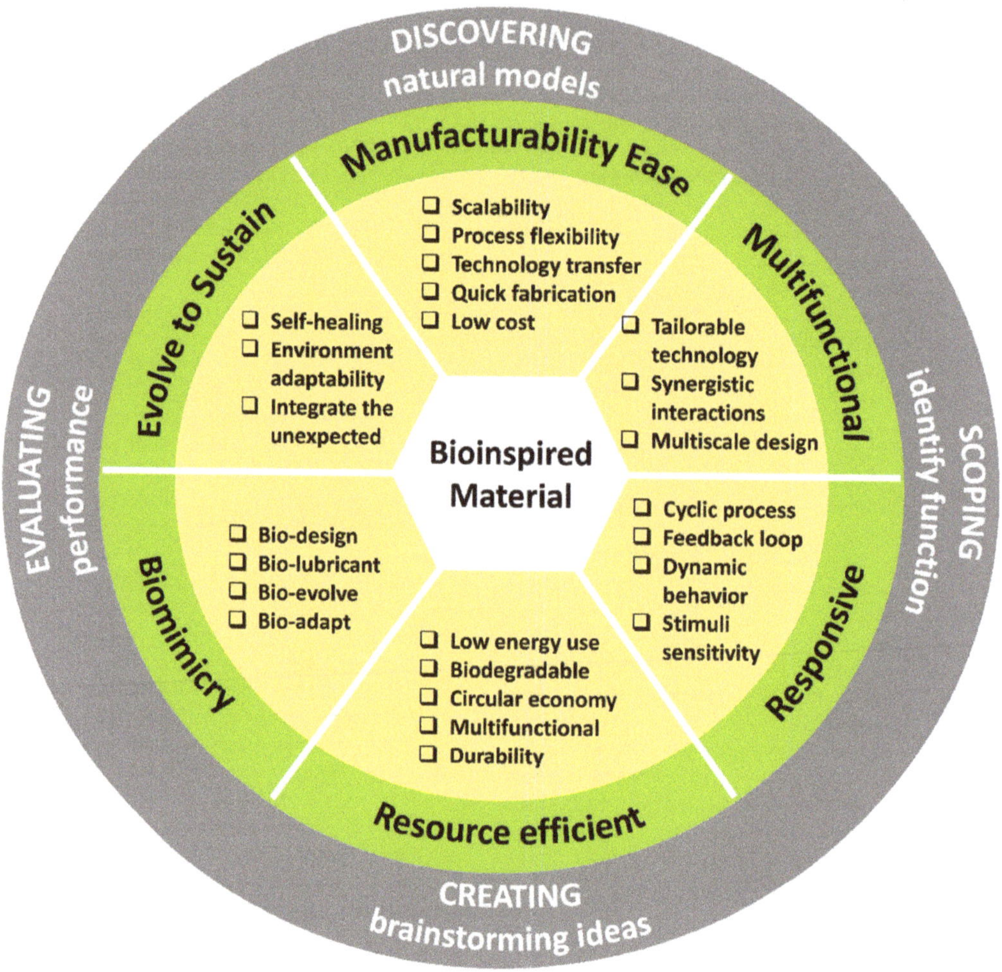

Figure 28. Circle of a bioinspired tribological material. Six key necessities (green) of a bioinspired tribological material are shown, i.e., biomimicry, evolve to sustain, manufacturability ease, multifunctionality, responsive, and resource-efficient. Under these necessities are their key drivers (yellowish). The outer periphery (gray) marks the guiding principle behind a typical biomimetic approach, i.e., discover natural models, identify function within material, create materialistic approach, and evaluate material's performance.

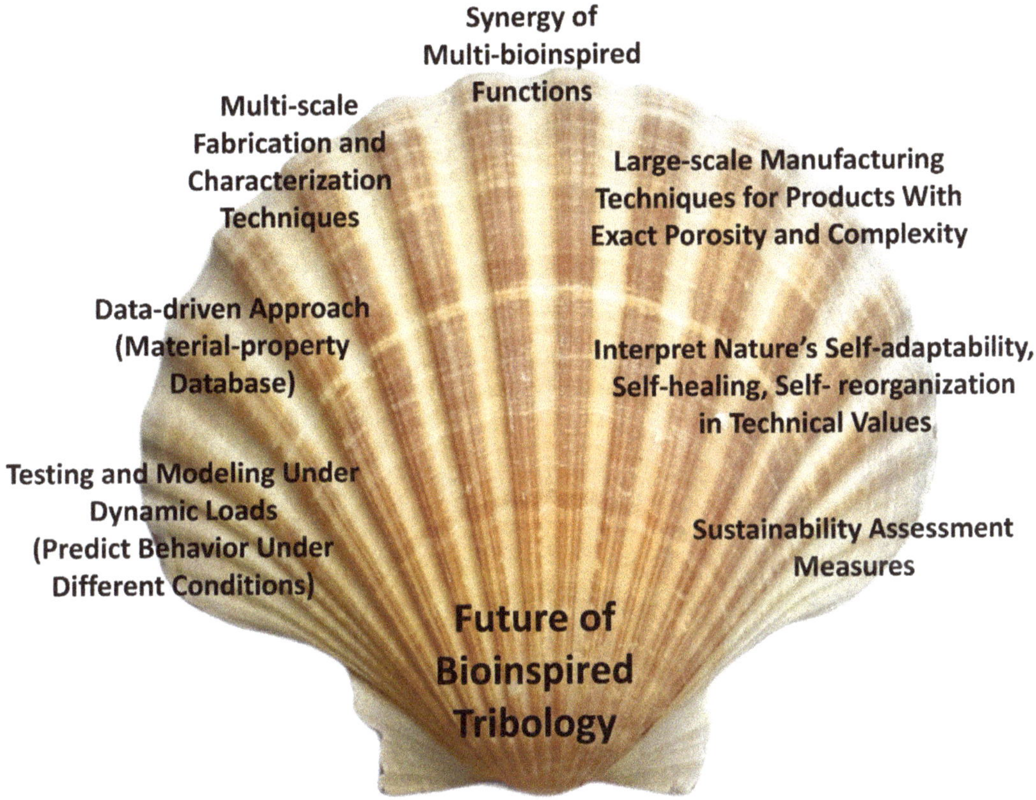

Figure 29. The future of bioinspired tribology advancement demands integrating key aspects of modeling and testing, data-driven approaches, novel multiscale fabrication and characterization techniques, synergy of multi-bioinspired functionalities, new manufacturing methods to fabricate intricate geometries of bioinspired materials, approaches towards interpreting nature's science, and measures towards sustainability assessment.

10. Summary

Drawing from the virtually limitless wellspring of bioinspiration, natural materials exhibit remarkable mechanical efficiency, offering a wealth of features, such as self-optimized microstructures, exceptional load-bearing ability, hierarchical arrangements, and environmental adaptability, etc. These attributes, combined with wear resistance and the ability to modulate stiffness and friction in response to external stimuli, provide a rich source of inspiration for developing advanced sustainable tribological materials. As bioinspired tribology evolves, a multidimensional approach encompassing design, manufacturing processes, and performance evaluation becomes imperative. By examining successful examples from nature, distinct patterns in microstructural organization become evident, leading to the classification of design elements crucial for wear resistance in biological materials under various conditions, such as sliding, erosion, impact, or machining. Moreover, the current study objectifies and delves into the common traits of biological materials essential for specific tribological applications, considering factors like self-organization of microstructures, self-cleaning, flexibility, hydrophobicity, and antibacterial properties, etc. Looking ahead, advancement regarding bioinspired wear-resistant materials hinges on understanding structure–composition–property relationships, establishing comprehensive databases for customizable design and optimization, and employing advanced manufacturing techniques to meet intricate design needs. Innovations in additive manufacturing, including 4D printing, offer opportunities for integrating biological materials into synthetic components, paving the way for novel bioinspired metamaterials with superior mechanical and tribological properties. Additionally, progress in testing and modeling of bioinspired materials under dynamic loads, as well as investigating the synergistic effects of multiple structural elements and understanding the role of self-healing and structural hierarchy, are crucial for future advancements. Optimal utilization of materials, porosity, and fluids for superior lubrication and heat transfer, alongside the creation of efficient databases and computational models, will facilitate technology transfer and collaboration, expanding the scope of knowledge in this field. By addressing these key areas, bioinspired tribology can significantly progress, leading to the development of durable, efficient, and sustainable materials and structures with enhanced tribological properties.

Author Contributions: Conceptualization, R.K. and I.H.; methodology, R.K., M.R. and I.H.; investigation, R.K., M.R., R.R., H.S.M. and N.K.; writing—original draft preparation, R.K.; writing—review and editing, R.K., M.R., R.R., H.S.M., N.K. and I.H.; resources and funding, I.H.; supervision, I.H. All authors have read and agreed to the published version of the manuscript.

Funding: This work was supported by Estonian Research Council grants PRG643, "BiLaTEX" MNHA23020, PUTJD1169, and "DuplexCER" MNHA22040. Additionally, Kamboj would like to acknowledge the MALAMA project, which is co-funded by the EU and implemented by the Pori Unit of the Turku School of Economics (TSE Pori) and the Department of Mechanical and Materials Engineering of the University of Turku (project code A80056). The funding authority is the Regional Council of Satakunta from the EU Regional and Structural Policy Programme "Innovation and Skills in Finland 2021–2027".

Institutional Review Board Statement: Not applicable.

Data Availability Statement: Some or all of the data, and models that support the findings of this study are available from the corresponding author upon request.

Conflicts of Interest: The authors declare no conflict of interest.

References

1. Tzanakis, I.; Hadfield, M.; Thomas, B.; Noya, S.; Henshaw, I.; Austen, S. Future perspectives on sustainable tribology. *Renew. Sustain. Energy Rev.* **2012**, *16*, 4126–4140. [CrossRef]
2. Woydt, M. The importance of tribology for reducing CO_2 emissions and for sustainability. *Wear* **2021**, *474–475*, 203768. [CrossRef]
3. Holmberg, K.; Erdemir, A. Influence of tribology on global energy consumption, costs and emissions. *Friction* **2017**, *5*, 263–284. [CrossRef]
4. Meng, Y.; Xu, J.; Jin, Z.; Prakash, B.; Hu, Y. A review of recent advances in tribology. *Friction* **2020**, *8*, 221–300. [CrossRef]
5. Kumar, R.; Kariminejad, A.; Antonov, M.; Goljandin, D.; Klimczyk, P.; Hussainova, I. Progress in Sustainable Recycling and Circular Economy of Tungsten Carbide Hard Metal Scraps for Industry 5.0 and Onwards. *Sustainability* **2023**, *15*, 12249. [CrossRef]
6. Allibone, T.E.; Meek, J.M. Relation between friction and adhesion. *Proc. R. Soc. London Ser. A Math. Phys. Sci.* **1950**, *202*, 244–253. [CrossRef]
7. Kumar, R.; Liu, L.; Antonov, M.; Ivanov, R.; Hussainova, I. Hot Sliding Wear of 88 wt.% TiB–Ti Composite from SHS Produced Powders. *Materials* **2021**, *14*, 1242. [CrossRef] [PubMed]
8. Kumar, R.; Hussainova, I.; Rahmani, R.; Antonov, M. Solid Lubrication at High-Temperatures—A Review. *Materials* **2022**, *15*, 1695. [CrossRef] [PubMed]
9. Kumar, R.; Antonov, M.; Klimczyk, P.; Mikli, V.; Gomon, D. Effect of cBN content and additives on sliding and surface fatigue wear of spark plasma sintered Al_2O_3-cBN composites. *Wear* **2022**, *494–495*, 204250. [CrossRef]
10. Wood, R.J.K. Multifunctional Materials for Tribological Applications. p. 379. Available online: https://books.google.com/books/about/Multifunctional_Materials_for_Tribologic.html?id=qZm9BwAAQBAJ (accessed on 30 September 2023).
11. Ibrahim, M.Z.; Sarhan, A.A.; Yusuf, F.; Hamdi, M. Biomedical materials and techniques to improve the tribological, mechanical and biomedical properties of orthopedic implants—A review article. *J. Alloys Compd.* **2017**, *714*, 636–667. [CrossRef]
12. Xu, Y.; Liu, W.; Zhang, G.; Li, Z.; Hu, H.; Wang, C.; Zeng, X.; Zhao, S.; Zhang, Y.; Ren, T. Friction stability and cellular behaviors on laser textured Ti–6Al–4V alloy implants with bioinspired micro-overlapping structures. *J. Mech. Behav. Biomed. Mater.* **2020**, *109*, 103823. [CrossRef] [PubMed]
13. Rezapourian, M.; Kamboj, N.; Jasiuk, I.; Hussainova, I. Biomimetic design of implants for long bone critical-sized defects. *J. Mech. Behav. Biomed. Mater.* **2022**, *134*, 105370. [CrossRef] [PubMed]

14. Kamboj, N.; Kazantseva, J.; Rahmani, R.; Rodríguez, M.A.; Hussainova, I. Selective laser sintered bio-inspired silicon-wollastonite scaffolds for bone tissue engineering. *Mater. Sci. Eng. C* **2020**, *116*, 111223. [CrossRef]
15. Zambrano-Mera, D.F.; Espinoza-González, R.; Rosenkranz, A.; Harvey, T.J.; Polcar, T.; Valenzuela, P.; Gacitúa, W. Enhanced erosion resistance of anti-reflective TiO_2/SiO_2 coatings induced by Zr-oxide doping. *Sol. Energy Mater. Sol. Cells* **2023**, *250*, 112079. [CrossRef]
16. Wiesinger, F.; Sutter, F.; Fernández-García, A.; Reinhold, J.; Pitz-Paal, R. Sand erosion on solar reflectors: Accelerated simulation and comparison with field data. *Sol. Energy Mater. Sol. Cells* **2016**, *145*, 303–313. [CrossRef]
17. Wang, X.; Xu, Q.; Wang, B.; Zhang, L.; Yang, H.; Peng, Z. Effect of surface waviness on the static performance of aerostatic journal bearings. *Tribol. Int.* **2016**, *103*, 394–405. [CrossRef]
18. Chen, S. Hydrodynamic Behaviour of Gliding Hydrofoil Crafts. Ph.D. Thesis, University of London, London, UK, 2013.
19. Li, M.S.; Melville, D.; Chung, E.; Stuart, H.S. Milliscale features increase friction of soft skin in lubricated contact. *IEEE Robot. Autom. Lett.* **2020**, *5*, 4781–4787. [CrossRef]
20. Majidi, C.; Shepherd, R.F.; Kramer, R.K.; Whitesides, G.M.; Wood, R.J. Influence of surface traction on soft robot undulation. *Int. J. Robot. Res.* **2013**, *32*, 1577–1584. [CrossRef]
21. Ben-Larbi, M.K.; Hensel, R.; Atzeni, G.; Arzt, E.; Stoll, E. Orbital debris removal using micropatterned dry adhesives: Review and recent advances. *Prog. Aerosp. Sci.* **2022**, *134*, 100850. [CrossRef]
22. Williamsen, J.E. *Vulnerability of Manned Spacecraft to Crew Loss from Orbital Debris Penetration*; NASA Marshall Space Flight Center: Huntsville, AL, USA, 1994.
23. Aïssa, B.; Tagziria, K.; Haddad, E.; Jamroz, W.; Loiseau, J.; Higgins, A.; Asgar-Khan, M.; Hoa, S.V.; Merle, P.G.; Therriault, D.; et al. The Self-Healing Capability of Carbon Fibre Composite Structures Subjected to Hypervelocity Impacts Simulating Orbital Space Debris. *ISRN Nanomater.* **2012**, *2012*, 16. [CrossRef]
24. Kumar, R.; Antonov, M.; Holovenko, Y.; Surzenkov, A. Erosive Wear Resistance of Nature-inspired Flexible Materials. *Tribol. Lett.* **2020**, *68*, 1–8. [CrossRef]
25. Koch, K.; Bhushan, B.; Barthlott, W. Multifunctional surface structures of plants: An inspiration for biomimetics. *Prog. Mater. Sci.* **2009**, *54*, 137–178. [CrossRef]
26. Bhushan, B. Lessons from nature for green science and technology: An overview and bioinspired superliquiphobic/philic surfaces. *Philos. Trans. R. Soc. A Math. Phys. Eng. Sci.* **2019**, *377*, 20180274. [CrossRef] [PubMed]
27. Nosonovsky, M.; Bhushan, B. Biologically Inspired Surfaces: Broadening the Scope of Roughness**. *Adv. Funct. Mater.* **2008**, *18*, 843–855. [CrossRef]
28. Ibrahim, M.; Jiang, J.; Wen, Z.; Sun, X. Surface Engineering for Enhanced Triboelectric Nanogenerator. *Nanoenergy Adv.* **2021**, *1*, 58–80. [CrossRef]
29. Liang, L.; Qing, Y. (Eds.) *Advanced Coating Materials*; Wiley: New York, NY, USA, 2018. [CrossRef]
30. Ma, S.; Scaraggi, M.; Yan, C.; Wang, X.; Gorb, S.N.; Dini, D.; Zhou, F. Bioinspired 3D Printed Locomotion Devices Based on Anisotropic Friction. *Small* **2019**, *15*, e1802931. [CrossRef]
31. Wu, Z.; Chen, Y.; Mukasa, D.; Pak, O.S.; Gao, W. Medical micro/nanorobots in complex media. *Chem. Soc. Rev.* **2020**, *49*, 8088–8112. [CrossRef] [PubMed]
32. Zhang, Y.; Zhang, Y.; Han, Y.; Gong, X. Micro/Nanorobots for Medical Diagnosis and Disease Treatment. *Micromachines* **2022**, *13*, 648. [CrossRef]
33. Fukuda, T.; Kajima, H.; Hasegawa, Y. Intelligent robots as artificial living creatures. *Artif. Life Robot.* **2004**, *8*, 101–110. [CrossRef]
34. Ge, J.Z.; Calderón, A.A.; Chang, L.; Pérez-Arancibia, N.O. An earthworm-inspired friction-controlled soft robot capable of bidirectional locomotion. *Bioinspir. Biomim.* **2019**, *14*, 036004. [CrossRef]
35. Zhang, Z.; Wang, X.; Tong, J.; Stephen, C. Innovative Design and Performance Evaluation of Bionic Imprinting Toothed Wheel. *Appl. Bionics Biomech.* **2018**, *2018*, 9806287. [CrossRef] [PubMed]
36. Tong, J.; Ji, W.; Jia, H.; Chen, D.; Yang, X. Design and tests of biomimetic blades for soil-rototilling and stubble-breaking. *J. Bionic Eng.* **2015**, *12*, 495–503. [CrossRef]
37. Zhu, D.; Zhang, X.; Ding, H. Tool wear characteristics in machining of nickel-based superalloys. *Int. J. Mach. Tools Manuf.* **2013**, *64*, 60–77. [CrossRef]
38. Mishnaevsky, L.; Hasager, C.B.; Bak, C.; Tilg, A.-M.; Bech, J.I.; Rad, S.D.; Fæster, S. Leading edge erosion of wind turbine blades: Understanding, prevention and protection. *Renew. Energy* **2021**, *169*, 953–969. [CrossRef]
39. Vadivel, H.S.; Somberg, J.; Kalin, M.; Emami, N. Tribological performance of aUHMWPE-based multiscale composite under different lubrication and loads. *Lubr. Sci.* **2022**, *34*, 480–492. [CrossRef]
40. Viswanathan, R.; Bakker, W. Materials for ultrasupercritical coal power plants—Turbine materials: Part II. *J. Mater. Eng. Perform.* **2001**, *10*, 96–101. [CrossRef]
41. Yang, X.; Xia, R.; Zhou, H.; Guo, L.; Zhang, L. Bionic surface design of cemented carbide drill bit. *Sci. China Technol. Sci.* **2016**, *59*, 175–182. [CrossRef]
42. Lu, J.; Wang, X.; Huang, Y.; Zhou, C.; Xu, B.; Fu, Q. Fabrication and cutting performance of bionic micro-serrated scalpels based on the miscanthus leaves. *Tribol. Int.* **2020**, *145*, 106162. [CrossRef]
43. Dutour, M.; Sikiri, S.; Antoniac, I.V.; Arango-Santander, S. Bioinspired Topographic Surface Modification of Biomaterials. *Materials* **2022**, *15*, 2383. [CrossRef]

44. Jiang, W. Bio-inspired self-sharpening cutting tool surface for finish hard turning of steel. *CIRP Ann.* **2014**, *63*, 517–520. [CrossRef]
45. Lv, J.; Song, Y.; Jiang, L.; Wang, J. Bio-inspired strategies for anti-icing. *ACS Nano* **2014**, *8*, 3152–3169. [CrossRef] [PubMed]
46. Fu, Y.F.; Yuan, C.Q.; Bai, X.Q. Marine drag reduction of shark skin inspired riblet surfaces. *Biosurf. Biotribol.* **2017**, *3*, 11–24. [CrossRef]
47. El-Atab, N.; Babatain, W.; Bahabry, R.; Alshanbari, R.; Shamsuddin, R.; Hussain, M.M. Ultraflexible Corrugated Monocrystalline Silicon Solar Cells with High Efficiency (19%), Improved Thermal Performance, and Reliability Using Low-Cost Laser Patterning. *ACS Appl. Mater. Interfaces* **2020**, *12*, 2269–2275. [CrossRef] [PubMed]
48. Luo, Y.; Yuan, L.; Li, J.; Wang, J. Boundary layer drag reduction research hypotheses derived from bio-inspired surface and recent advanced applications. *Micron* **2015**, *79*, 59–73. [CrossRef] [PubMed]
49. Zhang, D.; Guo, D.; Wang, F.; Dong, H.; Wang, Z. Low-velocity impact response of a novel bionic turtle shell back armor sandwich structure. *J. Mater. Res. Technol.* **2024**, *29*, 910–923. [CrossRef]
50. San Ha, N.; Lu, G. A review of recent research on bio-inspired structures and materials for energy absorption applications. *Compos. Part B Eng.* **2020**, *181*, 107496. [CrossRef]
51. Tilly, G.; Sage, W. The interaction of particle and material behaviour in erosion processes. *Wear* **1970**, *16*, 447–465. [CrossRef]
52. Zhou, L.; Zhang, H.; Ju, G.; Su, C.; Shen, H.; Zhao, X. Bio-inspired "rigid and flexible" structure design for robust superhydrophobic composite and its application. *Compos. Sci. Technol.* **2023**, *243*, 110244. [CrossRef]
53. Rahmani, R.; Karimi, J.; Kamboj, N.; Kumar, R.; Brojan, M.; Tchórz, A.; Skrabalak, G.; Lopes, S.I. Fabrication of localized diamond-filled copper structures via selective laser melting and spark plasma sintering. *Diam. Relat. Mater.* **2023**, *136*, 109916. [CrossRef]
54. Rahmani, R.; Kamboj, N.; Brojan, M.; Antonov, M.; Prashanth, K.G. Hybrid metal-ceramic biomaterials fabricated through powder bed fusion and powder metallurgy for improved impact resistance of craniofacial implants. *Materialia* **2022**, *24*, 101465. [CrossRef]
55. Han, Z.; Yin, W.; Zhang, J.; Jiang, J.; Niu, S.; Ren, L. Erosion-Resistant Surfaces Inspired by Tamarisk. *J. Bionic Eng.* **2013**, *10*, 479–487. [CrossRef]
56. Zidane, I.F.; Saqr, K.M.; Swadener, G.; Ma, X.; Shehadeh, M.F. On the role of surface roughness in the aerodynamic performance and energy conversion of horizontal wind turbine blades: A review. *Int. J. Energy Res.* **2016**, *40*, 2054–2077. [CrossRef]
57. Li, Y.; Yang, Y.; Li, R.; Tang, X.; Guo, D.; Qing, Y.; Qin, Y. Enhanced antibacterial properties of orthopedic implants by titanium nanotube surface modification: A review of current techniques. *Int. J. Nanomed.* **2019**, *14*, 7217–7236. [CrossRef] [PubMed]
58. Guo, Z.; Xie, X.; Yuan, C.; Bai, X. Study on influence of micro convex textures on tribological performances of UHMWPE material under the water-lubricated conditions. *Wear* **2019**, *426–427*, 1327–1335. [CrossRef]
59. Li, A.; Su, F.; Chu, P.K.; Sun, J. Articular cartilage inspired bilayer coating on Ti6Al4V alloy with low friction and high load-bearing properties. *Appl. Surf. Sci.* **2020**, *515*, 146065. [CrossRef]
60. Liu, G.; Zhang, P.; Liu, Y.; Zhang, D.; Chen, H. Self-Lubricanting Slippery Surface with Wettability Gradients for Anti-Sticking of Electrosurgical Scalpel. *Micromachines* **2018**, *9*, 591. [CrossRef] [PubMed]
61. Kumar, R.; Torres, H.; Aydinyan, S.; Antonov, M.; Varga, M.; Hussainova, I.; Ripoll, M.R. Tribological behavior of Ni-based self-lubricating claddings containing sulfide of nickel, copper, or bismuth at temperatures up to 600 °C. *Surf. Coat. Technol.* **2023**, *456*, 129270. [CrossRef]
62. Bhushan, B. Biomimetics: Lessons from nature—An overview. *Philos. Trans. R. Soc. A Math. Phys. Eng. Sci.* **2009**, *367*, 1445–1486. [CrossRef]
63. Katiyar, N.K.; Goel, G.; Hawi, S.; Goel, S. Nature-inspired materials: Emerging trends and prospects. *NPG Asia Mater.* **2021**, *13*, 56. [CrossRef]
64. Stute, F.; Mici, J.; Chamberlain, L.; Lipson, H. Digital Wood: 3D Internal Color Texture Mapping. *3D Print. Addit. Manuf.* **2018**, *5*, 285–291. [CrossRef]
65. Gibson, L.; Ashby, M.; Harley, B. *Cellular Materials in Nature and Medicine*; Cambridge University Press: Cambridge, UK, 2010.
66. Murphy, C.M.; Haugh, M.G.; O'Brien, F.J. The effect of mean pore size on cell attachment, proliferation and migration in collagen–glycosaminoglycan scaffolds for bone tissue engineering. *Biomaterials* **2010**, *31*, 461–466. [CrossRef] [PubMed]
67. Park, J.; Lee, Y.; Ha, M.; Cho, S.; Ko, H. Micro/nanostructured surfaces for self-powered and multifunctional electronic skins. *J. Mater. Chem. B* **2016**, *4*, 2999–3018. [CrossRef]
68. Wilson, D.J.; Lin, Z.; Bower, D.Q.; Deravi, L.F. Engineering color, pattern, and texture: From nature to materials. *Matter* **2021**, *4*, 2163–2171. [CrossRef]
69. Zhu, B.; Noack, M.; Merindol, R.; Barner-Kowollik, C.; Walther, A. Light-Adaptive Supramolecular Nacre-Mimetic Nanocomposites. *Nano Lett.* **2016**, *16*, 5176–5182. [CrossRef] [PubMed]
70. Pang, C.; Kwak, M.K.; Lee, C.; Jeong, H.E.; Bae, W.-G.; Suh, K.Y. Nano meets beetles from wing to tiptoe: Versatile tools for smart and reversible adhesions. *Nano Today* **2012**, *7*, 496–513. [CrossRef]
71. Mayer, G. Rigid biological systems as models for synthetic composites. *Science* **2005**, *310*, 1144–1147. [CrossRef] [PubMed]
72. Qu, J.; Teeple, C.B.; Oldham, K.R. Modeling Legged Microrobot Locomotion Based on Contact Dynamics and Vibration in Multiple Modes and Axes. *J. Vib. Acoust.* **2017**, *139*, 031013. [CrossRef]
73. Li, Z.; Guo, Z. Bioinspired surfaces with wettability for antifouling application. *Nanoscale* **2019**, *11*, 22636–22663. [CrossRef] [PubMed]

74. Jin, H.; Tian, L.; Bing, W.; Zhao, J.; Ren, L. Bioinspired marine antifouling coatings: Status, prospects, and future. *Prog. Mater. Sci.* **2022**, *124*, 100889. [CrossRef]
75. Zhang, J.; Wang, Y.; Zhou, S.; Wang, Y.; Wang, C.; Guo, W.; Lu, X.; Wang, L. Tailoring self-lubricating, wear-resistance, anticorrosion and antifouling properties of Ti/(Cu, MoS_2)-DLC coating in marine environment by controlling the content of Cu dopant. *Tribol. Int.* **2020**, *143*, 106029. [CrossRef]
76. Kumar, R.; Torres, H.; Aydinyan, S.; Antonov, M.; Varga, M.; Ripoll, M.R.; Hussainova, I. Microstructure and high temperature tribological behaviour of self-lubricating Ti-TiBx composite doped with Ni Bi. *Surf. Coat. Technol.* **2022**, *447*, 128827. [CrossRef]
77. Li, S.; An, Y.; Zhao, X.; Zhou, H.; Chen, J.; Hou, G. Bioinspired Smart Coating with Superior Tribological Performance. *ACS Appl. Mater. Interfaces* **2017**, *9*, 16745–16749. [CrossRef] [PubMed]
78. Fu, X.; Tian, L.; Fan, Y.; Ye, W.; Qiao, Z.-A.; Zhao, J.; Ren, L.; Ming, W. Stimuli-responsive self-healing anticorrosion coatings: From single triggering behavior to synergetic multiple protections. *Mater. Today Chem.* **2021**, *22*, 100575. [CrossRef]
79. Chopra, D.; Gulati, K.; Ivanovski, S. Bed of nails: Bioinspired nano-texturing towards antibacterial and bioactivity functions. *Mater. Today Adv.* **2021**, *12*, 100176. [CrossRef]
80. Parodi, A.; Molinaro, R.; Sushnitha, M.; Evangelopoulos, M.; Martinez, J.O.; Arrighetti, N.; Corbo, C.; Tasciotti, E. Bio-inspired engineering of cell- and virus-like nanoparticles for drug delivery. *Biomaterials* **2017**, *147*, 155–168. [CrossRef]
81. Martinez, J.O.; Brown, B.S.; Quattrocchi, N.; Evangelopoulos, M.; Ferrari, M.; Tasciotti, E. Multifunctional to multistage delivery systems: The evolution of nanoparticles for biomedical applications. *Chin. Sci. Bull.* **2012**, *57*, 3961–3971. [CrossRef]
82. Yoo, J.-W.; Irvine, D.J.; Discher, D.E.; Mitragotri, S. Bio-inspired, bioengineered and biomimetic drug delivery carriers. *Nat. Rev. Drug Discov.* **2011**, *10*, 521–535. [CrossRef] [PubMed]
83. Kim, H.N.; Jiao, A.; Hwang, N.S.; Kim, M.S.; Kang, D.H.; Kim, D.-H.; Suh, K.-Y. Nanotopography-guided tissue engineering and regenerative medicine. *Adv. Drug Deliv. Rev.* **2013**, *65*, 536–558. [CrossRef]
84. Walsh, L.A.; Allen, J.L.; Desai, T.A. Nanotopography applications in drug delivery. *Expert Opin. Drug Deliv.* **2015**, *12*, 1823–1827. [CrossRef]
85. Finbloom, J.A.; Huynh, C.; Huang, X.; Desai, T.A. Bioinspired nanotopographical design of drug delivery systems. *Nat. Rev. Bioeng.* **2023**, *1*, 139–152. [CrossRef]
86. Alvarez-Lorenzo, C.; Concheiro, A. Bioinspired drug delivery systems. *Curr. Opin. Biotechnol.* **2013**, *24*, 1167–1173. [CrossRef] [PubMed]
87. Yang, X.; Zhang, W.; Qin, X.; Cui, M.; Guo, Y.; Wang, T.; Wang, K.; Shi, Z.; Zhang, C.; Li, W.; et al. Recent Progress on Bioinspired Antibacterial Surfaces for Biomedical Application. *Biomimetics* **2022**, *7*, 88. [CrossRef] [PubMed]
88. Glinel, K.; Thebault, P.; Humblot, V.; Pradier, C.; Jouenne, T. Antibacterial surfaces developed from bio-inspired approaches. *Acta Biomater.* **2012**, *8*, 1670–1684. [CrossRef] [PubMed]
89. Han, Z.; Wang, Z.; Feng, X.; Li, B.; Mu, Z.; Zhang, J.; Niu, S.; Ren, L. Antireflective surface inspired from biology: A review. *Biosurf. Biotribol.* **2016**, *2*, 137–150. [CrossRef]
90. Borodich, F.M.; Savencu, O. Hierarchical Models of Engineering Rough Surfaces and Bio-Inspired Adhesives. In *Bio-Inspired Structured Adhesives*; Springer: Cham, Switzerland, 2017; pp. 179–219. [CrossRef]
91. The Biomimicry Institute—Nature-Inspired Innovation. Available online: https://biomimicry.org/ (accessed on 14 February 2024).
92. Hofmann, D.C.; Kolodziejska, J.; Roberts, S.; Otis, R.; Dillon, R.P.; Suh, J.-O.; Liu, Z.-K.; Borgonia, J.-P. Compositionally graded metals: A new frontier of additive manufacturing. *J. Mater. Res.* **2014**, *29*, 1899–1910. [CrossRef]
93. Dimas, L.S.; Bratzel, G.H.; Eylon, I.; Buehler, M.J. Tough Composites Inspired by Mineralized Natural Materials: Computation, 3D printing, and Testing. *Adv. Funct. Mater.* **2013**, *23*, 4629–4638. [CrossRef]
94. Yang, Y.; Song, X.; Li, X.; Chen, Z.; Zhou, C.; Zhou, Q.; Chen, Y. Recent Progress in Biomimetic Additive Manufacturing Technology: From Materials to Functional Structures. *Adv. Mater.* **2018**, *30*, e1706539. [CrossRef]
95. Bandyopadhyay, A.; Traxel, K.D.; Bose, S. Nature-inspired materials and structures using 3D Printing. *Mater. Sci. Eng. R Rep.* **2021**, *145*, 100609. [CrossRef] [PubMed]
96. Greco, L.; Buccino, F.; Xu, Z.; Vergani, L.; Berto, F.; Guagliano, M.; Razavi, N.; Bagherifard, S. Design and analysis of energy absorbent bioinspired lattice structures. *J. Bionic Eng.* **2023**, *20*, 1670–1686. [CrossRef]
97. Yang, J.; Gu, D.; Lin, K.; Zhang, Y.; Guo, M.; Yuan, L.; Zhang, H.; Zhang, H. Laser Additive Manufacturing of Bio-inspired Metallic Structures. *Chin. J. Mech. Eng. Addit. Manuf. Front.* **2022**, *1*, 100013. [CrossRef]
98. Velasco-Hogan, A.; Xu, J.; Meyers, M.A.; Velasco-Hogan, A.; Meyers, M.A.; Xu, J. Additive Manufacturing as a Method to Design and Optimize Bioinspired Structures. *Adv. Mater.* **2018**, *30*, 1800940. [CrossRef] [PubMed]
99. Kumar, R.; Antonov, M.; Beste, U.; Goljandin, D. Assessment of 3D printed steels and composites intended for wear applications in abrasive, dry or slurry erosive conditions. *Int. J. Refract. Met. Hard Mater.* **2020**, *86*, 105126. [CrossRef]
100. Li, P.; Dou, X.; Feng, C.; Schönherr, H. Enhanced cell adhesion on a bio-inspired hierarchically structured polyester modified with gelatin-methacrylate. *Biomater. Sci.* **2018**, *6*, 785–792. [CrossRef] [PubMed]
101. Yoon, E.-S.; Singh, R.; Kong, H.; Kim, B.; Kim, D.-H.; Jeong, H.; Suh, K. Tribological properties of bio-mimetic nano-patterned polymeric surfaces on silicon wafer. *Tribol. Lett.* **2006**, *21*, 31–37. [CrossRef]

102. Wang, C.; Shao, R.; Wang, G.; Sun, S. Hierarchical hydrophobic surfaces with controlled dual transition between rose petal effect and lotus effect via structure tailoring or chemical modification. *Colloids Surf. A Physicochem. Eng. Asp.* **2021**, *622*, 126661. [CrossRef]
103. Maurya, H.; Kollo, L.; Juhani, K.; Sergejev, F.; Prashanth, K. Effect of preheating and cooling of the powder bed by laser pulse shaping on the microstructure of the TiC based cermets. *Ceram. Int.* **2022**, *48*, 20612–20618. [CrossRef]
104. Li, W.; Yu, Y.; Huang, R.; Wang, X.; Lai, P.; Chen, K.; Shang, L.; Zhao, Y. Multi-Bioinspired Functional Conductive Hydrogel Patches for Wound Healing Management. *Adv. Sci.* **2023**, *10*, e2301479. [CrossRef]
105. Han, Z.; Fu, J.; Feng, X.; Niu, S.; Zhang, J.; Ren, L. Bionic anti-adhesive electrode coupled with maize leaf microstructures and TiO_2 coating. *RSC Adv.* **2017**, *7*, 45287–45293. [CrossRef]
106. Zhang, P.; Liu, G.; Zhang, D.; Chen, H. Liquid-Infused Surfaces on Electrosurgical Instruments with Exceptional Antiadhesion and Low-Damage Performances. *ACS Appl. Mater. Interfaces* **2018**, *10*, 33713–33720. [CrossRef]
107. Ashammakhi, N.; Ahadian, S.; Zengjie, F.; Suthiwanich, K.; Lorestani, F.; Orive, G.; Ostrovidov, S.; Khademhosseini, A. Advances and Future Perspectives in 4D Bioprinting. *Biotechnol. J.* **2018**, *13*, e1800148. [CrossRef]
108. YWang, Y.; Zheng, G.; Jiang, N.; Ying, G.; Li, Y.; Cai, X.; Meng, J.; Mai, L.; Guo, M.; Zhang, Y.S.; et al. Nature-inspired micropatterns. *Nat. Rev. Methods Prim.* **2023**, *3*, 68. [CrossRef]
109. Hazell, G.; Fisher, L.E.; Murray, W.A.; Nobbs, A.H.; Su, B. Bioinspired bactericidal surfaces with polymer nanocone arrays. *J. Colloid Interface Sci.* **2018**, *528*, 389–399. [CrossRef] [PubMed]
110. Sánchez, J.; Toro, A.; Estupiñán, H.; Leighton, G.; Endrino, J. Fabrication of bio-inspired deterministic surfaces by photochemical machining for tribological applications. *Tribol. Int.* **2020**, *150*, 106341. [CrossRef]
111. Skorb, E.V.; Möhwald, H. Ultrasonic approach for surface nanostructuring. *Ultrason. Sonochem.* **2016**, *29*, 589–603. [CrossRef] [PubMed]
112. Ren, L.; Zhou, X.; Liu, Q.; Liang, Y.; Song, Z.; Zhang, B.; Li, B. 3D magnetic printing of bio-inspired composites with tunable mechanical properties. *J. Mater. Sci.* **2018**, *53*, 14274–14286. [CrossRef]
113. May, R.M.; Hoffman, M.G.; Sogo, M.J.; Parker, A.E.; O'toole, G.A.; Brennan, A.B.; Reddy, S.T. Micro-patterned surfaces reduce bacterial colonization and biofilm formation in vitro: Potential for enhancing endotracheal tube designs. *Clin. Transl. Med.* **2014**, *3*, 8. [CrossRef] [PubMed]
114. Agarwal, K.; Sahay, R.; Baji, A.; Budiman, A.S. Impact-Resistant and Tough Helicoidally Aligned Ribbon Reinforced Composites with Tunable Mechanical Properties via Integrated Additive Manufacturing Methodologies. *ACS Appl. Polym. Mater.* **2020**, *2*, 3491–3504. [CrossRef]
115. Min, L.; Pan, H.; Chen, S.; Wang, C.; Wang, N.; Zhang, J.; Cao, Y.; Chen, X.; Hou, X. Recent progress in bio-inspired electrospun materials. *Compos. Commun.* **2019**, *11*, 12–20. [CrossRef]
116. del Campo, A.; Greiner, C.; Arzt, E. Contact Shape Controls Adhesion of Bioinspired Fibrillar Surfaces. *Langmuir* **2007**, *23*, 10235–10243. [CrossRef] [PubMed]
117. Kraus, T.; Brodoceanu, D.; Pazos-Perez, N.; Fery, A. Colloidal Surface Assemblies: Nanotechnology Meets Bioinspiration. *Adv. Funct. Mater.* **2013**, *23*, 4529–4541. [CrossRef]
118. Kandemir, S.; Yöyler, S.; Kumar, R.; Antonov, M.; Dieringa, H. Effect of Graphene Nanoplatelet Content on Mechanical and Elevated-Temperature Tribological Performance of Self-Lubricating ZE10 Magnesium Alloy Nanocomposites. *Lubricants* **2024**, *12*, 52. [CrossRef]
119. Rosenkranz, A.; Costa, H.L.; Baykara, M.Z.; Martini, A. Synergetic effects of surface texturing and solid lubricants to tailor friction and wear—A review. *Tribol. Int.* **2021**, *155*, 106792. [CrossRef]
120. Kumar, R.; Antonov, M.; Liu, L.; Hussainova, I. Sliding wear performance of in-situ spark plasma sintered Ti-TiBw composite at temperatures up to 900 °C. *Wear* **2021**, *476*, 203663. [CrossRef]
121. Abdel-Aal, H.A.; El Mansori, M.; Mezghani, S. Multi-scale investigation of surface topography of ball python (python regius) shed skin in comparison to human skin. *Tribol. Lett.* **2010**, *37*, 517–527. [CrossRef]
122. Filippov, A.E.; Gorb, S.N. Modelling of the frictional behaviour of the snake skin covered by anisotropic surface nanostructures. *Sci. Rep.* **2016**, *6*, 23539. [CrossRef] [PubMed]
123. Bapat, S.; Tiner, C.; Rajurkar, K.; Nath, S.; Atre, S.; Malshe, A. Understanding biologicalisation of the snake-skin inspired textures through additive manufacturing for mechanical traction. *CIRP Ann.* **2020**, *69*, 201–204. [CrossRef]
124. Schneider, J.; Djamiykov, V.; Greiner, C. Friction reduction through biologically inspired scale-like laser surface textures. *Beilstein J. Nanotechnol.* **2018**, *9*, 2561–2572. [CrossRef] [PubMed]
125. Sánchez, J.C.; Estupiñan, H.A.; Toro, A. Friction response of bioinspired AISI 52100 steel surfaces texturized by photochemical machining. *Surf. Topogr. Metrol. Prop.* **2021**, *9*, 014001. [CrossRef]
126. Greiner, C.; Schäfer, M. Bio-inspired scale-like surface textures and their tribological properties. *Bioinspir. Biomim.* **2015**, *10*, 044001. [CrossRef]
127. Mühlberger, M.; Rohn, M.; Danzberger, J.; Sonntag, E.; Rank, A.; Schumm, L.; Kirchner, R.; Forsich, C.; Gorb, S.; Einwögerer, B.; et al. UV-NIL fabricated bio-inspired inlays for injection molding to influence the friction behavior of ceramic surfaces. *Microelectron. Eng.* **2015**, *141*, 140–144. [CrossRef]
128. Kumar, M.; Ranjan, V.; Tyagi, R. Effect of Shape, Density, and an Array of Dimples on the Friction and Wear Performance of Laser Textured Bearing Steel under Dry Sliding. *J. Mater. Eng. Perform.* **2020**, *29*, 2827–2838. [CrossRef]

129. Khan, M.A.; Gupta, K. A study on machinability of nickel based superalloy using micro-textured tungsten carbide cutting tools. *Mater. Res. Express* **2020**, *7*, 016537. [CrossRef]
130. Thirunavukkarasu, N.; Peng, S.; Gunasekaran, H.B.; Yang, Z.; Wu, L.; Weng, Z. Responsive friction modulation of 3D printed elastomeric bioinspired structures. *Tribol. Int.* **2022**, *175*, 107823. [CrossRef]
131. Erdemir, A. Review of engineered tribological interfaces for improved boundary lubrication. *Tribol. Int.* **2005**, *38*, 249–256. [CrossRef]
132. Huang, W.; Wang, X. Biomimetic design of elastomer surface pattern for friction control under wet conditions. *Bioinspir. Biomim.* **2013**, *8*, 046001. [CrossRef] [PubMed]
133. Wu, J.; Yu, A.; Chen, Q.; Wu, M.; Sun, L.; Yuan, J. Tribological properties of bronze surface with dimple textures fabricated by the indentation method. *Proc. Inst. Mech. Eng. Part J J. Eng. Tribol.* **2020**, *234*, 1680–1694. [CrossRef]
134. Mzali, S.; Elwasli, F.; Mezlini, S.; Hajlaoui, K.; Alrasheedi, N.H. Tribological behavior of 3D printed biomimetic surfaces. *Tribol. Int.* **2024**, *193*, 109352. [CrossRef]
135. Wang, B.; Yang, W.; Sherman, V.R.; Meyers, M.A. Pangolin armor: Overlapping, structure, and mechanical properties of the keratinous scales. *Acta Biomater.* **2016**, *41*, 60–74. [CrossRef] [PubMed]
136. Tramsen, H.T.; Gorb, S.N.; Zhang, H.; Manoonpong, P.; Dai, Z.; Heepe, L. Inversion of friction anisotropy in a bio-inspired asymmetrically structured surface. *J. R. Soc. Interface* **2018**, *15*, 20170629. [CrossRef]
137. Huang, Q.; Shi, X.; Xue, Y.; Zhang, K.; Wu, C. Wear-triggered self-repairing behavior of bionic textured AISI 4140 steel filled with multi-solid lubricants. *Wear* **2022**, *504–505*, 204416. [CrossRef]
138. Huang, Q.; Shi, X.; Xue, Y.; Zhang, K.; Wu, C. Optimization of bionic textured parameter to improve the tribological performance of AISI 4140 self-lubricating composite through response surface methodology. *Tribol. Int.* **2021**, *161*, 107104. [CrossRef]
139. Varenberg, M.; Gorb, S.N. Hexagonal Surface Micropattern for Dry and Wet Friction. *Adv. Mater.* **2009**, *21*, 483–486. [CrossRef]
140. Drotlef, D.; Stepien, L.; Kappl, M.; Barnes, W.J.P.; Butt, H.; del Campo, A. Insights into the Adhesive Mechanisms of Tree Frogs using Artificial Mimics. *Adv. Funct. Mater.* **2013**, *23*, 1137–1146. [CrossRef]
141. Federle, W.; Barnes, W.; Baumgartner, W.; Drechsler, P.; Smith, J. Wet but not slippery: Boundary friction in tree frog adhesive toe pads. *J. R. Soc. Interface* **2006**, *3*, 689–697. [CrossRef] [PubMed]
142. Lu, G.; Lu, W.; Shi, X.; Zhai, W.; Zhang, J.; Yang, Z.; Chen, W. Tribological properties and self-compensating lubrication mechanisms of Ni_3Al matrix bio-inspired shell-like composite structure. *Appl. Surf. Sci.* **2022**, *573*, 151462. [CrossRef]
143. Wang, H.; Qiu, H.; Liu, W.; Tian, L.; Zheng, J.; Zhang, Z.; Zhou, Z. Anti-wear properties of multi-bioinspired textures with hierarchical and superhydrophobic structures under water lubrication. *Wear* **2023**, *532*, 205124. [CrossRef]
144. Kumar, R.; Antonov, M.; Varga, M.; Hussainova, I.; Ripoll, M.R. Synergistic effect of Ag and MoS_2 on high-temperature tribology of self-lubricating NiCrBSi composite coatings by laser metal deposition. *Wear* **2023**, *532–533*, 205114. [CrossRef]
145. Kumar, R.; Aydinyan, S.; Ivanov, R.; Liu, L.; Antonov, M.; Hussainova, I. High-Temperature Wear Performance of hBN-Added Ni-W Composites Produced from Combustion-Synthesized Powders. *Materials* **2022**, *15*, 1252. [CrossRef]
146. He, Y.; Zhang, Z.; Wang, Y.; Liu, M.; Yuan, J.; Li, P.; Yang, M.; Liu, W. Inspired with fish scale to manufacture biomimetic MXene derivative for the reinforcement on tribological and mechanical properties of PPS fabric/phenolic composites. *Tribol. Int.* **2023**, *179*, 108136. [CrossRef]
147. Liu, Y.; Zhang, H.; Dai, S.; Dong, G. Designing a bioinspired scaly textured surface for improving the tribological behaviors of starved lubrication. *Tribol. Int.* **2022**, *173*, 107594. [CrossRef]
148. Hoskins, J.K.; Zou, M. Designing a Bioinspired Surface for Improved Wear Resistance and Friction Reduction. *J. Tribol.* **2021**, *143*, 051107. [CrossRef]
149. Tong, J.; Ma, Y.-H.; Ren, L.-Q.; Li, J.-Q. Tribological characteristics of pangolin scales in dry sliding. *J. Mater. Sci. Lett.* **2000**, *19*, 569–572. [CrossRef]
150. Singh, R.A.; Pham, D.-C.; Kim, J.; Yang, S.; Yoon, E.-S. Bio-inspired dual surface modification to improve tribological properties at small-scale. *Appl. Surf. Sci.* **2009**, *255*, 4821–4828. [CrossRef]
151. TNarayana, T.; Saleem, S.S. Enhancing fretting wear behavior of Ti64 alloy: The impact of surface textures and CrN-MoS_2-Ag composite coating. *Tribol. Int.* **2024**, *193*, 109346. [CrossRef]
152. Jiao, Y.; Liu, S.; Sun, Y.; Yue, W.; Zhang, H. Bioinspired Surface Functionalization of Nanodiamonds for Enhanced Lubrication. *Langmuir* **2018**, *34*, 12436–12444. [CrossRef]
153. Liu, Y.; Wu, G.; de Groot, K. Biomimetic coatings for bone tissue engineering of critical-sized defects. *J. R. Soc. Interface* **2010**, *7*, S631–S647. [CrossRef]
154. Bandyopadhyay, A.; Dittrick, S.; Gualtieri, T.; Wu, J.; Bose, S. Calcium phosphate–titanium composites for articulating surfaces of load-bearing implants. *J. Mech. Behav. Biomed. Mater.* **2016**, *57*, 280–288. [CrossRef] [PubMed]
155. Li, W.; Thian, E.S.; Wang, M.; Wang, Z.; Ren, L. Surface Design for Antibacterial Materials: From Fundamentals to Advanced Strategies. *Adv. Sci.* **2021**, *8*, 2100368. [CrossRef]
156. Nodoushan, E.J.; Ebrahimi, N.G.; Ayazi, M. An anti-bacterial approach to nanoscale roughening of biomimetic rice-like pattern PP by thermal annealing. *Appl. Surf. Sci.* **2017**, *423*, 1054–1061. [CrossRef]
157. Liu, M.; Li, M.-T.; Xu, S.; Yang, H.; Sun, H.-B. Bioinspired Superhydrophobic Surfaces via Laser-Structuring. *Front. Chem.* **2020**, *8*, 835. [CrossRef]

158. Jia, Z.; Xiu, P.; Li, M.; Xu, X.; Shi, Y.; Cheng, Y.; Wei, S.; Zheng, Y.; Xi, T.; Cai, H.; et al. Bioinspired anchoring AgNPs onto micro-nanoporous TiO_2 orthopedic coatings: Trap-killing of bacteria, surface-regulated osteoblast functions and host responses. *Biomaterials* **2016**, *75*, 203–222. [CrossRef]
159. Yan, H.; Wu, Q.; Yu, C.; Zhao, T.; Liu, M. Recent Progress of Biomimetic Antifouling Surfaces in Marine. *Adv. Mater. Interfaces* **2020**, *7*, 2000966. [CrossRef]
160. Liu, Q.; Meng, F.; Wang, X.; Yang, B.; Tan, D.; Li, Q.; Shi, Z.; Shi, K.; Chen, W.; Liu, S.; et al. Tree Frog-Inspired Micropillar Arrays with Nanopits on the Surface for Enhanced Adhesion under Wet Conditions. *ACS Appl. Mater. Interfaces* **2020**, *12*, 19116–19122. [CrossRef]
161. Watson, G.S.; Green, D.W.; Schwarzkopf, L.; Li, X.; Cribb, B.W.; Myhra, S.; Watson, J.A. A gecko skin micro/nano structure—A low adhesion, superhydrophobic, anti-wetting, self-cleaning, biocompatible, antibacterial surface. *Acta Biomater.* **2015**, *21*, 109–122. [CrossRef]
162. Ibrahim, U.H.; Devnarain, N.; Govender, T. Biomimetic strategies for enhancing synthesis and delivery of antibacterial nanosystems. *Int. J. Pharm.* **2021**, *596*, 120276. [CrossRef]
163. Bandara, C.D.; Singh, S.; Afara, I.O.; Wolff, A.; Tesfamichael, T.; Ostrikov, K.; Oloyede, A. Bactericidal Effects of Natural Nanotopography of Dragonfly Wing on *Escherichia coli*. *ACS Appl. Mater. Interfaces* **2017**, *9*, 6746–6760. [CrossRef] [PubMed]
164. Bhushan, B.; Nosonovsky, M. The rose petal effect and the modes of superhydrophobicity. *Philos. Trans. R. Soc. A Math. Phys. Eng. Sci.* **2010**, *368*, 4713–4728. [CrossRef] [PubMed]
165. Pu, X.; Li, G.; Huang, H. Preparation, anti-biofouling and drag-reduction properties of a biomimetic shark skin surface. *Biol. Open* **2016**, *5*, 389–396. [CrossRef]
166. Lang, A.W.; Bradshaw, M.T.; Smith, J.A.; Wheelus, J.N.; Motta, P.J.; Habegger, M.L.; Hueter, R.E. Movable shark scales act as a passive dynamic micro-roughness to control flow separation. *Bioinspir. Biomim.* **2014**, *9*, 036017. [CrossRef]
167. Negrini, N.C.; Toffoletto, N.; Farè, S.; Altomare, L. Plant Tissues as 3D Natural Scaffolds for Adipose, Bone and Tendon Tissue Regeneration. *Front. Bioeng. Biotechnol.* **2020**, *8*, 723. [CrossRef]
168. Anselme, K.; Ploux, L.; Ponche, A. Cell/Material Interfaces: Influence of Surface Chemistry and Surface Topography on Cell Adhesion. *J. Adhes. Sci. Technol.* **2010**, *24*, 831–852. [CrossRef]
169. Asha, A.B.; Chen, Y.; Zhang, H.; Ghaemi, S.; Ishihara, K.; Liu, Y.; Narain, R. Rapid Mussel-Inspired Surface Zwitteration for Enhanced Antifouling and Antibacterial Properties. *Langmuir* **2019**, *35*, 1621–1630. [CrossRef] [PubMed]
170. Rasmussen, M.H.; Holler, K.R.; Baio, J.E.; Jaye, C.; Fischer, D.A.; Gorb, S.N.; Weidner, T. Evidence that gecko setae are coated with an ordered nanometre-thin lipid film. *Biol. Lett.* **2022**, *18*, 20220093. [CrossRef] [PubMed]
171. Román-Kustas, J.; Hoffman, J.B.; Alonso, D.; Reed, J.H.; Gonsalves, A.E.; Oh, J.; Hong, S.; Jo, K.D.; Dana, C.E.; Alleyne, M.; et al. Analysis of cicada wing surface constituents by comprehensive multidimensional gas chromatography for species differentiation. *Microchem. J.* **2020**, *158*, 105089. [CrossRef]
172. Guo, Z.; Liu, W.; Su, B.-L. Superhydrophobic surfaces: From natural to biomimetic to functional. *J. Colloid Interface Sci.* **2011**, *353*, 335–355. [CrossRef]
173. Shahali, H.; Hasan, J.; Mathews, A.; Wang, H.; Yan, C.; Tesfamichael, T.; Yarlagadda, P.K.D.V. Multi-biofunctional properties of three species of cicada wings and biomimetic fabrication of nanopatterned titanium pillars. *J. Mater. Chem. B* **2019**, *7*, 1300–1310. [CrossRef] [PubMed]
174. Kelleher, S.M.; Habimana, O.; Lawler, J.; O'reilly, B.; Daniels, S.; Casey, E.; Cowley, A. Cicada Wing Surface Topography: An Investigation into the Bactericidal Properties of Nanostructural Features. *ACS Appl. Mater. Interfaces* **2016**, *8*, 14966–14974. [CrossRef] [PubMed]
175. Gao, A.; Yan, Y.; Li, T.; Liu, F. Biomimetic urchin-like surface based on poly (lactic acid) membrane for robust anti-wetting and anti-bacteria properties. *Mater. Lett.* **2019**, *237*, 240–244. [CrossRef]
176. Li, W.; Zheng, Y.; Pang, W.; Lai, P. Bio-inspired adhesive hydrogel for wound healing. *Biomed. Technol.* **2023**, *1*, 65–72. [CrossRef]
177. Shi, T.; Hou, X.; Guo, S.; Zhang, L.; Wei, C.; Peng, T.; Hu, X. Nanohole-boosted electron transport between nanomaterials and bacteria as a concept for nano–bio interactions. *Nat. Commun.* **2021**, *12*, 493. [CrossRef]
178. Gao, Q.; Feng, T.; Huang, D.; Liu, P.; Lin, P.; Wu, Y.; Ye, Z.; Ji, J.; Huang, W. Antibacterial and hydroxyapatite-forming coating for biomedical implants based on polypeptide-functionalized titania nanospikes. *Biomater. Sci.* **2019**, *8*, 278–289. [CrossRef] [PubMed]
179. Dou, X.Q.; Zhang, D.; Feng, C.; Jiang, L. Bioinspired Hierarchical Surface Structures with Tunable Wettability for Regulating Bacteria Adhesion. *ACS Nano* **2015**, *9*, 10664–10672. [CrossRef] [PubMed]
180. Lin, W.; Klein, J. Recent Progress in Cartilage Lubrication. *Adv. Mater.* **2021**, *33*, 2005513. [CrossRef] [PubMed]
181. Lin, P.; Zhang, R.; Wang, X.; Cai, M.; Yang, J.; Yu, B.; Zhou, F. Articular Cartilage Inspired Bilayer Tough Hydrogel Prepared by Interfacial Modulated Polymerization Showing Excellent Combination of High Load-Bearing and Low Friction Performance. *ACS Macro Lett.* **2016**, *5*, 1191–1195. [CrossRef] [PubMed]
182. Lin, P.; Fu, D.; Zhang, T.; Ma, S.; Zhou, F. Microgel-Modified Bilayered Hydrogels Dramatically Boosting Load-Bearing and Lubrication. *ACS Macro Lett.* **2023**, *12*, 1450–1456. [CrossRef] [PubMed]
183. Maksimkin, A.; Chukov, D.; Senatov, F.; Salimon, A. Comparative analysis of the tribological characteristics of canine joint cartilage and UHMWPE-based biomimetic materials. *Mater. Lett.* **2017**, *191*, 105–107. [CrossRef]

184. Sun, J.; Li, A.; Su, F. Excellent Lubricating Ability of Functionalization Graphene Dispersed in Perfluoropolyether for Titanium Alloy. *ACS Appl. Nano Mater.* **2019**, *2*, 1391–1401. [CrossRef]
185. Zhiwu, H.; Junqiu, Z.; Chao, G.; Li, W.; Ren, L. Erosion Resistance of Bionic Functional Surfaces Inspired from Desert Scorpions. *Langmuir* **2012**, *28*, 2914–2921. [CrossRef] [PubMed]
186. Zhang, J.; Chen, W.; Yang, M.; Chen, S.; Zhu, B.; Niu, S.; Han, Z.; Wang, H. The Ingenious Structure of Scorpion Armor Inspires Sand-Resistant Surfaces. *Tribol. Lett.* **2017**, *65*, 110. [CrossRef]
187. Harrington, E. Bumps and Grooves Protect Surface: Sahara Scorpion—Asknature. *Asknature*. Available online: https://asknature.org/strategy/bumps-and-grooves-protect-surface/ (accessed on 13 February 2024).
188. Han, Z.; Zhu, B.; Yang, M.; Niu, S.; Song, H.; Zhang, J. The effect of the micro-structures on the scorpion surface for improving the anti-erosion performance. *Surf. Coat. Technol.* **2017**, *313*, 143–150. [CrossRef]
189. Han, Z.; Feng, H.; Yin, W.; Niu, S.; Zhang, J.; Chen, D. An Efficient Bionic Anti-Erosion Functional Surface Inspired by Desert Scorpion Carapace. *Tribol. Trans.* **2015**, *58*, 357–364. [CrossRef]
190. Han, Z.W.; Yang, M.K.; Yin, W.; Zhang, J.Q. Biomimetic Design and Manufacturing of Anti-Erosion Functional Surfaces Inspired from Desert Scorpion. *Adv. Sci. Technol.* **2017**, *100*, 187–196. [CrossRef]
191. Wang, Z.; Zhang, Z.; Sun, Y.; Gao, K.; Liang, Y.; Li, X.; Ren, L. Wear behavior of bionic impregnated diamond bits. *Tribol. Int.* **2016**, *94*, 217–222. [CrossRef]
192. Han, Z.; Yin, W.; Zhang, J.; Niu, S.; Ren, L. Active Anti-erosion Protection Strategy in Tamarisk (*Tamarix aphylla*). *Sci. Rep.* **2013**, *3*, 3429. [CrossRef] [PubMed]
193. Zhang, S.; Zhang, J.; Yu, H.; Niu, S.; Lian, Z.; Xu, J.; Han, Z.; Ren, L. Large-area preparation strategy and anti-erosion mechanism for morphology-material coupled biomimetic anti-erosion functional surface. *Wear* **2023**, *530*, 205008. [CrossRef]
194. Jung, S.; Yang, E.; Jung, W.; Kim, H.-Y. Anti-erosive mechanism of a grooved surface against impact of particle-laden flow. *Wear* **2018**, *406–407*, 166–172. [CrossRef]
195. Paswan, S.K.; Singh, A.K. Analysis of surface finishing mechanism in a newly developed rotational magnetorheological honing process for its productivity improvement. *Wear* **2019**, *426–427*, 68–82. [CrossRef]
196. Scarabaeus-Wikipedia. Available online: https://en.wikipedia.org/wiki/Scarabaeus (accessed on 13 February 2024).
197. Fan, S.; Han, X.; Tang, Y.; Wang, Y.; Kong, X. Shark Skin—An Inspiration for the Development of a Novel and Simple Biomimetic Turbulent Drag Reduction Topology. *Sustainability* **2022**, *14*, 16662. [CrossRef]
198. Martin, S.; Bhushan, B. Modeling and optimization of shark-inspired riblet geometries for low drag applications. *J. Colloid Interface Sci.* **2016**, *474*, 206–215. [CrossRef]
199. Xiao, G.; He, Y.; Huang, Y.; Li, Q. Shark-Skin-Inspired Micro-Riblets Forming Mechanism of TC17 Titanium Alloy with Belt Grinding. *IEEE Access* **2019**, *7*, 107635–107647. [CrossRef]
200. CBüttner, C.C.; Schulz, U. Shark skin inspired riblet structures as aerodynamically optimized high temperature coatings for blades of aeroengines. *Smart Mater. Struct.* **2011**, *20*, 094016. [CrossRef]
201. Bechert, D.W.; Bruse, M.; Hage, W.; VAN DER Hoeven, J.G.T.; Hoppe, G. Experiments on drag-reducing surfaces and their optimization with an adjustable geometry. *J. Fluid Mech.* **1997**, *338*, 59–87. [CrossRef]
202. Lang, A.; Habegger, M.L.; Motta, P. Shark Skin Drag Reduction. *Encycl. Nanotechnol.* **2015**, 1–8. [CrossRef]
203. Jo, S.; Ahn, S.; Lee, H.; Jung, C.-M.; Song, S.; Kim, D.R. Water-repellent Hybrid Nanowire and Micro-scale Denticle Structures on Flexible Substrates of Effective Air Retention. *Sci. Rep.* **2018**, *8*, 16631. [CrossRef]
204. Han, Z.; Mu, Z.; Yin, W.; Li, W.; Niu, S.; Zhang, J.; Ren, L. Biomimetic multifunctional surfaces inspired from animals. *Adv. Colloid Interface Sci.* **2016**, *234*, 27–50. [CrossRef]
205. Kleis, I.; Kulu, P. *Solid Particle Erosion: Occurrence, Prediction and Control*; Springer: Tallinn, Estonia, 2007. Available online: https://books.google.com/books?hl=en&lr=&id=NAWBg1jJJv4C&oi=fnd&pg=PR9&dq=particle+erosion+wear&ots=dVdCbdWFeK&sig=AEDbx5TXNC4Yb0mTbyUGEVOzr-Y (accessed on 4 February 2024).
206. Finnie, I.; Stevick, G.; Ridgely, J. The influence of impingement angle on the erosion of ductile metals by angular abrasive particles. *Wear* **1992**, *152*, 91–98. [CrossRef]
207. Finnie, I. Erosion of surfaces by solid particles. *Wear* **1960**, *3*, 87–103. [CrossRef]
208. Oka, Y.; Matsumura, M.; Kawabata, T. Relationship between surface hardness and erosion damage caused by solid particle impact. *Wear* **1993**, *162–164*, 688–695. [CrossRef]
209. Oka, Y.; Okamura, K.; Yoshida, T. Practical estimation of erosion damage caused by solid particle impact: Part 1: Effects of impact parameters on a predictive equation. *Wear* **2005**, *259*, 95–101. [CrossRef]
210. Sun, F.; Xu, H. A review of biomimetic research for erosion wear resistance. *Bio-Design Manuf.* **2020**, *3*, 331–347. [CrossRef]
211. Sınmazçelik, T.; Fidan, S.; Ürgün, S. Effects of 3D printed surface texture on erosive wear. *Tribol. Int.* **2020**, *144*, 106110. [CrossRef]
212. Han, Z.; Zhang, J.; Ge, C.; Lü, Y.; Jiang, J.; Liu, Q.; Ren, L. Anti-Erosion Function in Animals and its Biomimetic Application. *J. Bionic Eng.* **2010**, *7*, S50–S58. [CrossRef]
213. Jia, Z.; Yu, Y.; Hou, S.; Wang, L. Biomimetic architected materials with improved dynamic performance. *J. Mech. Phys. Solids* **2019**, *125*, 178–197. [CrossRef]
214. Porter, M.M.; Adriaens, D.; Hatton, R.L.; Meyers, M.A.; McKittrick, J. Why the seahorse tail is square. *Science* **2015**, *349*, 6243. [CrossRef]

215. Zheng, B.; Liu, Y.; Liu, J.; Yin, S.; Xu, J. Novel mechanical behaviors of DNA-inspired helical structures with chirality. *Int. J. Mech. Sci.* **2019**, *161–162*, 105025. [CrossRef]
216. Han, S.; Chen, F.; Ye, H.; Zheng, Z.; Chen, L.; Wang, G. Bamboo-Inspired Renewable, High-Strength, Vibration-Damping Composites for Structural Applications. *ACS Sustain. Chem. Eng.* **2023**, *11*, 1146–1156. [CrossRef]
217. Ha, N.S.; Lu, G.; Xiang, X. High energy absorption efficiency of thin-walled conical corrugation tubes mimicking coconut tree configuration. *Int. J. Mech. Sci.* **2018**, *148*, 409–421. [CrossRef]
218. Zhang, W.; Yin, S.; Yu, T.; Xu, J. Crushing resistance and energy absorption of pomelo peel inspired hierarchical honeycomb. *Int. J. Impact Eng.* **2019**, *125*, 163–172. [CrossRef]
219. Yu, W.; Zhu, K.; Aman, Y.; Guo, Z.; Xiong, S. Bio-inspired design of SiCf-reinforced multi-layered Ti-intermetallic composite. *Mater. Des.* **2016**, *101*, 102–108. [CrossRef]
220. Lin, C.; Han, Y.; Guo, C.; Chang, Y.; Han, X.; Lan, L.; Jiang, F. Synthesis and mechanical properties of novel Ti-(SiCf/Al_3Ti) ceramic-fiber-reinforced metal-intermetallic-laminated (CFR-MIL) composites. *J. Alloys Compd.* **2017**, *722*, 427–437. [CrossRef]
221. Damiens, R.; Rhee, H.; Hwang, Y.; Park, S.; Hammi, Y.; Lim, H.; Horstemeyer, M. Compressive behavior of a turtle's shell: Experiment, modeling, and simulation. *J. Mech. Behav. Biomed. Mater.* **2012**, *6*, 106–112. [CrossRef]
222. Balani, K.; Patel, R.R.; Keshri, A.K.; Lahiri, D.; Agarwal, A. Multi-scale hierarchy of Chelydra serpentina: Microstructure and mechanical properties of turtle shell. *J. Mech. Behav. Biomed. Mater.* **2011**, *4*, 1440–1451. [CrossRef]
223. Achrai, B.; Wagner, H.D. Micro-structure and mechanical properties of the turtle carapace as a biological composite shield. *Acta Biomater.* **2013**, *9*, 5890–5902. [CrossRef]
224. Achrai, B.; Bar-On, B.; Wagner, H.D. Biological armors under impact—Effect of keratin coating, and synthetic bio-inspired analogues. *Bioinspir. Biomim.* **2015**, *10*, 016009. [CrossRef] [PubMed]
225. Chen, Q.; Pugno, N.M. Bio-mimetic mechanisms of natural hierarchical materials: A review. *J. Mech. Behav. Biomed. Mater.* **2013**, *19*, 3–33. [CrossRef]
226. Suresh Kumar, S.; Sankar, P.; S, V.K. Effect ofs bio-inspired surface pattern (Pangolin's scales) and grooved mechanisms on the high velocity ballistic performance of aluminum 6061-T6 targets. *Mech. Adv. Mater. Struct.* **2022**, *29*, 4151–4169. [CrossRef]
227. Ha, N.S.; Le, V.T.; Goo, N.S. Investigation of punch resistance of the Allomyrira dichtoloma beetle forewing. *J. Bionic Eng.* **2018**, *15*, 57–68. [CrossRef]
228. Zhang, W.; Xu, J.; Yu, T. Dynamic behaviors of bio-inspired structures: Design, mechanisms, and models. *Eng. Struct.* **2022**, *265*, 114490. [CrossRef]
229. Hao, P.; Du, J. Mechanical properties of bio-mimetic energy-absorbing materials under impact loading. *J. Mater. Sci.* **2018**, *53*, 3189–3197. [CrossRef]
230. Xiang, J.; Du, J.; Li, D.; Scarpa, F. Numerical analysis of the impact resistance in aluminum alloy bi-tubular thin-walled structures designs inspired by beetle elytra. *J. Mater. Sci.* **2017**, *52*, 13247–13260. [CrossRef]
231. San Ha, N.; Lu, G.; Xiang, X. Energy absorption of a bio-inspired honeycomb sandwich panel. *J. Mater. Sci.* **2019**, *54*, 6286–6300.
232. Behera, R.P.; Le Ferrand, H. Impact-resistant materials inspired by the mantis shrimp's dactyl club. *Matter* **2021**, *4*, 2831–2849. [CrossRef]
233. De Falco, P.; Ongaro, F.; Pugno, N. Multi-hit damage and perforation of plates inspired by the attacks of the mantis shrimp. *Mech. Mater.* **2023**, *181*, 104587. [CrossRef]
234. Yang, J.; Gu, D.; Lin, K.; Yuan, L.; Guo, M.; Zhang, H.; Liu, H. Laser powder bed fusion of mechanically efficient helicoidal structure inspired by mantis shrimp. *Int. J. Mech. Sci.* **2022**, *231*, 107573. [CrossRef]
235. Taylor, J.R.A.; Scott, N.I.; Rouse, G.W. Evolution of mantis shrimp telson armour and its role in ritualized fighting. *J. R. Soc. Interface* **2019**, *16*, 20190203. [CrossRef] [PubMed]
236. Yang, B.; Chen, W.; Xin, R.; Zhou, X.; Tan, D.; Ding, C.; Wu, Y.; Yin, L.; Chen, C.; Wang, S.; et al. Pomelo Peel-Inspired 3D-Printed Porous Structure for Efficient Absorption of Compressive Strain Energy. *J. Bionic Eng.* **2022**, *19*, 448–457. [CrossRef]
237. Chen, Y.; Jin, Z.; Kang, W.; Liu, Z.; Yang, W.; Li, Y. 3D printed bio-inspired self-similar carbon fiber reinforced composite sandwich structures for energy absorption. *Compos. Sci. Technol.* **2024**, *248*, 110453. [CrossRef]
238. Prakash, V.A.; Bourchak, M.; Alshahrani, H.; Juhany, K.A. Development of cashew nut shell lignin-acrylonitrile butadiene styrene 3D printed core and industrial hemp/aluminized glass fiber epoxy biocomposite for morphing wing and unmanned aerial vehicle applications. *Int. J. Biol. Macromol.* **2023**, *253*, 127068. [CrossRef]
239. Xiang, J.; Du, J. Energy absorption characteristics of bio-inspired honeycomb structure under axial impact loading. *Mater. Sci. Eng. A* **2017**, *696*, 283–289. [CrossRef]
240. Zou, M.; Yu, Y.J.; Zhang, R.R.; Wei, C.G.; Wang, H.X. Simulation analysis of energy-absorption properties of thin-wall tube based on horn structure. *Jilin Daxue Xuebao (Gongxueban)/J. Jilin Univ. Eng. Technol. Ed.* **2015**, *45*, 1863–1868. [CrossRef]
241. Zhou, J.; Liu, S.; Guo, Z.; Xu, S.; Song, J.; Zou, M. Study on the energy absorption performance of bionic tube inspired by yak horn. *Mech. Adv. Mater. Struct.* **2022**, *29*, 7246–7258. [CrossRef]
242. Zhang, P.; Heyne, M.A.; To, A.C. Biomimetic staggered composites with highly enhanced energy dissipation: Modeling, 3D printing, and testing. *J. Mech. Phys. Solids* **2015**, *83*, 285–300. [CrossRef]
243. Ko, K.; Jin, S.; Lee, S.E.; Hong, J.-W. Impact resistance of nacre-like composites diversely patterned by 3D printing. *Compos. Struct.* **2020**, *238*, 111951. [CrossRef]

244. Gu, G.X.; Takaffoli, M.; Buehler, M.J. Hierarchically Enhanced Impact Resistance of Bioinspired Composites. *Adv. Mater.* **2017**, *29*, 1700060. [CrossRef] [PubMed]
245. Meng, L.; Zhao, J.; Lan, X.; Yang, H.; Wang, Z. Multi-objective optimisation of bio-inspired lightweight sandwich structures based on selective laser melting. *Virtual Phys. Prototyp.* **2020**, *15*, 106–119. [CrossRef]
246. Yu, X.; Pan, L.; Chen, J.; Zhang, X.; Wei, P. Experimental and numerical study on the energy absorption abilities of trabecular–honeycomb biomimetic structures inspired by beetle elytra. *J. Mater. Sci.* **2019**, *54*, 2193–2204. [CrossRef]
247. Zhang, X.; Cai, Z.-B.; Li, W.; Zhu, M.-H. Understanding hydration effects on mechanical and impacting properties of turtle shell. *J. Mech. Behav. Biomed. Mater.* **2018**, *78*, 116–123. [CrossRef] [PubMed]
248. Tasdemirci, A.; Akbulut, E.F.; Guzel, E.; Tuzgel, F.; Yucesoy, A.; Sahin, S.; Guden, M. Crushing behavior and energy absorption performance of a bio-inspired metallic structure: Experimental and numerical study. *Thin-Walled Struct.* **2018**, *131*, 547–555. [CrossRef]
249. Xiao, Y.; Yin, H.; Fang, H.; Wen, G. Crashworthiness design of horsetail-bionic thin-walled structures under axial dynamic loading. *Int. J. Mech. Mater. Des.* **2016**, *12*, 563–576. [CrossRef]
250. Bührig-Polaczek, A.; Fleck, C.; Speck, T.; Schüler, P.; Fischer, S.F.; Caliaro, M.; Thielen, M. Biomimetic cellular metals—Using hierarchical structuring for energy absorption. *Bioinspir. Biomim.* **2016**, *11*, 045002. [CrossRef]
251. Lazarus, B.S.; Velasco-Hogan, A.; Río, T.G.-D.; Meyers, M.A.; Jasiuk, I. A review of impact resistant biological and bioinspired materials and structures. *J. Mater. Res. Technol.* **2020**, *9*, 15705–15738. [CrossRef]
252. Edelsten, L.; Jeffrey, J.E.; Burgin, L.V.; Aspden, R.M. Viscoelastic deformation of articular cartilage during impact loading. *Soft Matter* **2010**, *6*, 5206–5212. [CrossRef]
253. Zhu, Z.D.; Ma, G.J.; Wu, C.W.; Chen, Z. Numerical study of the impact response of woodpecker's head. *AIP Adv.* **2012**, *2*, 42173. [CrossRef]
254. Holmberg, K.; Laukkanen, A.; Ghabchi, A.; Rombouts, M.; Turunen, E.; Waudby, R.; Suhonen, T.; Valtonen, K.; Sarlin, E. Computational modelling based wear resistance analysis of thick composite coatings. *Tribol. Int.* **2014**, *72*, 13–30. [CrossRef]
255. Spongey Cranium Absorbs Impact—Biological Strategy—AskNature. Available online: https://asknature.org/strategy/spongey-cranium-absorbs-impact/#jp-carousel-7917 (accessed on 27 February 2024).
256. Van Wassenbergh, S.; Ortlieb, E.J.; Mielke, M.; Böhmer, C.; Shadwick, R.E.; Abourachid, A. Woodpeckers minimize cranial absorption of shocks. *Curr. Biol.* **2022**, *32*, 3189–3194.e4. [CrossRef]
257. Yoon, S.-H.; Park, S. A mechanical analysis of woodpecker drumming and its application to shock-absorbing systems. *Bioinspir. Biomim.* **2011**, *6*, 016003. [CrossRef] [PubMed]
258. Antonov, M.; Zahavi, A.; Kumar, R.; Tamre, M.; Klimczyk, P. Performance of Al_2O_3-cBN materials and the perspective of using hyperspectral imaging during cutting tests. *Proc. Estonian Acad. Sci.* **2021**, *70*, 524–532. [CrossRef]
259. Yu, H.; Han, Z.; Zhang, J.; Zhang, S. Bionic design of tools in cutting: Reducing adhesion, abrasion or friction. *Wear* **2021**, *482–483*, 203955. [CrossRef]
260. Lu, Y.; Liu, Y.; Li, X.; Kang, Y. A new method of drilling long boreholes in low permeability coal by improving its permeability. *Int. J. Coal Geol.* **2010**, *84*, 94–102. [CrossRef]
261. Wang, Z.; Gao, K.; Sun, Y.; Zhang, Z.; Zhang, S.; Liang, Y.; Li, X.; Ren, L. Effects of bionic units in different scales on the wear behavior of bionic impregnated diamond bits. *J. Bionic Eng.* **2016**, *13*, 659–668. [CrossRef]
262. Zhang, W.; Xiang, J.; Huang, R.; Liu, H. A review of bio-inspired geotechnics-perspectives from geomaterials, geo-components, and drilling & excavation strategies. *Biogeotechnics* **2023**, *1*, 100025. [CrossRef]
263. Hu, Y.; Chen, X.; Chen, J.; Zhang, C.; Fu, W. The influence of crescent texture parameters on the axial force when drilling bone. *Med. Eng. Phys.* **2021**, *87*, 87–94. [CrossRef]
264. Wang, C. Bionic design and test of polycrystalline diamond compact bit for hard rock drilling in coal mine. *Adv. Mech. Eng.* **2020**, *12*, 1687814020923180. [CrossRef]
265. Banken, E.; Oeffner, J. Biomimetics for innovative and future-oriented space applications—A review. *Front. Space Technol.* **2023**, *3*, 1000788. [CrossRef]
266. Zhang, Z.; Jia, H.; Sun, J. Bioinspired design of a ridging shovel with anti-adhesive and drag reducing. *Adv. Mech. Eng.* **2015**, *7*, 3. [CrossRef]
267. Qaisrani, A.; Tong, J.; Ren, L.; Chen, B. The Effects of Unsmoothed Surfaces on Soil Adhesion and Draft of Bulldozing Plates. 1993. Available online: https://www.cabidigitallibrary.org/doi/full/10.5555/19932457271 (accessed on 10 February 2024).
268. El Salem, A.; Zhang, G.; Wang, H.; Salem, H.M.; Abdalla, M.A.; Ghazy, A.A. The effect of integrating a bio-inspired convex structure with a low-surface energy polymer on soil adhesion and friction. *J. Terramech.* **2023**, *109*, 93–100. [CrossRef]
269. Wang, L.; Zhou, Q. Surface hydrophobicity of slippery zones in the pitchers of two Nepenthes species and a hybrid. *Sci. Rep.* **2016**, *6*, 19907. [CrossRef]
270. Zhang, P.; Chen, H.; Zhang, L.; Ran, T.; Zhang, D. Transparent self-cleaning lubricant-infused surfaces made with large-area breath figure patterns. *Appl. Surf. Sci.* **2015**, *355*, 1083–1090. [CrossRef]
271. Espinosa, H.D.; Zaheri, A.; Nguyen, H.; Restrepo, D.; Daly, M.; Frank, M.; McKittrick, J. In situ Wear Study Reveals Role of Microstructure on Self-Sharpening Mechanism in Sea Urchin Teeth. *Matter* **2019**, *1*, 1246–1261. [CrossRef]
272. Killian, C.E.; Metzler, R.A.; Gong, Y.; Churchill, T.H.; Olson, I.C.; Trubetskoy, V.; Christensen, M.B.; Fournelle, J.H.; De Carlo, F.; Cohen, S.; et al. Self-Sharpening Mechanism of the Sea Urchin Tooth. *Adv. Funct. Mater.* **2011**, *21*, 682–690. [CrossRef]

273. Wang, R.Z.; Addadi, L.; Weiner, S. Design strategies of sea urchin teeth: Structure, composition and micromechanical relations to function. *Philos. Trans. R. Soc. B Biol. Sci.* **1997**, *352*, 469–480. [CrossRef]
274. Foronda, E.; Ramírez-Gil, F.J.; Delgado-Mejía, Á.; Ballesteros, L.M.; Rudas, J.; Olmos-Villalba, L.C. Thermal enhancement of heat sinks with bio-inspired textured surfaces. *Therm. Sci. Eng. Prog.* **2023**, *46*, 102170. [CrossRef]
275. RKumar, R.; Malaval, B.; Antonov, M.; Zhao, G. Performance of polyimide and PTFE based composites under sliding, erosive and high stress abrasive conditions. *Tribol. Int.* **2020**, *147*, 106282. [CrossRef]
276. Yin, W.; Han, Z.; Feng, H.; Zhang, J.; Cao, H.; Tian, Y. Gas–Solid Erosive Wear of Biomimetic Pattern Surface Inspired from Plant. *Tribol. Trans.* **2017**, *60*, 159–165. [CrossRef]
277. Liu, Q.; Mao, H.; Niu, L.; Chen, F.; Ma, P. Excellent flexibility and stab-resistance on pangolin-inspired scale-like structure composite for versatile protection. *Compos. Commun.* **2022**, *35*, 101266. [CrossRef]
278. Su, Y.; Li, Z.; Li, L.; Wang, J.; Gao, H.; Wang, G. Cutting performance of micro-textured polycrystalline diamond tool in dry cutting. *J. Manuf. Process.* **2017**, *27*, 1–7. [CrossRef]
279. Sharma, V.; Pandey, P.M. Recent advances in turning with textured cutting tools: A review. *J. Clean. Prod.* **2016**, *137*, 701–715. [CrossRef]
280. Gundiah, N.; Jaddivada, S. Making the cut: Mechanics of cutting and steering of insect probes. *Curr. Opin. Insect Sci.* **2020**, *42*, 84–89. [CrossRef]
281. Zhang, W.; Huang, R.; Xiang, J.; Zhang, N. Recent advances in bio-inspired geotechnics: From burrowing strategy to underground structures. *Gondwana Res.* **2024**, *130*, 1–17. [CrossRef]
282. Ji, W.; Chen, D.; Jia, H.; Tong, J. Experimental Investigation into Soil-Cutting Performance of the Claws of Mole Rat (*Scaptochirus moschatus*). *J. Bionic Eng.* **2010**, *7*, S166–S171. [CrossRef]
283. Morris, J.; Wang, Y.; Backeljau, T.; Chapelle, G. Biomimetic and bio-inspired uses of mollusc shells. *Mar. Genom.* **2016**, *27*, 85–90. [CrossRef]
284. Bermejo, R. "Toward seashells under stress": Bioinspired concepts to design tough layered ceramic composites. *J. Eur. Ceram. Soc.* **2017**, *37*, 3823–3839. [CrossRef]
285. Gao, K.; Li, M.; Dong, B.; Sun, Y.; Sun, Y.; Liu, J. Bionic coupling polycrystalline diamond composite bit. *Pet. Explor. Dev.* **2014**, *41*, 533–537. [CrossRef]
286. Malshe, A.; Rajurkar, K.; Samant, A.; Hansen, H.N.; Bapat, S.; Jiang, W. Bio-inspired functional surfaces for advanced applications. *CIRP Ann.* **2013**, *62*, 607–628. [CrossRef]
287. Yu, H.; Zhang, J.; Zhang, S.; Han, Z. Bionic structures and materials inspired by plant leaves: A comprehensive review for innovative problem-solving. *Prog. Mater. Sci.* **2023**, *139*, 101181. [CrossRef]
288. Qi, H.; Ma, Z.; Xu, Z.; Wang, S.; Ma, Y.; Wu, S.; Guo, M. The Design and Experimental Validation of a Biomimetic Stubble-Cutting Device Inspired by a Leaf-Cutting Ant's Mandibles. *Biomimetics* **2023**, *8*, 555. [CrossRef] [PubMed]
289. Jiang, Z.; Sun, J.; Xiong, Q.; Wang, L.; Li, J. Structural design of groove and micro-blade of the end mill in aluminum alloys machining based on bionics. *Int. J. Adv. Manuf. Technol.* **2017**, *88*, 3343–3356. [CrossRef]
290. Drew, J.; Philipp, C.; Westneat, M.W. Shark Tooth Weapons from the 19th Century Reflect Shifting Baselines in Central Pacific Predator Assemblies. *PLoS ONE* **2013**, *8*, e59855. [CrossRef]
291. Sea Urchin Teeth | California Academy of Sciences. Available online: https://www.calacademy.org/explore-science/sea-urchin-teeth (accessed on 24 February 2024).
292. HMaurya, H.; Vikram, R.; Kumar, R.; Rahmani, R.; Juhani, K.; Sergejev, F.; Prashanth, K. EBSD investigation of microstructure and microtexture evolution on additively manufactured TiC-Fe based cermets—Influence of multiple laser scanning. *Micron* **2024**, *180*, 103613. [CrossRef]
293. Xie, F.; Lei, X.B. The Influence of Bionic Micro-Texture's Surface on Tool's Cutting Performance. *Key Eng. Mater.* **2016**, *693*, 1155–1162. [CrossRef]
294. Fatima, A.; Mativenga, P.T. On the comparative cutting performance of nature-inspired structured cutting tool in dry cutting of AISI/SAE 4140. *Proc. Inst. Mech. Eng. Part B J. Eng. Manuf.* **2015**, *231*, 1941–1948. [CrossRef]
295. Wei, H.; Chen, G.; Chen, Z.; Yu, Z.; Huang, J. Progress on Bionic Textured Cutting Tools: A Review and Prospects. *J. Bionic Eng.* **2023**, *21*, 19–55. [CrossRef]
296. Yu, Q.; Zhang, X.; Miao, X.; Liu, X.; Zhang, L. Performances of concave and convex microtexture tools in turning of Ti_6Al_4V with lubrication. *Int. J. Adv. Manuf. Technol.* **2020**, *109*, 1071–1092. [CrossRef]
297. Siju, A.S.; Waigaonkar, S.D. Effects of rake surface texture geometries on the performance of single-point cutting tools in hard turning of titanium alloy. *J. Manuf. Process.* **2021**, *69*, 235–252. [CrossRef]
298. Li, X.; Deng, J.; Lu, Y.; Duan, R.; Ge, D. Machinability Investigation and Sustainability Assessment of Dry Cutting AISI1045 Steel Using Tools Configured with Shark-Skin-Inspired Structures and WS2/C Coatings. *Int. J. Precis. Eng. Manuf.-Green Technol.* **2022**, *9*, 83–106. [CrossRef]
299. Liu, Y.; Deng, J.; Wu, F.; Duan, R.; Zhang, X.; Hou, Y. Wear resistance of carbide tools with textured flank-face in dry cutting of green alumina ceramics. *Wear* **2017**, *372–373*, 91–103. [CrossRef]
300. Gao, K.; Sun, Y.-H.; Ren, L.-Q.; Cao, P.-L.; Li, W.-T.; Fan, H.-K. Design and Analysis of Ternary Coupling Bionic Bits. *J. Bionic Eng.* **2008**, *5*, 53–59. [CrossRef]

301. Zhang, P.; Zhang, L.; Chen, H.; Dong, Z.; Zhang, D. Surfaces Inspired by the *Nepenthes* Peristome for Unidirectional Liquid Transport. *Adv. Mater.* **2017**, *29*, 1702995. [CrossRef] [PubMed]
302. Chen, H.; Zhang, P.; Zhang, L.; Liu, H.; Jiang, Y.; Zhang, D.; Han, Z.; Jiang, L. Continuous directional water transport on the peristome surface of Nepenthes alata. *Nature* **2016**, *532*, 85–89. [CrossRef] [PubMed]
303. Sun, J.; Wang, Y.; Ma, Y.; Tong, J.; Zhang, Z. DEM simulation of bionic subsoilers (tillage depth >40 cm) with drag reduction and lower soil disturbance characteristics. *Adv. Eng. Softw.* **2018**, *119*, 30–37. [CrossRef]
304. Tang, Q.; Guo, W.; Gao, K.; Gao, R.; Zhao, Y.; Sun, Y.; Zhou, Y. Design and test of a self-adaptive bionic polycrystalline diamond compact bit inspired by cat claw. *Adv. Mech. Eng.* **2018**, *10*, 1687814018810249. [CrossRef]
305. Feng, Y.; Yuan, P.; Wang, L.; Zhang, J.; Zhang, J.; Zhou, X. Experimental investigation of different morphology textured ceramic tools by in-situ formed for the dry cutting. *Int. J. Appl. Ceram. Technol.* **2020**, *17*, 1108–1118. [CrossRef]
306. Zhang, W.; Zhang, L.; Wang, B.; Wang, S. Finite element simulation analysis of bionic ball-end milling cutter. *Int. J. Adv. Manuf. Technol.* **2019**, *103*, 3151–3161. [CrossRef]
307. Jia, H.; Li, C.; Zhang, Z.; Wang, G. Design of Bionic Saw Blade for Corn Stalk Cutting. *J. Bionic Eng.* **2013**, *10*, 497–505. [CrossRef]
308. Jaggessar, A.; Shahali, H.; Mathew, A.; Yarlagadda, P.K.D.V. Bio-mimicking nano and micro-structured surface fabrication for antibacterial properties in medical implants. *J. Nanobiotechnol.* **2017**, *15*, 64. [CrossRef]
309. Luo, Z.; Weiss, D.E.; Liu, Q.; Tian, B. Biomimetic approaches toward smart bio-hybrid systems. *Nano Res.* **2018**, *11*, 3009–3030. [CrossRef]
310. Bjørge, I.M.; Correia, C.R.; Mano, J.F. Hipster microcarriers: Exploring geometrical and topographical cues of non-spherical microcarriers in biomedical applications. *Mater. Horiz.* **2022**, *9*, 908–933. [CrossRef]
311. Kamboj, N.; Ressler, A.; Hussainova, I. Bioactive Ceramic Scaffolds for Bone Tissue Engineering by Powder Bed Selective Laser Processing: A Review. *Materials* **2021**, *14*, 5338. [CrossRef]
312. Nierstrasz, V.A. Textile-based drug release systems. *Smart Text. Med. Healthc. Mater. Syst. Appl.* **2007**, 50–73. [CrossRef]
313. JZimmermann, J.; Reifler, F.A.; Fortunato, G.; Gerhardt, L.; Seeger, S. A Simple, One-Step Approach to Durable and Robust Superhydrophobic Textiles. *Adv. Funct. Mater.* **2008**, *18*, 3662–3669. [CrossRef]
314. Gerhardt, L.-C.; Lottenbach, R.; Rossi, R.; Derler, S. Tribological investigation of a functional medical textile with lubricating drug-delivery finishing. *Colloids Surf. B Biointerfaces* **2013**, *108*, 103–109. [CrossRef]
315. Guerra, C.; Schwartz, C.J. Investigation of the influence of textiles and surface treatments on blistering using a novel simulant. *Ski. Res. Technol.* **2012**, *18*, 94–100. [CrossRef] [PubMed]
316. Wollina, U.; Heide, M.; Müller-Litz, W.; Obenauf, D.; Ash, J. Functional textiles in prevention of chronic wounds, wound healing and tissue engineering. *Curr. Probl. Dermatol.* **2003**, *31*, 82–97. [CrossRef]
317. Tanner, T.; Marks, R. Delivering drugs by the transdermal route: Review and comment. *Ski. Res. Technol.* **2008**, *14*, 249–260. [CrossRef] [PubMed]
318. Marian, M.; Shah, R.; Gashi, B.; Zhang, S.; Bhavnani, K.; Wartzack, S.; Rosenkranz, A. Exploring the Lubrication Mechanisms of Synovial Fluids for Joint Longevity—A Perspective. *Colloids Surf. B Biointerfaces* **2021**, *206*, 111926. [CrossRef] [PubMed]
319. Sterner, O.; Karageorgaki, C.; Zürcher, M.; Zürcher, S.; Scales, C.W.; Fadli, Z.; Spencer, N.D.; Tosatti, S.G.P. Reducing Friction in the Eye: A Comparative Study of Lubrication by Surface-Anchored Synthetic and Natural Ocular Mucin Analogues. *ACS Appl. Mater. Interfaces* **2017**, *9*, 20150–22016. [CrossRef] [PubMed]
320. Černohlávek, M.; Brandejsová, M.; Štěpán, P.; Vagnerová, H.; Hermannová, M.; Kopecká, K.; Kulhánek, J.; Nečas, D.; Vrbka, M.; Velebný, V.; et al. Insight into the Lubrication and Adhesion Properties of Hyaluronan for Ocular Drug Delivery. *Biomolecules* **2021**, *11*, 1431. [CrossRef] [PubMed]
321. Yamasaki, K.; Drolle, E.; Nakagawa, H.; Hisamura, R.; Ngo, W.; Jones, L. Impact of a low molecular weight hyaluronic acid derivative on contact lens wettability. *Contact Lens Anterior Eye* **2021**, *44*, 101334. [CrossRef] [PubMed]
322. Kathuria, A.; Shamloo, K.; Jhanji, V.; Sharma, A. Categorization of Marketed Artificial Tear Formulations Based on Their Ingredients: A Rational Approach for Their Use. *J. Clin. Med.* **2021**, *10*, 1289. [CrossRef] [PubMed]
323. Morrison, S.B.; Sullivan, D.A.; Sullivan, B.D.; Sheardown, H.; Schmidt, T.A. Dose-Dependent and Synergistic Effects of Proteoglycan 4 on Boundary Lubrication at a Human Cornea–Polydimethylsiloxane Biointerface. *Eye Contact Lens Sci. Clin. Pract.* **2012**, *38*, 27–35. [CrossRef]
324. Markoulli, M.; Kolanu, S. Contact lens wear and dry eyes: Challenges and solutions. *Clin. Optom.* **2017**, *9*, 41–48. [CrossRef]
325. Kim, S.; Laschi, C.; Trimmer, B. Soft robotics: A bioinspired evolution in robotics. *Trends Biotechnol.* **2013**, *31*, 287–294. [CrossRef]
326. Lamping, F.; Gorb, S.N.; de Payrebrune, K.M. Frictional Properties of a Novel Artificial Snakeskin for Soft Robotics. *Biotribology* **2022**, *30*, 100210. [CrossRef]
327. Hao, T.; Xiao, H.; Liu, S.; Liu, Y. Fingerprint-inspired surface texture for the enhanced tip pinch performance of a soft robotic hand in lubricated conditions. *Friction* **2023**, *11*, 1349–1358. [CrossRef]
328. Peng, Y.; Serfass, C.M.; Kawazoe, A.; Shao, Y.; Gutierrez, K.; Hill, C.N.; Santos, V.J.; Visell, Y.; Hsiao, L.C. Elastohydrodynamic friction of robotic and human fingers on soft micropatterned substrates. *Nat. Mater.* **2021**, *20*, 1707–1711. [CrossRef]
329. Shen, Z.; Chen, F.; Zhu, X.; Yong, K.-T.; Gu, G. Stimuli-responsive functional materials for soft robotics. *J. Mater. Chem. B* **2020**, *8*, 8972–8991. [CrossRef]
330. Gao, X.; Jiang, L. Water-repellent legs of water striders. *Nature* **2004**, *432*, 36. [CrossRef]
331. Vyas, L.; Aquino, D.; Kuo, C.; Dai, J.S.; Dasgupta, P. Flexible Robotics. *BJU Int.* **2011**, *107*, 187–189. [CrossRef] [PubMed]

332. Pfeifer, R.; Lungarella, M.; Iida, F. The challenges ahead for bio-inspired 'soft' robotics. *Commun. ACM* **2012**, *55*, 76–87. [CrossRef]
333. Cianchetti, M.; Laschi, C.; Menciassi, A.; Dario, P. Biomedical applications of soft robotics. *Nat. Rev. Mater.* **2018**, *3*, 143–153. [CrossRef]
334. Yoshida, K. Achievements in space robotics. *IEEE Robot. Autom. Mag.* **2009**, *16*, 20–28. [CrossRef]
335. Li, X.; Chen, X.; Zhang, C.; Luo, J. Preparation of self-lubricating NiTi alloy and its self-adaptive behavior. *Tribol. Int.* **2019**, *130*, 43–51. [CrossRef]
336. Yang, C.; Zhao, Y.; Tian, Y.; Wang, F.; Liu, X.; Jing, X. Fabrication and stability investigation of bio-inspired superhydrophobic surface on nitinol alloy. *Colloids Surf. A Physicochem. Eng. Asp.* **2019**, *567*, 16–26. [CrossRef]
337. Tang, G.; Ho, J.K.; Dong, G.; Hua, M. Fabrication self-recovery bulge textures on TiNi shape memory alloy and its tribological properties in lubricated sliding. *Tribol. Int.* **2016**, *96*, 11–22. [CrossRef]
338. NASA. New Commercial Robot Copies Geckos' Toes. Available online: https://www.nasa.gov/technology/tech-transfer-spinoffs/new-commercial-robot-copies-geckos-toes/ (accessed on 23 February 2024).
339. Li, H.; Li, Y.; Wu, J.; Jia, X.; Yang, J.; Shao, D.; Feng, L.; Wang, S.; Song, H. Bio-Inspired Hollow Carbon Microtubes for Multifunctional Photothermal Protective Coatings. *ACS Appl. Mater. Interfaces* **2022**, *14*, 29302–29314. [CrossRef] [PubMed]
340. Dalili, N.; Edrisy, A.; Carriveau, R. A review of surface engineering issues critical to wind turbine performance. *Renew. Sustain. Energy Rev.* **2009**, *13*, 428–438. [CrossRef]
341. Kim, M.S.; Lee, J.H.; Kwak, M.K. Review: Surface Texturing Methods for Solar Cell Efficiency Enhancement. *Int. J. Precis. Eng. Manuf.* **2020**, *21*, 1389–1398. [CrossRef]
342. Verma, A.S.; Jiang, Z.; Ren, Z.; Caboni, M.; Verhoef, H.; van der Mijle-Meijer, H.; Castro, S.G.; Teuwen, J.J. A probabilistic long-term framework for site-specific erosion analysis of wind turbine blades: A case study of 31 Dutch sites. *Wind. Energy* **2021**, *24*, 1315–1336. [CrossRef]
343. Niu, Y.; Hao, X.; Xia, A.; Wang, L.; Liu, Q.; Li, L.; He, N. Effects of textured surfaces on the properties of hydrodynamic bearing. *Int. J. Adv. Manuf. Technol.* **2022**, *118*, 1589–1596. [CrossRef]
344. Das, L.; Hocker, S.J.A.; Wiesner, V.L.; Wohl, C.J. Bio-Inspired Surface Structures to Mitigate Interfacial Particle Adhesion and Erosion—A Review. In Proceedings of the 44th Annual Meeting of the Adhesion Society, Virtual, 22–25 February 2021; Langley Research Center: Hampton, VA, USA.
345. Sun, X.; Jia, X.; Yang, J.; Wang, S.; Li, Y.; Shao, D.; Song, H. Bamboo fiber-reinforced chitosan sponge as a robust photothermal evaporator for efficient solar vapor generation. *J. Mater. Chem. A* **2021**, *9*, 23891–23901. [CrossRef]
346. Kant, K.; Pitchumani, R. Erosion wear analysis of heat exchange surfaces in a falling particle-based concentrating solar power system. *Sol. Energy Mater. Sol. Cells* **2024**, *266*, 112629. [CrossRef]
347. Mishnaevsky, L.; Branner, K.; Petersen, H.N.; Beauson, J.; McGugan, M.; Sørensen, B.F. Materials for Wind Turbine Blades: An Overview. *Materials* **2017**, *10*, 1285. [CrossRef] [PubMed]
348. Dashtkar, A.; Hadavinia, H.; Sahinkaya, M.N.; Williams, N.A.; Vahid, S.; Ismail, F.; Turner, M. Rain erosion-resistant coatings for wind turbine blades: A review. *Polym. Polym. Compos.* **2019**, *27*, 443–475. [CrossRef]
349. Wind Turbine Tech Inspired by Owl Feathers and Maple Seeds—Innovation—AskNature. Available online: https://asknature.org/innovation/wind-turbine-tech-inspired-by-owl-feathers-and-maple-seeds/ (accessed on 20 February 2024).
350. Chen, L.; Zhang, Y.; Cui, Y.; Wang, J.; Wang, M. Effects of Snake-Bioinspired Surface Texture on the Finger-Sealing Performance under Varied Working Conditions. *Machines* **2022**, *10*, 569. [CrossRef]
351. CKhatri, C.B.; Yadav, S.K.; Thakre, G.D.; Rajput, A.K. Design optimization of vein-bionic textured hydrodynamic journal bearing using genetic algorithm. *Acta Mech.* **2023**, *235*, 167–190. [CrossRef]
352. Vadivel, H.S.; Golchin, A.; Emami, N. Tribological behaviour of carbon filled hybrid UHMWPE composites in water. *Tribol. Int.* **2018**, *124*, 169–177. [CrossRef]
353. Li, X.; Guo, Z.; Huang, Q.; Yuan, C. Application of Bionic Tribology in Water-Lubricated Bearing: A Review. *J. Bionic Eng.* **2022**, *19*, 902–934. [CrossRef]
354. Pattnayak, M.R.; Pandey, R.K.; Dutt, J.K. Performance improvement of an oil-lubricated journal bearing using bionic-textures fused micro-pockets. *J. Tribol.* **2022**, *144*, 041804. [CrossRef]
355. Ji, M.; Zhang, H.; Xu, J.; Li, C.; Yu, D.; Chen, M.; El Mansori, M. Toward the mechanisms of surface texturing on the wear behavior of dental zirconia ceramics under dry and saliva lubricated conditions. *Wear* **2021**, *484–485*, 203845. [CrossRef]
356. Xie, Z.; Zhang, Y.; Zhou, J.; Zhu, W. Theoretical and experimental research on the micro interface lubrication regime of water lubricated bearing. *Mech. Syst. Signal Process.* **2021**, *151*, 107422. [CrossRef]
357. Metamaterial-Wikipedia. Available online: https://en.wikipedia.org/wiki/Metamaterial (accessed on 22 March 2024).
358. Garland, A.P.; Adstedt, K.M.; Casias, Z.J.; White, B.C.; Mook, W.M.; Kaehr, B.; Jared, B.H.; Lester, B.T.; Leathe, N.S.; Schwaller, E.; et al. Coulombic friction in metamaterials to dissipate mechanical energy. *Extreme Mech. Lett.* **2020**, *40*, 100847. [CrossRef]
359. Lehn, J.-M. Toward Self-Organization and Complex Matter. *Science* **2002**, *295*, 2400–2403. [CrossRef] [PubMed]
360. Fox-Rabinovich, G.S.; Gershman, I.S.; Veldhuis, S. Thin-Film PVD Coating Metamaterials Exhibiting Similarities to Natural Processes under Extreme Tribological Conditions. *Nanomaterials* **2020**, *10*, 1720. [CrossRef] [PubMed]
361. Liefferink, R.W.; Weber, B.; Coulais, C.; Bonn, D. Geometric control of sliding friction. *Extreme Mech. Lett.* **2021**, *49*, 101475. [CrossRef]

362. Rafsanjani, A.; Zhang, Y.; Liu, B.; Rubinstein, S.M.; Bertoldi, K. Kirigami skins make a simple soft actuator crawl. *Sci. Robot.* **2018**, *3*, 15. [CrossRef] [PubMed]
363. Marvi, H.; Cook, J.P.; Streator, J.L.; Hu, D.L. Snakes move their scales to increase friction. *Biotribology* **2016**, *5*, 52–60. [CrossRef]
364. Denny, M. Locomotion: The Cost of Gastropod Crawling. *Science* **1980**, *208*, 1288–1290. [CrossRef] [PubMed]
365. Ashby, M.F. The properties of foams and lattices. *Philos. Trans. R. Soc. A Math. Phys. Eng. Sci.* **2006**, *364*, 15–30. [CrossRef]
366. Al-Ketan, O.; Lee, D.-W.; Rowshan, R.; Abu Al-Rub, R.K. Functionally graded and multi-morphology sheet TPMS lattices: Design, manufacturing, and mechanical properties. *J. Mech. Behav. Biomed. Mater.* **2020**, *102*, 103520. [CrossRef] [PubMed]
367. Gao, T.; Liu, K.; Wang, X.; Wei, K.; Wang, Z. Multi-level mechanism of biomimetic TPMS hybridizations with tailorable global homogeneity and heterogeneity. *Extreme Mech. Lett.* **2024**, *68*, 102136. [CrossRef]
368. Xi, H.; Zhou, Z.; Zhang, H.; Huang, S.; Xiao, H. Multi-morphology TPMS structures with multi-stage yield stress platform and multi-level energy absorption: Design, manufacturing, and mechanical properties. *Eng. Struct.* **2023**, *294*, 116733. [CrossRef]
369. Fan, H.; Tian, Y.; Yang, L.; Hu, D.; Liu, P. Multistable mechanical metamaterials with highly tunable strength and energy absorption performance. *Mech. Adv. Mater. Struct.* **2022**, *29*, 1637–1649. [CrossRef]
370. Rezapourian, M.; Jasiuk, I.; Saarna, M.; Hussainova, I. Selective laser melted Ti6Al4V split-P TPMS lattices for bone tissue engineering. *Int. J. Mech. Sci.* **2023**, *251*, 108353. [CrossRef]
371. Rezapourian, M.; Hussainova, I. Optimal mechanical properties of Hydroxyapatite gradient Voronoi porous scaffolds for bone applications—A numerical study. *J. Mech. Behav. Biomed. Mater.* **2023**, *148*, 106232. [CrossRef]
372. Al-Ketan, O.; Abu Al-Rub, R.K. Multifunctional Mechanical Metamaterials Based on Triply Periodic Minimal Surface Lattices. *Adv. Eng. Mater.* **2019**, *21*, 1900524. [CrossRef]
373. Wong, T.-S.; Kang, S.H.; Tang, S.K.Y.; Smythe, E.J.; Hatton, B.D.; Grinthal, A.; Aizenberg, J. Bioinspired self-repairing slippery surfaces with pressure-stable omniphobicity. *Nature* **2011**, *477*, 443–447. [CrossRef] [PubMed]
374. Broeckhoven, C.; du Plessis, A. Escaping the Labyrinth of Bioinspiration: Biodiversity as Key to Successful Product Innovation. *Adv. Funct. Mater.* **2022**, *32*, 2110235. [CrossRef]
375. Gaspar, V.M.; Lavrador, P.; Borges, J.; Oliveira, M.B.; Mano, J.F. Advanced Bottom-Up Engineering of Living Architectures. *Adv. Mater.* **2020**, *32*, e1903975. [CrossRef] [PubMed]
376. Perricone, V.; Santulli, C.; Rendina, F.; Langella, C. Organismal Design and Biomimetics: A Problem of Scale. *Biomimetics* **2021**, *6*, 56. [CrossRef] [PubMed]
377. Cavaleri, L.; Asteris, P.G.; Psyllaki, P.P.; Douvika, M.G.; Skentou, A.D.; Vaxevanidis, N.M. Prediction of Surface Treatment Effects on the Tribological Performance of Tool Steels Using Artificial Neural Networks. *Appl. Sci.* **2019**, *9*, 2788. [CrossRef]
378. Mahapatra, S.; Patnaik, A. Study on mechanical and erosion wear behavior of hybrid composites using Taguchi experimental design. *Mater. Des.* **2009**, *30*, 2791–2801. [CrossRef]
379. Grützmacher, P.G.; Profito, F.J.; Rosenkranz, A. Multi-Scale Surface Texturing in Tribology—Current Knowledge and Future Perspectives. *Lubricants* **2019**, *7*, 95. [CrossRef]
380. Gachot, C.; Rosenkranz, A.; Hsu, S.M.; Costa, H.L. A critical assessment of surface texturing for friction and wear improvement. *Wear* **2017**, *372–373*, 21–41. [CrossRef]
381. Rosenkranz, A.; Marian, M.; Profito, F.J.; Aragon, N.; Shah, R. The Use of Artificial Intelligence in Tribology—A Perspective. *Lubricants* **2020**, *9*, 2. [CrossRef]
382. Marian, M.; Tremmel, S. Current Trends and Applications of Machine Learning in Tribology—A Review. *Lubricants* **2021**, *9*, 86. [CrossRef]
383. Curtarolo, S.; Setyawan, W.; Hart, G.L.; Jahnatek, M.; Chepulskii, R.V.; Taylor, R.H.; Wang, S.; Xue, J.; Yang, K.; Levy, O.; et al. AFLOW: An automatic framework for high-throughput materials discovery. *Comput. Mater. Sci.* **2012**, *58*, 218–226. [CrossRef]
384. Jia, D.; Duan, H.; Zhan, S.; Jin, Y.; Cheng, B.; Li, J. Design and Development of Lubricating Material Database and Research on Performance Prediction Method of Machine Learning. *Sci. Rep.* **2019**, *9*, 20277. [CrossRef]
385. Zhao, Y.; Mei, H.; Chang, P.; Chen, C.; Cheng, L.; Dassios, K.G. Infinite Approaching Superlubricity by Three-Dimensional Printed Structures. *ACS Nano* **2021**, *15*, 240–257. [CrossRef] [PubMed]
386. Zafar, M.Q.; Zhao, H. 4D Printing: Future Insight in Additive Manufacturing. *Met. Mater. Int.* **2019**, *26*, 564–585. [CrossRef]
387. Gharde, S.; Surendren, A.; Korde, J.M.; Saini, S.; Deoray, N.; Goud, R.; Nimje, S.; Kandasubramanian, B. Recent advances in additive manufacturing of bio-inspired materials. *Biomanufacturing* **2019**, 35–68. [CrossRef]
388. Liu, J.; Wang, X.; Liu, Y.; Liu, X.; Fan, K. Bioinspired three-dimensional and multiple adsorption effects toward high lubricity of solvent-free graphene-based nanofluid. *Carbon* **2022**, *188*, 166–176. [CrossRef]
389. Zhu, Y.; Yang, F.; Guo, Z. Bioinspired surfaces with special micro-structures and wettability for drag reduction: Which surface design will be a better choice? *Nanoscale* **2021**, *13*, 3463–3482. [CrossRef] [PubMed]

MDPI AG
Grosspeteranlage 5
4052 Basel
Switzerland
Tel.: +41 61 683 77 34

Biomimetics Editorial Office
E-mail: biomimetics@mdpi.com
www.mdpi.com/journal/biomimetics

www.ingramcontent.com/pod-product-compliance
Lightning Source LLC
Chambersburg PA
CBHW061435210926
51850CB00008B/1519
* 9 7 8 3 7 2 5 8 7 7 6 1 4 *